Audio Effects

Audio effects are used pervasively in music performance and production, and the creation of new digital audio tools is a significant industry. They are also commonly used in game production, broadcasting, and film and television.

This book is intended as an educational textbook on audio effects, with relevance to audio signal processing, music informatics, sound engineering, and related topics.

It provides fundamental background information on digital signal processing, focusing on audio-specific aspects. This constitutes the building block on which audio effects are developed. It integrates theory and practice, relating technical implementation to musical implications.

- This book can be used to gain an understanding of the operation of existing audio effects or to create new ones.
- It includes detailed coverage of common audio effects (and plenty of unusual ones).
- It includes discussion of current digital audio standards, like VST. Accompanying source code is provided in C/C++ for audio plugin development using JUCE.
- Each section of this book also has examples, exercises, sound samples, lecture slides, and additional information on related topics.

This second edition includes revised and expanded chapters, with new content on equalization, distortion, phase vocoders, and reverberation. It also has new chapters on immersive audio and on advanced concepts in audio programming, revised source code using the latest version of JUCE, and corrections and improvements throughout.

AUDIO ENGINEERING SOCIETY PRESENTS…

www.aes.org

Editorial Board

Chair: Francis Rumsey, Logophon Ltd, Hyun Kook Lee, University of Huddersfield, Natanya Ford, University of West England, Kyle Snyder, University of Michigan

Digital Audio Forensics Fundamentals: From Capture to Courtroom
James Zjalic

Drum Sound and Drum Tuning: Bridging Science and Creativity
Rob Toulson

Sound and Recording, 8th Edition: Applications and Theory
Francis Rumsey with Tim McCormick

Performing Electronic Music Live Kirsten Hermes Working with the Web Audio API
Joshua Reiss

Modern Recording Techniques, 10th Edition: A Practical Guide to Modern Music Production
David Miles Huber and Emiliano Caballero

Recording Orchestra and Other Classical Music Ensembles, 2nd Edition
Richard King

Sound Reproduction, 4th Edition: The Acoustics and Psychoacoustics of Loudspeakers, Rooms, and Headphones
Floyd E. Toole, Sean Olive, and Todd Welti

Classical Recording, 2nd Edition: A Practical Guide in the Decca Tradition
Caroline Haigh, John Dunkerley, and Mark Rogers

Handbook for Sound Engineers, 6th Edition
Edited by Glen Ballou and Doug Jones

Audio Effects: Theory, Implementation and Application
Joshua D. Reiss and Andrew P. McPherson

For more information about this series, please visit: www.routledge.com/Audio-Engineering-Society-Presents/book-series/AES

Audio Effects
Theory, Implementation and Application

2nd Edition

Joshua D. Reiss and Andrew P. McPherson

LONDON AND NEW YORK

Designed cover image: Shutterstock

MATLAB® and Simulink® are trademarks of The MathWorks, Inc. and are used with permission. The MathWorks does not warrant the accuracy of the text or exercises in this book. This book's use or discussion of MATLAB® or Simulink® software or related products does not constitute endorsement or sponsorship by The MathWorks of a particular pedagogical approach or particular use of the MATLAB® and Simulink® software.

Second edition published 2026
by Routledge
605 Third Avenue, New York, NY 10158

and by Routledge
4 Park Square, Milton Park, Abingdon, Oxon, OX14 4RN

CRC Press is an imprint of Taylor & Francis Group, LLC

© 2026 Joshua D. Reiss and Andrew P. McPherson

First edition published by CRC Press 2014

Reasonable efforts have been made to publish reliable data and information, but the author and publisher cannot assume responsibility for the validity of all materials or the consequences of their use. The authors and publishers have attempted to trace the copyright holders of all material reproduced in this publication and apologize to copyright holders if permission to publish in this form has not been obtained. If any copyright material has not been acknowledged please write and let us know so we may rectify in any future reprint.

Except as permitted under U.S. Copyright Law, no part of this book may be reprinted, reproduced, transmitted, or utilized in any form by any electronic, mechanical, or other means, now known or hereafter invented, including photocopying, microfilming, and recording, or in any information storage or retrieval system, without written permission from the publishers.

For permission to photocopy or use material electronically from this work, access www.copyright.com or contact the Copyright Clearance Center, Inc. (CCC), 222 Rosewood Drive, Danvers, MA 01923, 978-750-8400. For works that are not available on CCC please contact mpkbookspermissions@tandf.co.uk

For Product Safety Concerns and Information please contact our EU representative GPSR@taylorandfrancis.com. Taylor & Francis Verlag GmbH, Kaufingerstraße 24, 80331 München, Germany.

Trademark notice: Product or corporate names may be trademarks or registered trademarks and are used only for identification and explanation without intent to infringe.

ISBN: 9781032974286 (hbk)
ISBN: 9781032974859 (pbk)
ISBN: 9781003593942 (ebk)

DOI: 10.1201/9781003593942

Typeset in Palatino
by codeMantra

Access the Support Material: www.routledge.com/9781032974286

Contents

Preface

Audio effects are used in broadcasting, television, film, games, and music production. Once they were used primarily to enhance a recording and to correct artifacts in the production process, but now they are used creatively and pervasively.

This book aims to describe the theory behind the effects, to explain how they can be implemented, and to illustrate many ways in which they can be used. The concepts covered in this book have relevance to sound engineering, digital signal processing, acoustics, audio signal processing, music informatics, and related topics.

Both authors have taught courses on this subject. We were aware of excellent texts on the use of audio effects, especially for mixing and music production. We also knew excellent reference material for audio signal processing and for audio effect research. But it was still challenging to find the right material that teaches the reader, from the ground up, how and why to create audio effects and how they are used.

That is the purpose of this book. It provides students and researchers with knowledge of how to use the tools and the basics of how they work, as well as how to create them. It is primarily educational and geared toward undergraduate and master's level students, though it can also serve as a reference for practitioners and researchers. It explains how sounds can be processed and modified by mathematical or computer algorithms. It teaches the theory and principles behind the full range of audio effects and provides the reader with an understanding of how to analyze, implement, and use them.

We chose not to shy away from presenting the math and science behind the implementations and applications. Thus, it is one of the few resources for use in the classroom with a mathematical and technical approach to audio effects. It provides a detailed overview of audio effects and includes example questions to aid in learning and understanding. It has a special focus on programming and implementation with industry standards and provides source code for generating plug-in versions of many of the effects.

Chapter 1 begins by covering some fundamental concepts used often in later chapters. It also introduces the notation that we use throughout. Here, we describe some essential concepts from digital signal processing, thus allowing the subject matter to be mostly self-contained, without the reader needing to consult other texts.

Chapter 2 is about how to build the audio effects as software plug-ins. We focused on the C++ VST format, which is probably the most popular standard and available for most platforms and hosts. This chapter (and to some extent, Chapters 13 and 14) may be read at any point or independently of the others. It makes reference to the effects discussed previously, but the chapter is focused on practical implementation. It complements the supplementary material, which includes source code that may be used to build VST plug-ins for a large number of effects described in this book.

In Chapter 3, we introduce delay lines and related effects such as delay, vibrato, chorus, and flanging. These are some of the most basic effects, and the concept of delay lines is useful for understanding implementations of the effects introduced in later sections.

Chapter 4 then covers filter fundamentals. We chose a quite general approach here and introduce techniques that allow the reader to construct a wide variety of high-order filters. Attention is also paid to some additional filters often used in other effects, such as the all-pass filter and the exponential moving average.

In Chapter 5, we explore filters in more detail, covering effects that have filters as their essential components. These include the graphic and parametric equalizer, wah-wah, and phaser.

We then move on to nonlinear effects. Chapter 6 discusses modulation, focusing primarily on tremolo and ring modulation. Chapter 7 goes into detail on dynamics processing, especially the dynamic range compressor and the noise gate. Here, much emphasis is given on correct implementation and perceptual qualities of these effects. Chapter 8 then covers distortion effects. These are concerned with the sounds that result from highly nonlinear processing beyond the dynamics processors of the previous chapter.

Having introduced the important signal-processing concepts, we can now move on to the phase vocoder and introduce several effects that do their processing in the frequency domain. This is the focus of Chapter 9.

Up to this point, none of the effects attempted to recreate how a natural sound might be perceived by a human listener in a real acoustic space. The next three chapters deal with spatial sound reproduction and spatial sound phenomena. In Chapter 10, we cover reverberation, paying particular emphasis on both algorithmic and convolutional approaches to artificial reverberation. Though grouped together with the other chapters concerned with spatial sound, the reverberation approaches described here do not necessarily require the processing of two or more channels of audio.

Chapter 11 covers many of the main spatialization techniques, starting with panning and precedence as can be used in stereo positioning and then moving on to techniques requiring more and more channels: vector-based amplitude panning, ambisonics, and wavefield synthesis. The final technique describes binaural sound reproduction using HRTFs, for listening with headphones.

Chapter 12 covers the rapidly growing field of immersive audio. Immersive audio is not well-defined, but often allows both the listener and all sources to be moving in any direction from arbitrary positions, with arbitrary orientations, as well as to define directional properties of the sources, how sound levels decay with distance and how the spatial sound is finally rendered. It aims to fully immerse the listener in a sonic environment. Here, an immersive audio effect is constructed, where sounds are rendered, given full knowledge of the environment. We also show how the Doppler effect, a physical phenomenon associated with moving sources or moving listeners, is derived and simulated.

Chapter 13 is about audio production. This is of course a very broad area, so we focus on the architecture of mixing consoles and digital audio workstations, and how the effects we've described may be used in these devices. We then discuss how to order and combine the audio effects in order to accomplish various production challenges.

Chapter 14 picks up where Chapter 2 left off in exploring principles and practical strategies for implementing audio effects in code. This chapter examines several aspects of evaluating and improving the computational performance of audio effect plug-ins, including computational efficiency, memory footprint, and latency. It also introduces why, where, and how to use multi-threading in audio code, where some functions run asynchronously to the main audio processing thread. Finally, this chapter introduces the implementation of audio effects on embedded systems such as microcontrollers and single-board computers.

Authors

Joshua D. Reiss is Professor with the Centre for Digital Music at Queen Mary University of London. He has published more than 200 scientific papers and co-authored three books. He is a past President and Fellow of the Audio Engineering Society. He has served as an expert witness in major cases related to audio technology and patent litigation. He is also Entrepreneur-in-Resident at QMUL, having co-founded four spin-out companies: LandR, Waveshaper AI, RoEx, and Nemisindo. His primary focus of research is on the use of state-of-the-art signal processing and machine learning techniques for sound design and audio production.

Andrew P. McPherson is Professor in the Dyson School of Design Engineering, Imperial College London. He leads the Augmented Instruments Laboratory, a research team designing and exploring new musical instruments. He has published over 200 papers and has held research fellowships from ERC, UKRI, and the Royal Academy of Engineering and has spun out two companies including Bela.io, which makes high-performance embedded audio hardware.

Changes in the Second Edition

The first edition of this book was published in 2015. Much has changed in the approximately 10 years from that first edition to the writing of this new preface in 2025. The field has changed massively. At the same time, we also identified some typos in the first edition and some better explanations for certain concepts. So a new edition was needed.

There have been small changes throughout, but some significant changes are as follows:

- Building Audio Effect Plug-ins was moved from the last chapter to the second chapter, right after Fundamentals. This way, the basics of plug-ins are introduced before the reader explores any of the plug-ins in the online material and before any code examples of audio effects are given. This change also required a rewrite of this chapter.
- Some explanations have been significantly rewritten in the chapters on Delay, Filter Design, Filter Effects, Distortion, and Phase Vocoder. This was sometimes done to simplify the math (without omitting important details) and at other times to provide a better explanation of a theoretical concept.
- Several concepts were mostly overlooked in the first edition and are now discussed, at least briefly, in this edition. This includes Feedback Delay Networks in the chapter on Reverberation, and databases of measured HRTFs for binaural rendering in the Spatial Audio chapter.
- The Doppler chapter has now been replaced with a chapter on Immersive Audio. This chapter includes the previous material on the Doppler Effect but also goes much further in the rendering of a virtual environment.
- This book now ends with a chapter on Advanced Concepts in Audio Programming. It addresses aspects of latency and multithreading, for instance, that become more important when one moves beyond simple demonstrations of audio effects.
- Each chapter now includes more references, especially to recent publications. The chapters also end with a Further Reading section, referring the reader to other useful works.

- The accompanying source code has been rewritten. It has been improved in places and made to work with current versions of JUCE, Visual Studio, and Xcode.
- Accompanying videos have been made available on a YouTube channel.

Acknowledgments

The text of the first edition benefited greatly from the comments of expert reviewers, most notably Dr. Pedro Duarte Pestana. We are also deeply indebted to Brecht De Man, who revised the audio effects source code and contributed several implementations.

Since the first edition, two people have rewritten the accompanying code for later versions of JUCE: Juan Gil and Shane Dunne. Though their implementations are not entirely compatible with the very latest version of JUCE, their brilliant work served as the basis for the new accompanying code for the second edition.

This book would also not have been possible without all of the excellent work that has gone on before. We are indebted to various people whose work is frequently cited throughout the text: Vesa Valimaki, Julius Smith, Roey Izhaki, Udo Zoelzer, Ville Pulkki, Sophocles Orfanidis, to name just a few. The errors and omissions are ours, whereas the best explanations are found in the work of the cited authors.

Finally, the first author (JR) dedicates this book to his family: brothers Alex and Ben (and their families), his parents Chris and Judith, his daughters Eliza and Laura, and his wife Sabrina. The second author (AM) dedicates his contributions to Barry Vercoe (1937-2025), a computer music pioneer and AM's Master's thesis advisor at the MIT Media Lab.

Resources

The main website accompanying this book is:
https://joshreiss.github.io/Audio_Effects-Theory_Implementation_and_Application/

It includes the following:

- YouTube playlist with videos related to the book:
  - https://tinyurl.com/yk4fxfkc
- GitHub Repository
  - https://github.com/joshreiss/Audio_Effects-Theory_Implementation_and_Application

For each chapter, the GitHub repository includes

- C++ Source code and JUCE projects
- Powerpoint presentations
- MATLAB files for generating images
- Sound samples

1

Fundamentals

In digital audio signal processing and digital audio effects, we are primarily concerned with systems that take a discrete, uniformly sampled audio signal, process it, and produce a discrete, uniformly sampled output audio signal. Therefore, we will start by introducing some fundamental properties of sound that are frequently used, then how we represent sound as a digital signal, and then move on to how we describe the systems that act on and modify such signals. This is not meant to give a detailed overview of digital signal processing, which would involve discussion of continuous time signals, infinite signals, and mathematical relationships. Rather, we intend to focus on just the type of signals and systems that are encountered in audio effects, and on the most useful properties and representations. Having said that, this is also intended to be self-contained. Very little prior knowledge is assumed, and it should not be necessary to refer to more detailed discussions in other texts in order to understand these concepts.

Understanding Sound and Digital Audio

Fundamentally, all audio comprises waveforms. Vibrating objects create pressure waves in the air; when these waves reach our ears, we perceive them as sound. With the invention of the telephone in the 19th century, audio was first encoded as an electric signal, with the changes in electric voltage representing the changes in pressure over time. Until the late 20th century, electric recording and transmission were all analog: sound was represented by a continuous waveform over time.

In this book, we will work almost exclusively with digital audio. Rather than representing audio as a continuous voltage, as in analog, the waveform will be composed of discrete samples over time. These samples can be stored, processed, and ultimately reconstructed as sound we can hear. Digital audio systems generally begin with an *Analog to Digital Converter* (ADC), which captures periodic snapshots of the electrical voltage on an audio transmission line and represents these snapshots as discrete numbers. By capturing the voltage many thousands of times per second, one can achieve a very close approximation of the original audio signal. This

DOI: 10.1201/9781003593942-1

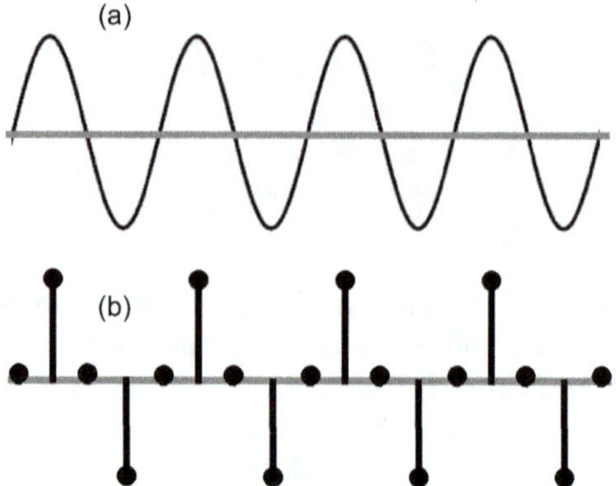

FIGURE 1.1
A continuous time signal (a), and its digital representation, found by sampling the signal uniformly in time (b).

encoding method is known as *pulse code modulation*, and is the encoding format used in the WAV and AIFF audio formats. Pulse code modulation is also one of the most popular forms of ADC, and certainly one of the simplest to explain.

Thus, a continuous time audio signal, such as captured from a microphone, is represented as a digital signal with uniform timing between samples (see Figure 1.1). But digital audio signals need not be derived from analog, nor even represent any physical sound. It can be completely synthetic and generated using digital signal processing techniques. We will touch on this later in the text when discussing low-frequency oscillators (Chapter 3), phase vocoders (Chapter 9), and other concepts. It is important to note that, unless additional information is stored, there is no distinction between those digital audio signals that were generated from the conversion of analog signals and those that were generated from digital sound synthesis techniques (though, of course, 'real world' signals are likely to have more noise and more complex phenomena).

There are three important characteristics of almost any digital audio data: sample rate, bit depth, and number of channels.

Sample rate is the rate at which the samples are captured or played back. It is typically measured in Hertz (Hz), or cycles per second. In this case, one cycle represents one sample. An audio CD has a sample rate of 44,100 Hz, or 44.1 kHz. Higher sampling rates allow a digital recording to accurately record higher frequencies of sound, or to provide a safety margin in case of additional noise or artifacts introduced in the recording, processing, or playback. 48 kHz is often used in audiovisual production, and sample rates of 96 or 192 kHz are used in high-resolution audio, such as in DVD-Audio, or in professional audio production.

The *bit depth* specifies how many bits are used to represent each audio sample. The most common choices in audio are 16 and 24 bit. The bit depth also determines the theoretical dynamic range of the audio signal. In digital audio, amplitude is often expressed as a unitless number, representing a ratio between the current intensity and the highest (or lowest) possible intensity that can be represented. The maximum absolute value for this ratio is known as *dynamic range*. In an ideal ADC, the dynamic range, in decibels (see below), is very roughly 6.02 times the number of bits. Thus, 16-bit audio could represent signals whose loudness ranges over 96 dB, e.g., from a quiet whisper to a loud rock concert.

The *number of channels* actually refers to the fact that audio content often consists of several different channels, each representing its own signal. This is most often the case in stereo or surround sound, where each channel may represent the sound sent to each loudspeaker. Monaural audio, however, is typically encoded as a single channel. We will return to these concepts in Chapter 11.

Digital audio may be encoded with or without *data compression*. When data compression is used, sophisticated algorithms are used to encode and re-represent the data such that it takes up much less space. Hence, a decoder must be used to convert the data back into time domain samples before playback. The compression can either be *lossless* (the decoded data is identical to the original data before compression) or *lossy*. Modern lossy audio compression techniques use knowledge of psychoacoustics to minimize the perceived degradation of audio that occurs when a substantial amount of the information contained in the original signal is discarded.

Data compression also introduces one more characteristic of audio data, the *bit rate*. This is the number of bits per unit of time. For lossless signals, this is simply the bit depth times the sample rate times the number of channels. For instance, CD audio would typically have a bit rate of 1,411.2 kbps (kilobits per second),

$$16\frac{\text{bits}}{\text{sample}}\cdot 44,100\frac{\text{samples}}{\text{second}}\cdot 2\,(\#\,\text{channels}) = 1,411,200\frac{\text{bits}}{\text{second}}. \quad (1.1)$$

For audio signals that have undergone lossy compression, the bit rate is usually greatly reduced. Most compression schemes, including mp3 and aac, transmit audio with a bit rate between 30 and 500 kbps.

It should be noted that there is a lot of fine detail regarding quantization, sampling, dynamic range, and lossy compression of audio data that has been omitted here. For the purpose of this text, it is sufficient to know the format and general meaning of these concepts, but the reader is also encouraged to refer to signal processing texts for more detailed discussion [1–4].

WHY 44.1 KHZ?

44.1 kHz, or 44,100 samples per second, is perhaps the most popular sample rate used in digital audio, especially for music content. The short answer as to why it is so popular is simple; it was the sample rate chosen for the Compact Disc, and thus is the sample rate of much audio taken from CDs, and the default sample rate of much audio workstation software.

As to why it was chosen as the sample rate for the Compact Disc, the answer is a bit more interesting. In the 1970s, when digital recording was still in its infancy, many different sample rates were used, including 3 and 50 kHz in Soundstream's recordings. In the late 70s, Philips and Sony collaborated on the Compact Disc, and there was much debate between the two companies regarding sample rate. In the end, 44.1 kHz was chosen for a number of reasons.

According to the Nyquist theorem, 44.1 kHz allows reproduction of all frequency content below 22.05 kHz. This covers all frequencies heard by a normal person. Though there is still debate about the perception of high-frequency content, it is generally agreed that few people can hear tones above 20 kHz.

44.1 kHz also allowed the creators of the CD format to fit at least 80 min of music (more than on a vinyl LP record) on a 120 mm disc, which was considered a strong selling point.

But 44,100 is a rather special number. $44{,}100 = 2^2 \times 3^2 \times 5^2 \times 7^2$, and hence 44.1 kHz is actually an easy number to work with for many calculations.

Working with Decibels

We often deal with quantities that can cover a very wide range of values, from very large to very small. The *decibel scale* is a useful way to represent such quantities. The decibel (dB) is a logarithmic representation of the ratio between two values. Typically, both values represent power, and hence, the decibel is unitless. One of these values is usually a reference, so that the decibel scale can represent absolute levels. The decibel representation of a level is then ten times the logarithm to base 10 of the ratio of the two power quantities. Since power is usually the square of a magnitude, we can write a value in decibels in terms of the magnitudes or powers as,

$$x_{\text{dB}} = 10\log_{10}\left(x^2/x_0^2\right) = 20\log_{10}\left(|x|/|x_0|\right). \tag{1.2}$$

If not specified, x_0 is usually assumed to be 1. So for example, one million is 60 dB, and 0.001 is –30 dB. Whether a decibel or linear scale is used often depends just on which one best conveys the relevant information.

Level Measurements

The sound pressure measured from a source is inversely proportional to the distance from the source. In other words, the measured sound pressure p is proportional to $1/r$, where r represents the distance between the source and a listener. Suppose a sound pressure p_1 is measured at distance r_1 from the source; then the sound pressure p_2 at distance r_2 can be calculated as:

$$p_2 = \frac{p_1 r_1}{r_2}. \tag{1.3}$$

Not all sources radiate uniformly in every direction. For example, a violin radiates more sound upward from the top of the instrument than from the sides or back. Measurements at different angles may therefore give different results.

The intensity I of a sound is a physical measure of its acoustical energy flow, and is measured in watts per unit area, W/m^2. Whereas pressure is proportional to the distance to the sound source, the intensity is proportional to the square of the distance to the sound source, giving a $1/r^2$ relationship.

A decibel scale is used to represent the very wide range of sound intensities that can be perceived. The sound intensity level, L_I, given in dB, is the log ratio of a given intensity I to a reference. The reference level is usually set to $I_0 = 10^{-12}$ W/m^2, which is considered to be roughly the threshold of hearing at 1 kHz.

$$L_I = 10\log_{10}(I/I_0) = 120 + 10\log_{10} I. \tag{1.4}$$

Exact measurement of intensity is difficult, and the intensity values will fluctuate over time. So sound pressure level (SPL) is often used instead. The sound pressure level or sound level L_p is also given in decibels (dB) above a standard reference level, p_{ref} of 2×10^{-5} N/m^2 = 20 μPa, the sound pressure threshold of human hearing. That is,

$$L_p = 10\log_{10}\left(p_{\text{rms}}^2 / p_{\text{ref}}^2\right) = 20\log_{10}\left(p_{\text{rms}} / p_{\text{ref}}\right), \tag{1.5}$$

where p_{ref} is the reference sound pressure and p_{rms} is the rms (root mean squared, or square root of the average value of the squared signal) sound pressure being measured. In this text, we do not often refer to SPLs, since we will mostly deal with the processing of digital signals, where the physical sound level is not known.

For digital signals, level measurements are also given in a decibel representation. But now, decibels are measured relative to full scale, denoted dBFS. This is possible since most digital systems have a defined maximum available peak level.

Zero dBFS represents the maximum possible digital level. For example, if the maximum signal amplitude on a linear scale is 1 and the actual amplitude is 0.5, then the signal level would be defined as $20\log_{10}(0.5/1) = -3.01$ dBFS, or 3 decibels below peak level. However, if RMS (root mean squared) measurements are used, then the definition may be ambiguous. Different conventions are used for RMS measurements. Some RMS-based level measurements set the reference level so that peak and RMS measurements of a square wave will produce the same result, and all dBFS measurements will be negative, and the maximum sine wave that can be produced without clipping will have value –3.01dBFS.

An alternative (though much less common) definition gives the reference level so that peak and RMS measurements of a sine wave will produce the same result. A full-scale sine wave would be at 0 dBFS, but a full-scale square wave would exceed this, at +3dBFS. In audio production (see Chapter 13), meters are provided so that the user will know if maximum levels are exceeded and clipping will occur.

In Chapter 7, we will discuss some methods of estimating the levels of digital signals for use in dynamics processing. Though given on a decibel scale, these estimates are tailored to the audio effect and may be different from the dBFS value described here.

Representing and Understanding Digital Signals

The time between samples may be given as T_s so that the sampling frequency is given as $f_s = 1/T_s$. Therefore, the digital input signal may be represented as a discrete sampling of a continuous signal, $x(0)$, $x(T_s)$, $x(2T_s)$, etc. If we only consider the sample number, then a finite signal consisting of N samples can be represented as $x[0]$, $x[1]$, … $x[N-1]$.[1]

Representing Complex Numbers

Almost always, we will have real-valued signals, and our audio effects produce real-valued results. But a lot of the maths and analysis of signals and effects is actually easier to do when generalizing the discussion to complex numbers. So it is extremely useful to keep a few properties of complex numbers in mind. First off, any complex number can be written in several ways;

$$x = a + bj = re^{j\theta}. \tag{1.6}$$

Here, a is the real part and b is the imaginary part, and j is defined to be $\sqrt{-1}$. The complex number can be plotted as a point on a plane where the x direction is the real component and the y direction is the imaginary one. In which case, r is the magnitude of the vector from (0,0) to (a,b), and θ is

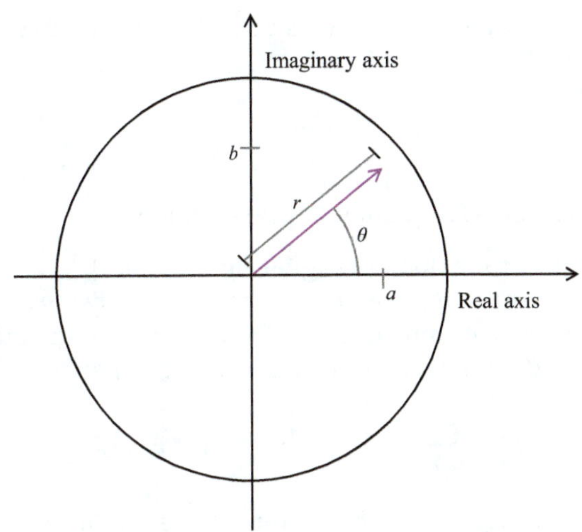

FIGURE 1.2
A complex number $a + jb = re^{j\theta}$ depicted in the complex plane. The unit circle, $\cos\theta + j\sin\theta = e^{j\theta}$ is also depicted.

the angle between the x-axis and the vector, known as the phase. This is depicted in Figure 1.2.

Using a form of Euler's identity, $e^{j\theta} = \cos\theta + j\sin\theta$, we have,

$$\begin{aligned} a &= r\cos\theta \\ b &= r\sin\theta \\ r &= \sqrt{a^2 + b^2}. \end{aligned} \tag{1.7}$$

It's a bit tricky to derive the phase θ from a and b. It's not just the arctan(b/a), since this doesn't distinguish between the case when b is positive and a negative, and when b is negative and a positive. That is, the arc tangent function has a range of $-\pi/2$ to $\pi/2$, but the phase ranges from $-\pi$ to π. So we use the following;

$$\theta = a\tan 2(b,a) = \begin{cases} \arctan(b/a) & a \ge 0, b \ne 0 \\ \arctan(b/a) + \pi\,\mathrm{sgn}(b) & a < 0 \\ \text{undefined} & a, b = 0 \end{cases}. \tag{1.8}$$

The conjugate of a complex number is simply defined as the same number, but with the sign changed on the complex component. Denoting complex conjugation by *, we then have, from (1.6),

$$x^* = a - bj = re^{-j\theta}. \tag{1.9}$$

The square magnitude of a number is given by that number times its complex conjugate,

$$|x|^2 = x \cdot x^* = a^2 + b^2 = r^2. \tag{1.10}$$

Frequency and Time-Frequency Representations

Let's return to our time domain digital signal, $x[0]$, $x[1]$, … $x[N-1]$. There are many other ways to represent this signal. Probably the most important is the Discrete Fourier Transform (DFT), which is intended to represent a finite, discrete signal in terms of its frequency components.

$$X[k] = \sum_{n=0}^{N-1} x[n] e^{-jnk2\pi/N} \quad 0 \le k, n \le N-1. \tag{1.11}$$

This converts the signal from being represented in terms of a real value at sample number n to representation in terms of a complex value at frequency bin k. In the same sense that time domain sample n corresponds to a discrete sample at time nT_s, frequency bin k corresponds to frequency kf_s/N. Now notice that the output involves complex numbers. More precisely, $X[k]$ gives a phase and amplitude for the frequency content from $(k-1/2)f_s/N$ to $(k+1/2)f_s/N$.

This transformation has a large number of properties, but for understanding digital signals, the following are most important.

$$\begin{aligned} y[n] &= x_1[n] + x_2[n] \leftrightarrow Y[k] = X_1[k] + X_2[k] \\ y[n] &= ax[n] \leftrightarrow Y[k] = aX[k]. \end{aligned} \tag{1.12}$$

In other words, if we add two signals together, we add their discrete Fourier transforms together, and if we multiply a signal by a constant, we multiply its discrete Fourier transforms by the same constant.

Now suppose our signal x is a complex sinusoid, $x[n] = ae^{jnl2\pi/N}, n = 0, 1, \dots N-1$, where a is some constant. Then,

$$X[k] = \sum_{n=0}^{N-1} ae^{j2\pi(l-k)n/N} = \begin{cases} a\sum_{n=0}^{N-1} 1 = aN & l = k \\ a\dfrac{1-e^{j2\pi(l-k)}}{1-e^{j2\pi(l-k)/N}} = 0 & l \ne k \end{cases}, \tag{1.13}$$

where we used a well-known identity $\sum_{n=0}^{N-1} x^n = \dfrac{1-x^N}{1-x}$. So each frequency bin in a Discrete Fourier Transform represents the magnitude of a complex sinusoid. That implies that any finite signal can be represented as a sum of weighted sinusoids.

Finally, we can convert from the frequency representation back to the time domain using the Inverse Discrete Fourier Transform (IDFT),

$$x[n]=\frac{1}{N}\sum_{k=0}^{N-1}X[k]e^{j2\pi nk/N}. \tag{1.14}$$

The DFT allows us to represent the signal as complex-valued frequency components. Each of these has a magnitude and phase. Thus, we can plot the magnitude and phase as a function of frequency for any signal. More common than magnitude plots, however, is the power spectrum, which is given as the power in each frequency bin,

$$P(k)=|X(k)|^2/N^2, \tag{1.15}$$

as a function of frequency.

The Discrete Fourier Transform and its inverse are very powerful tools, though very computationally intensive. But there is a lot of redundancy in the calculation. An implementation known as the Fast Fourier Transform (FFT) is commonly used. However, in its standard implementation, it requires that the number of samples be a power of two.

Even with the FFT, it is still quite slow to compute over a large number of samples. Furthermore, the incoming signal may be very long or infinite, yet one would like to know the frequency content at any given time. Thus, the Short Time Fourier Transform (STFT) is used;

$$STFT\{x[n]\}\equiv X[m,k]=\sum_{n=mR}^{N+mR-1}x[n]e^{-j(n-mR)k2\pi/N}\quad 0\le k\le N-1. \tag{1.16}$$

This provides estimates of the frequency content at times mR, where R is the hop size, in samples, between successive DFTs. And we often plot the spectrogram, $S[m,k]\equiv|X[m,k]|^2$. The relationship between spectrogram and STFT is completely analogous to the relationship between power spectrum and DFT.

Figure 1.3 depicts the waveform, power spectrum, and spectrogram of an excerpt of a solo guitar performance.

Aliasing

An important concept, and one that will feature heavily in dealing with nonlinear processing, is *aliasing*. Suppose $x[n]=\cos(nl2\pi/N)=\left[e^{jnl2\pi/N}+e^{-jnl2\pi/N}\right]/2, n=0,1,\ldots N-1$. Then,

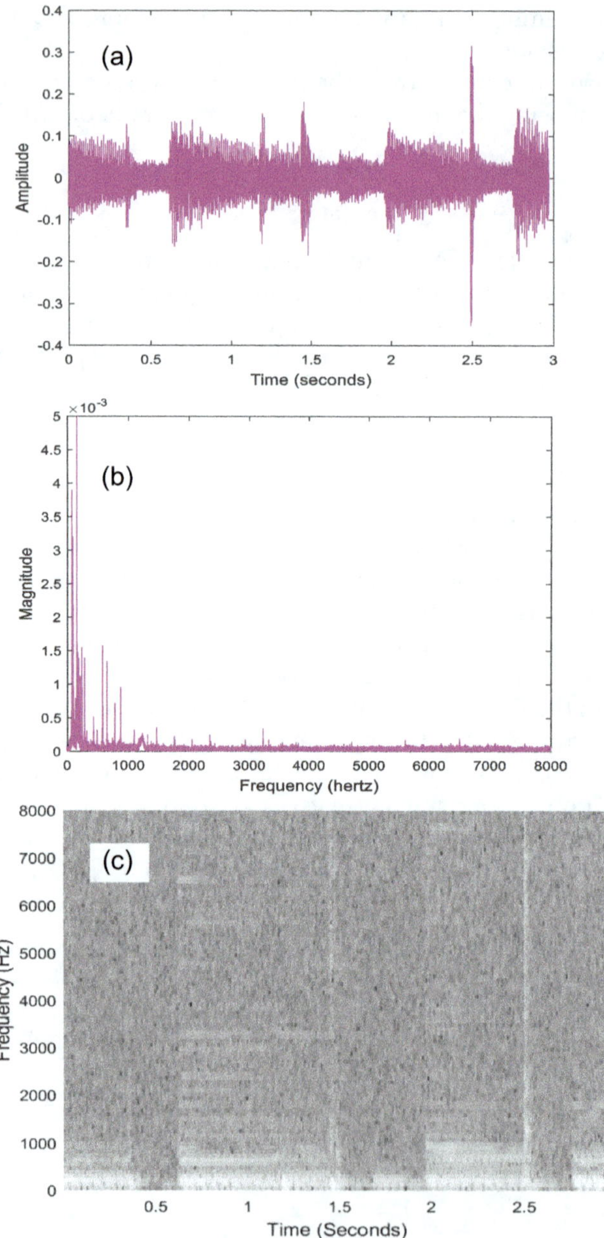

FIGURE 1.3
The time domain waveform (a), frequency domain power spectrum (b), and spectrogram of a 3 s excerpt of solo guitar performance.

$$X[k]=\sum_{n=0}^{N-1}\left[e^{jnl2\pi/N}+e^{-jnl2\pi/N}\right]e^{-j2\pi kn/N}/2=\begin{cases} N/2 & l=k \\ N/2 & l=N-k \\ 0 & l\neq k,N-k \end{cases}, \tag{1.17}$$

So a real-valued sinusoid will give two nonzero frequency components, one at k and one at $N-k$. This implies that, for real valued input signals, the frequency spectrum from $N/2$ to N is the mirror image of the spectrum from 0 to $N/2$. We can also easily show that the spectrum for an input frequency sinusoid with frequency $f+f_s$ is the same as for a sinusoid with frequency f. So if a continuous signal is sampled at a frequency f_s, then the sampled signal cannot reproduce frequencies above $f_s/2$.

In fact, when sampling at a frequency f_s, any signal at a frequency f_c+Nf_s or Nf_s-f_c, for any integer N, will be indistinguishable from a signal at frequency f_c. As an example of this, consider Figure 1.4. A sinusoid of frequency $0.7f_s$ is sampled at frequency f_s. The samples that result could equally well represent a sinusoid of frequency $0.3f_s$. This property is known as *aliasing*, since these signals are aliases of each other.

For this reason, signals should, in general, be bandlimited before sampling, so that they contain no frequency components greater than $f_s/2$. $f_s/2$ is known as the *Nyquist frequency*. This is also a consequence of the Nyquist-Shannon sampling theorem, which states that such bandlimited signals can be (in theory) completely reconstructed when sampled at a frequency of at least f_s.

Modifying and Processing Digital Signals

Up to now, we've been looking at how to represent and analyze signals. But how do we modify them? We start by introducing the difference equation, a formula for computing the nth output sample based on current and previous input samples and previous output samples. A linear, time-invariant digital filter may be given as a difference equation,

$$y[n]=b_0x[n]+b_1x[n-1]+...b_Nx[n-N]-a_1y[n-1]-...a_My[n-M], \tag{1.18}$$

where x is some input signal and y is the output signal. The constants b_0, ... b_N and a_0, ... a_N are known as coefficients or multipliers. Figure 1.5 shows this as a block diagram, with the input signal entering at the top left, and the output signal appearing on the right.

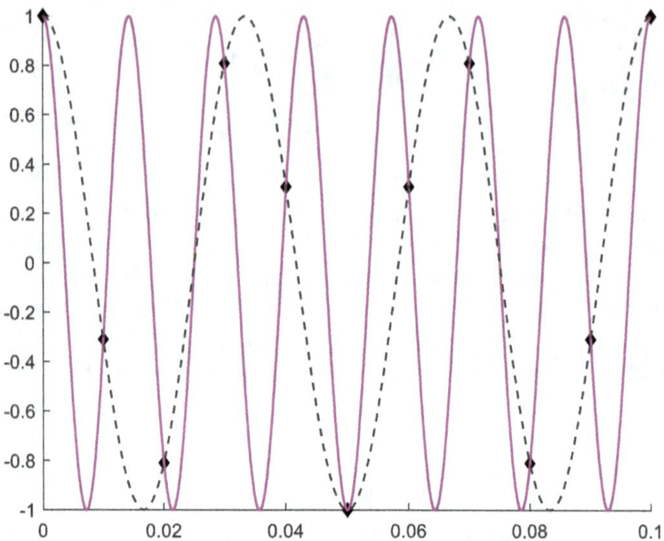

FIGURE 1.4
Two sampled sinusoids, one with frequency $0.3f_s$ and the other with frequency $0.7\,f_s$. They appear identical when sampled at a frequency f_s.

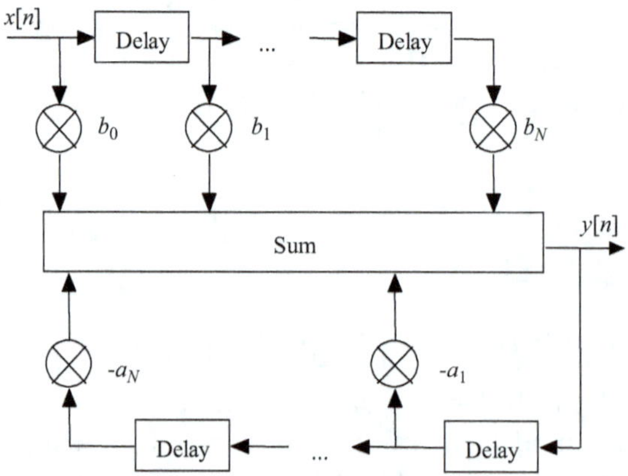

FIGURE 1.5
A time domain block diagram of a digital linear filter.

There are many different ways to represent this system. One of the most important is the impulse response, which describes the output when the input is just a single pulse.

$$h[n] = y[n], x[n] = \delta[n] \equiv \begin{cases} 1 & n = 0 \\ 0 & n \neq 0 \end{cases}. \tag{1.19}$$

Let's quickly look at two cases, $y[n] = x[n] + 0.5y[n-1]$ and $y[n] = x[n] + 2y[n-1]$. They have the impulse responses $h[n] = 1, 0.5, 0.25, 0.125\ldots$ and $h[n] = 1, 2, 4, 8\ldots$ In the first case, the output exponentially approaches zero, but in the second case, the output keeps growing even though the input has stopped. This first case is an example of a stable system, and whenever the impulse response does not exponentially converge to zero, the system is said to be unstable.

Any discrete time signal can be represented as a weighted sum of delayed impulses. Since the filter is linear, the response to that signal is a weighted sum of delayed impulse responses. So the impulse response fully characterizes the system.

The *Z* Transform and Filter Representation

Another important transformation of a discrete sequence is the Z transform,

$$Z\{x[n]\} = X(z) = \sum_{n=0}^{\infty} x[n]z^{-n}, z = e^{j\omega} = e^{j2\pi f/f_s}. \tag{1.20}$$

This converts the signal from being represented in terms of sample number n to representation in terms of complex number z. Note that this is related to the Fourier domain transform, of which we gave the discrete version previously. Note also that we use a normalized frequency, $\omega = 2\pi f/f_s$, so that if the signal has frequency components between 0 and f_s, then ω is between 0 and π, and z is between +1 and −1. This notation for normalized frequency will be used throughout this text.

This transformation has a large number of properties, but for understanding digital filters, the following are most important.

$$\begin{aligned}
&Z\{x_1[n]+x_2[n]\} = Z\{x_1[n]\}+Z\{x_2[n]\} = X_1(z)+X_2(z)\\
&Z\{ax[n]\} = aZ\{x[n]\} = aX(z)\\
&Z\{x[n-1]\} = \sum_{n=-\infty}^{\infty} x[n-1]z^{-n} = z^{-1}\sum_{n=-\infty}^{\infty} x[n-1]z^{-(n-1)} = z^{-1}X(z).
\end{aligned} \tag{1.21}$$

In other words, if we add two signals together, we add their Z transforms together. If we multiply a signal by a constant, we multiply its Z transform by the same constant, and if we delay a signal by one sample, then we multiply its z domain transform by $z-^{1}$. Similarly, if we delay a signal by n samples, we multiply its z domain transform by z^{-n}.

The Z domain representation allows us to redraw the block diagram as shown in Figure 1.6.

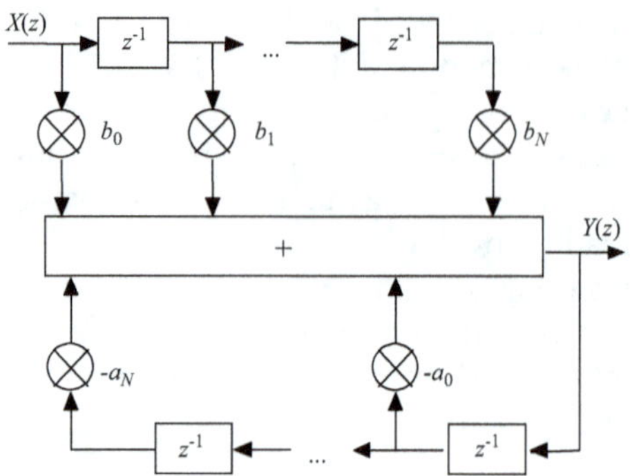

FIGURE 1.6
A z-domain block diagram of a digital linear filter.

And we can transform Eq. (1.18) into the Z domain,

$$Y(z) = b_0 X(z) + b_1 z^{-1} X(z) + \ldots b_N z^{-N} X(z) - a_1 z^{-1} Y(z) - \ldots a_M z^{-M} Y(z). \tag{1.22}$$

Now, we don't need to worry about what X and Y are. Equation (1.22) holds regardless of the exact nature of the input signal. We can use a little algebra to group terms involving X and terms involving Y.

$$\left[1 + a_1 z^{-1} + \ldots a_M z^{-M}\right] Y(z) = \left[b_0 + b_1 z^{-1} + \ldots b_N z^{-N}\right] X(z). \tag{1.23}$$

If $M < N$, we can always add a few zero coefficients to create additional delay terms. So we can now refer to N as the order of the filter. When N is large, we refer to this as a *high-order filter*. The *transfer function* is then defined as,

$$H(z) = \frac{Y(z)}{X(z)} = \frac{B(z)}{A(z)} = \frac{b_0 + b_1 z^{-1} + \ldots b_N z^{-N}}{1 + a_1 z^{-1} + \ldots a_N z^{-N}}. \tag{1.24}$$

We can rewrite this by expressing the polynomials in terms of positive powers of z (this notation is often used in this text) and factoring the polynomials.

$$H(z) = \frac{b_0 z^N + b_1 z^{N-1} + \ldots b_N}{z^N + a_1 z^{N-1} + \ldots a_M} = g \frac{(z - q_1)(z - q_2) \cdots (z - q_N)}{(z - p_1)(z - p_2) \cdots (z - p_N)}. \tag{1.25}$$

This gives rise to yet another way to define the filter; in terms of its poles p_1, p_2,...p_N, zeros q_1, q_2,...q_N and gain factor g. A *pole* is an infinite value of $H(z)$ obtained when z is set so that the denominator $A(z)$ equals zero, and a *zero* is a zero value of $H(z)$ obtained when z is set so that the numerator $B(z)$ has zero value. When poles and zeros are all plotted in the complex plane, this is known as a *pole zero plot.* When all the poles have magnitude less than 1 (they are inside the unit circle), then the filter will be stable. That is, the impulse response described in Eq. (1.19) will decay towards 0. If all the zeros are inside the unit circle, the filter is said to be *minimum phase,* and minimum phase filters produce the minimum delay for a given magnitude response.

An important distinction needs to be made here. When the filter has no feedback, i.e., the current output does not depend on previous outputs, then it can be written as;

$$
\begin{aligned}
y[n] &= b_0 x[n] + b_1 x[n-1] + \ldots b_N x[n-N] \\
H(z) &= b_0 + b_1 z^{-1} + \ldots b_N z^{-N} = g\frac{(z-q_1)(z-q_2)\cdots(z-q_N)}{z^N}.
\end{aligned}
\tag{1.26}
$$

Thus, all poles are located at $z = 0$ and the filter is inherently stable. This is known as a finite impulse response (FIR) filter. Filters containing a feedback path are known as infinite impulse response (IIR) filters.

A filter is characterized by its magnitude response, $|H(\omega)|$, and phase response, $\angle H(\omega)$, as a function of ω, where here we write the transfer function as a function of radial frequency ω rather than z.

To summarize, we have the following representations of a linear digital filter.

- Difference equation
- Block diagram
- Impulse response
- Transfer function
- Poles, zeros, and gain
- Magnitude and phase response.

There are, of course, many other ways to represent the filter, but these are probably the most important. It is also very useful to be able to go back and forth between the different representations, since it allows you to conceptualize the filter in different ways. In Chapter 4, we will look at how various filters are designed, and in Chapter 5, we will look at some audio effects based on filter design techniques.

Digital Filter Example

At this point, it is instructive to look at an example. Consider the following filter,

$$y[n]=4x[n]-4x[n-1]+x[n-2]-y[n-1]-y[n-2]/2. \tag{1.27}$$

This filter's current output is dependent on the current input and the two previous outputs and inputs. We can rewrite this in the Z domain as follows,

$$\begin{aligned} &Y(z)=4X(z)-4X(z)z^{-1}+X(z)z^{-2}-Y(z)z^{-1}-X(z)z^{-2}/2\rightarrow \\ &H(z)=\frac{Y(z)}{X(z)}=\frac{4z^2-4z+1}{z^2+z+1/2}. \end{aligned} \tag{1.28}$$

We can see that it is a second-order filter, since it can be expressed with a transfer function consisting of no more than second-order polynomials in the numerator and denominator,

$$H(z)=\frac{Y(z)}{X(z)}=\frac{4z^2-4z+1}{z^2+z+1/2}. \tag{1.29}$$

By factoring the polynomials, we can rewrite this transfer function in terms of zeros and poles,

$$H(z)=\frac{4\left(z-\frac{1}{2}\right)\left(z-\frac{1}{2}\right)}{\left(z-\frac{-1+j}{2}\right)\left(z-\frac{-1-j}{2}\right)}. \tag{1.30}$$

So we see that there are two zeros, both located at 1/2, and two poles, located at $(-1 \pm j)/2$. The poles are complex and can be written in polar form as $e^{\pm 3j\pi/4}$.

To find the impulse response of this filter, we simply enter a 1 as $x[0]$ in Eq. (1.27), set all previous inputs and future inputs to zero, and calculate the output values. To find the magnitude and phase response, plot $|H(\omega)|$ and $\angle H(\omega)$ as a function of ω. The pole zero plot, impulse response, magnitude (in decibels) response, and phase response for this filter are given in Figure 1.7.

It is clear from Figure 1.7 that this filter will attenuate low frequencies and boost high frequencies, with the maximum boost occurring at a normalized frequency of approximately $\omega = 0.785\pi$. For a sampling rate of 44.1 kHz, this corresponds to a frequency of approximately $f = 44.1 \times 0.785 \times 0.5 = 17.3$ kHz.

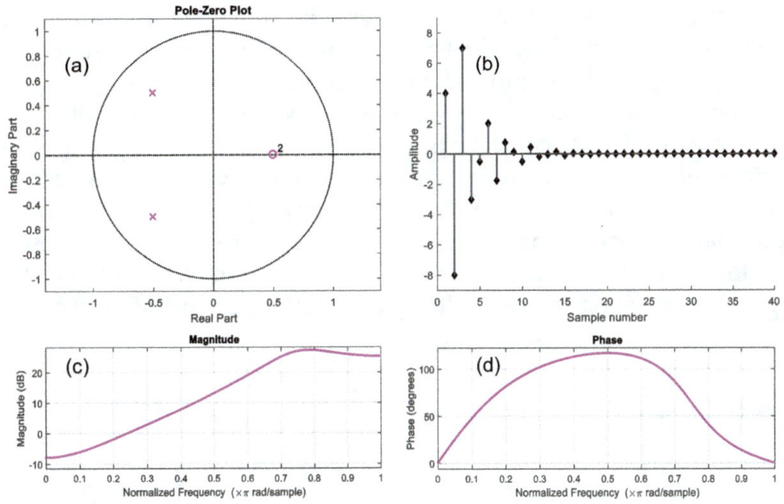

FIGURE 1.7
Pole zero plot (a), impulse response (b), magnitude response (c), and phase response (d) of a digital linear filter.

For very low and very high frequencies, this filter makes only small changes to the phase. However, for input sinusoids with normalized frequency $\omega = 0.5\pi$ (frequency $f = 11.025$ kHz for sampling frequency 44.1 kHz), the output signal will be a sinusoid shifted by about 116°.

Nonlinear and Time-Varying Effects

The filter described by Eq. (1.18) was linear and time invariant. That is, it is linear in the sense that summing two filter outputs gives the same result as applying the filter on the sum of two signals, and scaling the filter output gives the same result as scaling the filter input. And it is time invariant in that delaying the input signal is the same as delaying the output signal. Note that these same properties apply to the Z transformation, which is one reason why it is so useful for working with linear filters.

But many of the audio effects that we will encounter do not have these properties. Dynamic range compression (Chapter 7), for instance, is nonlinear, since applying a gain to the input signal is not the same as applying that same gain to the input signal. And many effects are not time invariant, since they incorporate a low-frequency oscillator (Chapter 3) with an explicit dependence on time.

However, this does not completely invalidate the linear representations. Most of the time varying systems that we will encounter have relatively slow time variation, at least compared to the sampling rate. So we can consider the transfer function and other representations for a snapshot of the system, where time is held constant. Similarly, we can sometimes consider how nonlinear effects act on a given signal level, and many nonlinear effects will have a range of input signal levels where they act like a linear filter. Where appropriate, we will also apply other analysis techniques. The representations for linear, time-invariant filters are simply some of the tools at our disposal for working with almost any audio effect that we will encounter.

Problems

1. Compute the z domain representation of an impulse $x(n)=\begin{cases}1 & n=0\\0 & n\neq 0\end{cases}$, and of a step function, $x(n)=\begin{cases}1 & n\geq 0\\0 & n<0\end{cases}$.
2. Show that IDFT of DFT gives the original signal.
3. Show that the spectrum for an input frequency sinusoid with frequency $f+f_s$, where f_s is the sample rate, is the same as for a sinusoid with frequency f.
4. Explain the difference between *FIR* and *IIR* digital filter designs. Which is computationally less expensive? Which usually exhibits the lowest overall latency? Which best preserves the phase relationships of the original signal? Why?

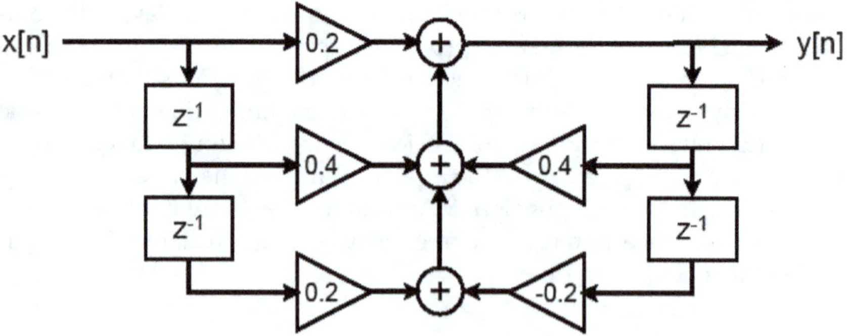

FIGURE 1.8
Example block diagram.

5.

  i. Write the transfer function corresponding to the block diagram in Figure 1.8. Use $y[n]$ to represent the output and $x[n]$ for the input. Is this an FIR or IIR filter? Why?
  ii. From the transfer function, write the frequency response $H(z) = Y(z)/X(z)$.
  iii. What are the values of $H(z)$ at $z = e^{j0} = 1$ and $e^{j\pi} = -1$)? Roughly, how does this filter affect the frequency content of the input signal?

Note

1 In this chapter, brackets are generally used when we refer to functions of discrete, integer samples, and parentheses are used for functions of continuous time.

2

Building Audio Effect Plug-ins

Most of the chapters in this book present the theory and implementation of the major types of audio effects, examining the mathematical principles behind each effect. This chapter describes how to put these principles into practice by creating an audio plug-in.

Audio plug-ins are the most common way of implementing audio effects in software. A typical plug-in is a self-contained block of code, which is compiled to run on a particular processor and operating system, and can be used within an audio software environment. This chapter will examine the process of creating plug-ins using the *JUCE* (*Jules' Utility Class Extensions*) programming framework, which can be used to create plug-ins for many different software platforms.

Audio Plug-in Basics

Audio plug-ins are usually designed for use within digital audio workstations (DAWs). To ensure compatibility across different DAWs, several industry-standard plug-in formats have been developed. These include Steinberg's *VST* (Virtual Studio Technology) format, widely supported in nearly all professional audio software; Apple's *AudioUnit* format, supported on most Mac programs; and the *AAX* (Avid Audio Extension) format by Digidesign/Avid. Each format provides similar functionality, typically including ways of passing audio into and out of the plug-in, negotiating sample rates and number of channels, and querying and setting user-adjustable parameters for the effect.

Programming Language

The most common programming language for writing audio plug-ins, by far, is C++, though other languages may be used. The code will be compiled to run on a specific processor and operating system. However, the same code can typically be compiled to run on any hardware and operating system, as long as the code does not use any OS-specific functionality.

DOI: 10.1201/9781003593942-2

This book is not intended to cover programming fundamentals or as an introduction to C++. Examples in the remainder of this chapter will be presented in C++ with the assumption that the reader has a basic familiarity (though not necessarily significant expertise) with the language.

Plug-in Properties

The essential task of an audio plug-in is to receive an input audio signal, apply an effect to it based on some control parameters, and produce an output audio signal. In some cases, including virtual instrument or synthesizer plug-ins, no audio input is used, and the output may be produced in response to *MIDI* (Musical Instrument Digital Interface) messages. However, the effects in this book all assume an input and an output.

Audio effect plug-ins process audio. They receive digital audio and process it through to their outputs. Their function is similar to that of hardware audio processors. They can also be chained together. That is, most hosts will allow the audio output from one effect to be used as input to another effect.

In contrast, virtual instrument plugins generate audio. They can act as standalone software synthesizers, samplers, or digital musical instruments. They typically use MIDI messages to control instrument parameters. Some effect plug-ins also accept MIDI input. In fact, the use of MIDI for control is not the distinguishing factor between effects and instruments. Think of the instruments as being similar to the effects, but without the need for digital audio input.

There are also other types of audio plug-ins. Monitoring effects are a special type of effect that provides feedback about an input signal without processing audio. Typically, they present a visualization of the input signal, like a level meter. Think of these as having digital audio input, but no need for any output signal.

Finally, it's worth mentioning MIDI effects, which do not really operate on audio at all. They have MIDI input and MIDI output. So they could, for instance, apply pitch shift on a MIDI data stream. The output MIDI messages could then be sent to other virtual instruments or hardware devices.

Our focus is on just the audio effect plug-ins. Important properties of such a plug-in include the *number of channels* it supports and the allowable *sample rates*. Many effects can operate with different numbers of channels, but others will require specific channel configurations. For example, a stereo panning effect would need at least two output channels, but it could take one or two input channels for stereo or mono input. Some effects may have restrictions on the sample rates they support, though it is useful wherever possible to write plug-ins that operate at any sample rate. Plug-ins will also define one or more user-adjustable *parameters,* which can typically be changed either through a standard interface provided by the DAW or by a custom GUI created by the plug-in author.

The Plug-in Host

A plug-in host is a software application or hardware device in which the plugins run. It typically presents plugin user interfaces and routes digital audio and MIDI to and from plugins.

The most well-known form of host for audio effect plug-ins is the digital audio workstation, or DAW. These are applications and devices used for a wide range of audio production tasks, such as recording, editing and rendering multitrack music or soundtracks. However, audio effect plug-ins might also be used in a stand-alone host which lacks some DAW features but has been optimized for use in live performance. They might also be used in video editing tools, or in game engines.

The JUCE Framework

The example code for this book uses the *JUCE* (Jules' Utility Class Extensions) environment, which provides a cross-platform, multi-format method for building audio plug-ins. JUCE was created in 2004 by Jules Storer. It is an open source cross-platform C++ codebase for developing desktop and mobile applications and plug-ins. It is used to write software so that it will run with the same user experience on many platforms (Windows, macOS, Linux, iOS, Android...) and in many formats. JUCE furthermore provides a cross-platform set of graphical user interface (GUI) controls and a very large library of useful C++ classes covering commonly used features by apps and plug-ins, such as graphics, audio, XML parsing, networking, cryptography, multi-threading... Thus, it also reduces the number of third-party libraries needed in a project.

But the real reason why we focus on JUCE here is that it is the *most widely used framework for audio application and plug-in development*. As mentioned, JUCE has a very large number of built-in libraries, but this is especially true for those related to audio functionality. JUCE has support for audio devices (CoreAudio, JACK, DirectSound) and MIDI playback, DSP building blocks, polyphonic synthesizers... and built-in readers for common audio file formats (WAV, AIFF, FLAC, MP3, Vorbis...). It also comes with wrapper classes for building all major audio plugin formats. Since all platform and format-specific code is contained in the wrapper, one can build for almost any plug-in format & platform from a single codebase.

JUCE is free for use in open-source projects. Documentation and download links can be found on its website: https://juce.com/.

VST – Virtual Studio Technology

The most popular audio plugin format is VST, which stands for Virtual Studio Technology. VST was created by Steinberg (now owned by Yamaha) in 1996. VST is an audio plug-in software interface. That is, it is an audio plugin standard that allows virtual instruments and effects (and more, as we will see) to be integrated into digital audio workstations. It is the main industry standard for audio plug-ins.

VST plug-ins usually run within digital audio workstation (DAW), though stand-alone VST plug-in hosts also exist. VST plug-ins are intended to provide additional functionality to the host, and they often use digital signal processing to simulate traditional recording studio hardware. VSTs usually have graphical user interfaces that display controls similar to physical switches and knobs on audio hardware.

Most VSTs are either instruments (VSTi) or effects (VSTfx), though other categories exist, such as the monitoring effects mentioned previously.

Theory of Operation

JUCE audio plug-ins are divided into two components: a *processor* that handles the audio calculations and an *editor* or GUI that lets the user interact with the plug-in. The processor provides several functions: a *callback function* that the DAW calls every time it needs a new block of audio samples; methods for getting and setting effect parameters; and initialization and cleanup routines. The editor provides graphical controls for the user to see and change the parameters. Most DAWs will provide a generic editor when the plug-in does not define its own.

Callback Function

The most important task for an audio plug-in is receiving and processing audio samples. How does the plug-in know how many samples to process, and when to process them? If the effect is operating in real time, it is clearly impossible to wait for the entire audio signal to arrive before applying the effect. Instead, audio needs to be processed in small *blocks* or *buffers* of samples as it comes in. To receive blocks of samples, plug-ins implement a *callback function*, a function which the DAW calls every time it has new audio to process. Thus, it is always the DAW, and *not* the plug-in, which determines how many audio samples to process and when. The advantage of this arrangement is that the plug-in author never needs to be concerned with where the audio samples come from, when they should arrive, or where they go after the effect has been applied. The author simply needs to write a callback function that processes as many samples as requested by the DAW.

When the DAW runs the callback function, it will provide several pieces of information. These include the *buffer size* (how many audio samples to process), the *sample rate*, the *number of input and output channels*, and a *buffer* (region of memory) containing the input audio. The host will also provide a buffer in which the audio output should be stored. In JUCE (as in many plug-in formats), the plug-in is expected to put its output in the same buffer where the input samples were found.

When the callback function finishes, it returns control to the DAW, which decides what should happen to the processed samples. The callback function will be called again when there are more samples to process. The buffer size used by the DAW partly determines the overall *latency* (delay) from input to output; smaller buffer sizes produce a low delay but increase the risk of *underruns* (gaps) if the computer cannot respond quickly enough. A typical conservative buffer size would be 512 samples; a high-performance buffer size might be as small as 32 samples. At a 44.1 kHz sample rate, a buffer size of 32 means that the callback function would run over 1,300 times per second!

Managing Parameters

Nearly every audio effect will have one or more user-adjustable parameters. For example, a parametric equalizer plug-in might let the user change the center frequency, the gain, and the Q of the filter. Most plug-ins provide a user interface to allow the user to adjust parameters. Every operating system and plug-in format provides different routines for managing a GUI, but there are several common requirements. In particular, the plug-in must provide a set of functions for the user to see or change the parameters, and there must be a way for the callback function to discover the current parameter values.

JUCE provides a convenient class for almost any parameter that the plug-in would expose to the DAW, the `audioProcessorParameter`. JUCE also provides several *methods* (C++ object functions) related to managing parameters, described in detail in the example in the next section. Two of the important methods are `getParameterTree()` and `setParameterTree()` within the audio processor object (see Figure 2.1). When the DAW calls `getParameterTree()`, the group of parameters managed by that audio processor is returned. Parameter values are most often in the form of floating-point numbers (C++ type `float`). It is up to the plug-in to define what they mean (for example, a parametric equalizer might have frequency as parameter 0, gain as parameter 1, and Q as parameter 2). With `setParameterTree()`, the group of parameters managed by the audio processor can be set. The host provides an index of the parameter and a floating-point value to which it should be set. The plug-in stores this information so the audio callback function can access it later. In particular, the callback function, `processBlock()` accesses the values of these variables to discover the current parameter settings.

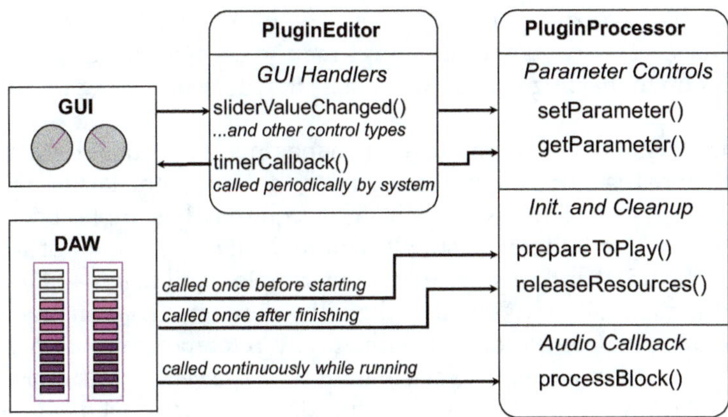

FIGURE 2.1
Basic components of a JUCE audio plugin and their relationship to the digital audio workstation (DAW).

Initialization and Cleanup

Before a plug-in can process audio, certain initialization tasks must be performed. For example, in a delay plug-in, memory for delay buffers might need to be allocated; for an equalizer plug-in, filter coefficients might need to be calculated and internal variables holding previous samples might need to be initialized to 0. For plug-ins with low-frequency oscillators, the phase of the oscillator may need to be initialized.

Every plug-in format will provide a method for initializing the plug-in. In JUCE, basic initialization can be performed in the *constructor* of the audio processor object, which runs once when the plug-in is first loaded. JUCE also provides a method called `prepareToPlay()`, which runs immediately before audio processing begins. By allocating resources just before the audio starts, the plug-in only uses resources while actively running.

All resources that are allocated by the plug-in will eventually need to be released. JUCE provides a method called `releaseResources()`, which runs immediately after the host stops processing audio. This method should be used to free any resources allocated in `prepareToPlay()`. It is safe to assume that the audio callback function will never run after a call to `releaseResources()`. Similarly, the counterpart to the C++ constructor is the *destructor*, which runs once when the user removes the plug-in from the host environment. Anything that is allocated in the constructor should be freed in the destructor.

Preserving State

Each time the DAW calls the plug-in's callback function, it will provide only a small block of input samples to be processed. The plug-in often depends on previous state information to know how to process these samples.

For example, phases of oscillators, pointers within circular buffers, and previous values of input and output samples may be needed to calculate the output. It is up to the plug-in to save any state that it needs during the callback function.

To understand the importance of managing state, consider a simple effect where the current output is equal to the sum of the last two inputs: $y[n]=x[n]+x[n-1]$. Suppose that the host requests 512 samples beginning at sample N. It will therefore supply a buffer containing 512 input samples $x[N]$ to $x[N+511]$ and require 512 output samples $y[N]$ to $y[N+511]$.

At first glance, it may seem as if the callback function has enough information to calculate the output without any reference to what has happened before. But what about calculating $y[N]$, the very first sample in the buffer? We have $y[N]=x[N]+x[N-1]$, but $x[N-1]$ was supplied *last time* the callback was run, and that value is no longer available. The plug-in must therefore use a separate instance variable to remember what has come before. Here, a single `float` for each channel would be needed to remember the previous sample. The float should be declared inside the class, so its value is preserved across calls to the callback function.

Setup

This section briefly covers how to set up the tools to begin audio plug-in development.

Required Software

To build the plug-ins that come with the book, the following software is required:

1. The *JUCE (Jules' Utility Class Extensions)* C++ library by Julian Storer. JUCE runs on nearly every platform, including Mac, Windows, and Linux.
2. An Integrated Development Environment, or IDE, with a C++ compiler. This is an application with lots of tools for software development, like a code editor and a debugger. Common ones include Visual Studio for Windows, Xcode for macOS and iOS, Android Studio, and Code::Blocks. They are all generally available as free downloads, sometimes after a registration process. Here, we will do everything in Windows with Visual Studio, but the steps should be similar with different development environments and operating systems.

3. To build VST plug-ins, the VST3 SDK (2.x versions are still supported but no longer recommended) is required. This is bundled with JUCE. Optionally on Mac, AudioUnits can be created instead. Whether you use VST or AudioUnits, your program in JUCE will be written identically.
4. A suitable digital audio workstation (DAW) or other host environment in which to test the plug-ins. JUCE ships with a VST plug-in host, which is sufficient for the sort of testing we need. On the Mac, the Xcode Audio Tools provide the *AU Lab* program, which is a simple, lightweight way of testing AudioUnit plug-ins. Several other cross-platform options are available, including Audacity and Reaper.

Download the Tools

If you do not have one already installed, you will need to download an IDE. When installing, you should be able to do a basic installation, needing only C++.

Now download JUCE from https://juce.com/download/. Unpack the JUCE folder and put it in some convenient location on your computer.

Setting Up the ProJucer

In JUCE, new projects are configured using a program called the *Projucer.* The Projucer is JUCE's tool for creating and managing JUCE projects. Once files and settings for a JUCE project are specified, it automatically generates third-party project files to allow the project to compile natively on each target platform. The Projucer also has a code editor, integrated GUI editor, and wizards for creating new projects and files.

Depending where you download JUCE from, it may or may not come with a pre-built version of the Projucer. If not, then the Projucer needs to be compiled and built on your system before anything further can take place. Within the main JUCE folder, Projucer project files for each of the major development environments (Visual Studio, Xcode, Linux Makefiles, etc.) can be found in extras/Projucer/Builds. Open the project in your development environment and build it. It should compile without errors if JUCE was correctly installed. Run the Projucer once it finishes compiling.

A common problem is that the JUCE libraries, or modules, are not easily found. The steps to set their location are shown in Figure 2.2, and are as follows. First, the Global paths may need to be set in the Projucer app. Navigate to the menu item *File -> Global Paths* on Windows or Linux, or *Projucer -> Global Paths* on macOS. Open global paths and set them to the right locations (1), if they are not set there by default. Now close the global paths window. Then, click in the *Modules* section on the left (2). Click

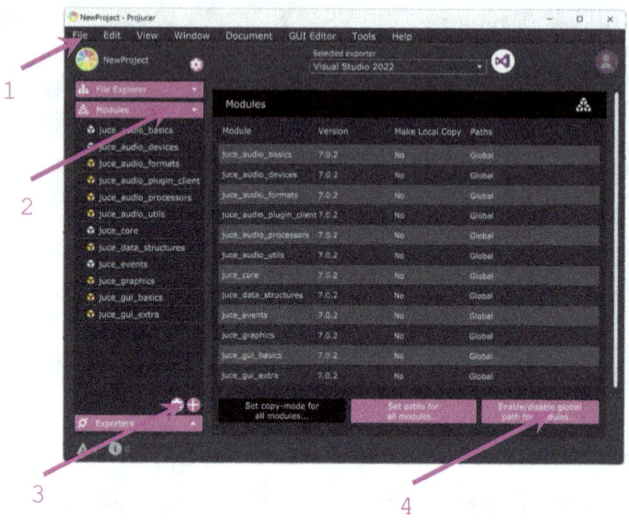

FIGURE 2.2
The steps in the Projucer for setting the global path for modules.

on the cog wheel and scroll down (3). Select *Enable/disable global path for modules...* and ensure that the Global Path is used for all modules (4).

The Projucer can now be used to create new JUCE projects, view tutorials, run examples, and much more. It is also possible to include the JUCE modules source code in an existing project directly, or build them into a static or dynamic library, which can be linked into a project.

Opening Example Plug-Ins

This book comes with many example audio plug-ins, which can be opened in the IDE or your choice. Within each plug-in directory, look in the `Builds` directory for a version specific to your IDE. If you do not find one, open the `.jucer` file in the Projucer and create a new target for your platform. Following the instructions in this chapter, you should be able to compile the plug-in and get it running in your audio host environment.

Example: Building a Gain Effect in JUCE

This section will describe the implementation of a simple plug-in in the JUCE environment. We will take as an example a basic gain effect, with user-adjustable control for a gain parameter that multiplies the incoming signal to produce the outgoing signal. It will demonstrate many of the concepts that have been discussed, including the management of parameters and preservation of state between callbacks.

The complete code for this effect can be found in the materials that accompany the book.

Creating a New Plug-in in JUCE

Once the Projucer is running, select *New Project* from the *File* menu, which brings up the *New Project* window. The Projucer will automatically generate different starter files depending on what you want to develop. Enter a project name (here, we call our project HelloWorld) and set the Project Type to Plug-In ->Basic (see Figure 2.3).

Click Create Project to make the JUCE project for your new plug-in. The Projucer displays a new screen now. On the left there is an accordion (stacked list user interface element), with three headings,

- File Explorer
- Modules
- Exporters.

The next step is to make an export target for your specific Integrated Development Environment. By default, the Projucer should create a target

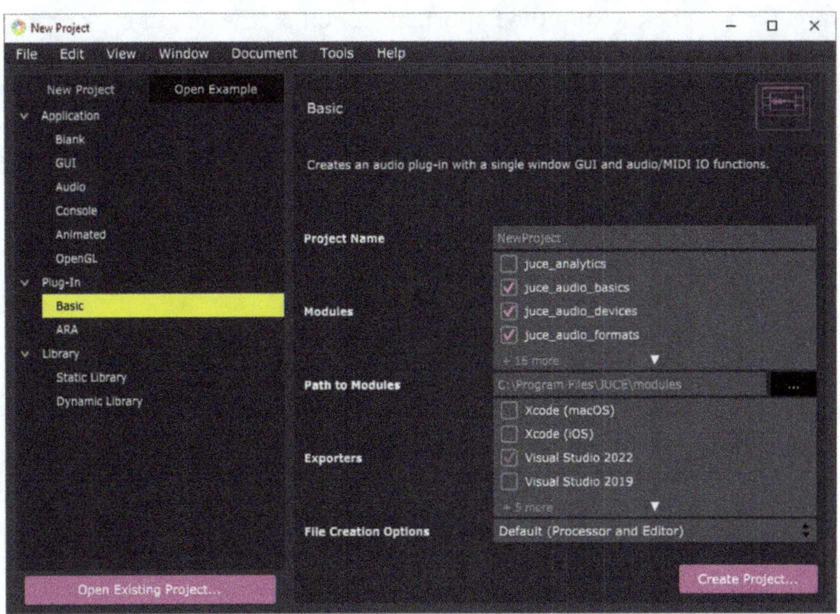

FIGURE 2.3
The new project window, for creating a new audio plug-in with the Projucer.

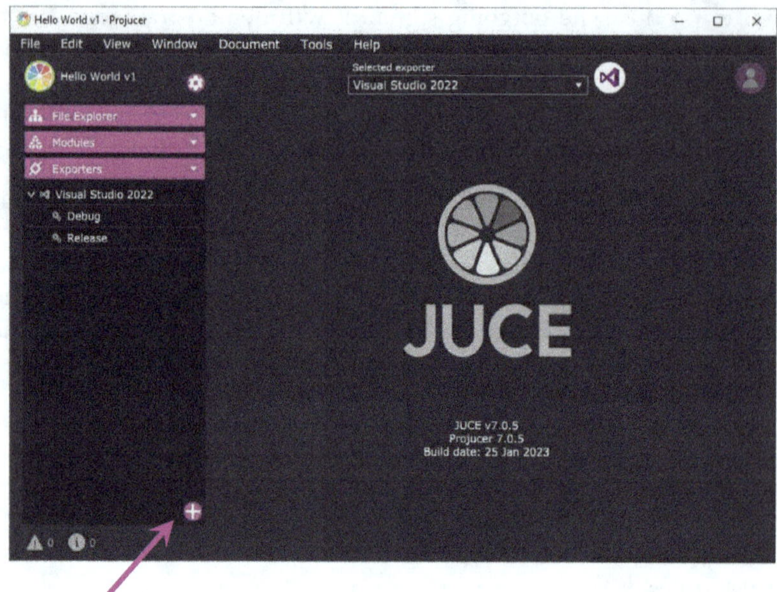

FIGURE 2.4
Adding an export target in the Project window.

for the IDE on your platform. If not, or if you want to make an additional target, right-click within the Exporters section on the left, or click on the plus symbol in that section, to create a new export target (Figure 2.4).

Selecting the export target from the list on the left-hand side of the window brings up settings specific to the platform. Most of these options can be left at their default values. Similarly, module settings can also usually be left at their defaults.

Now click on the cog wheel near the top left corner to get the *Project Settings* window where you can edit the details of the plug-in (Figure 2.5).

There are several fields worth mentioning here;

- **Project name**: the name of the project. This is used for naming various methods and files when the Projucer generates code. They can be renamed, but it's useful to pick a preferred name at this stage.
- **Project version**: the version number of your plug-in (the default value is fine for new plug-ins)
- **Company name**: the name of your company or organization, as you want it to appear in the audio host environment
- **Bundle identifier**: a unique identifier for the plug-in. The format looks like a reversed web address, which identifies the organization and the plug-in name.

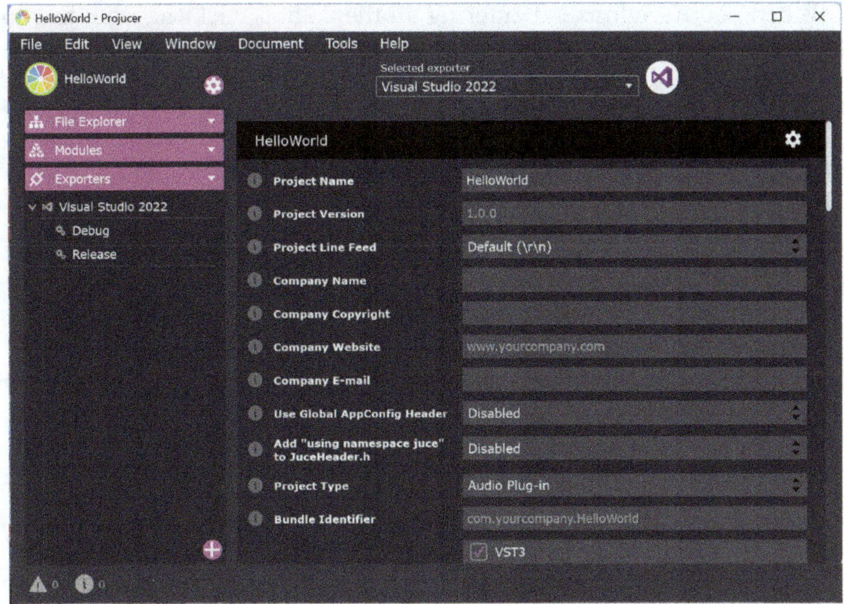

FIGURE 2.5
The Project Settings window.

- **Plugin formats**: these boxes select which format of plug-in to build. On most platforms, including Windows, VST3 is the most common format; on Mac, AudioUnit is the usual selection. Selecting those two checkboxes is a common option.
- **Plugin Name**: the name of your plug-in as you want it to appear in the host environment, usually the same as the Project Name
- **Plugin description**: a short description of your plug-in
- **Plugin manufacturer**: company name, similar to *Company Name* above
- **Plugin manufacturer code**: a four-letter code identifying your company or organization. Use a single code consistently across all the plug-ins you develop.
- **Plugin channel configurations**: pairs of numbers which identify valid configurations of input channels and output channels. For example, `{1,1}, {2,2}` indicates a plug-in that can take mono input and mono output, or stereo input and stereo output. The default is to leave this field blank.

Edit these settings as you wish, but make sure to give the plug-in an appropriate name and check VST3 and/or AU for the plug-in format. The other settings can usually be left at their defaults.

When you have finished entering all the details, click the icon next to *Selected exporter* at the top of the window. This should bring up the project within your IDE, where it can be built using the same procedure you normally use to compile projects. On the Mac, building the project will automatically copy the resulting plug-in to the system plug-ins directory (`~/Library/Audio/Plug-Ins/Components/`) where it will be found by audio software including AU Lab (which can be downloaded from Apple at https://www.apple.com/apple-music/apple-digital-masters/docs/au_lab.zip). On Windows, your DAW may need to be manually set to look for your plug-in. In many cases, restarting the audio host program (AU Lab, Audacity, etc.) is necessary every time the plug-in is recompiled.

File Overview

Juce plug-ins contain four source files by default:

- **PluginProcessor.h**: the header file for PluginProcessor.cpp; defines the C++ object whose code is filled out in PluginProcessor.cpp. Includes declaration of methods and object variables.
- **PluginProcessor.cpp**: the main file where the audio processing is done.
- **PluginEditor.h**: the header file for PluginEditor.cpp, containing the definition of the C++ object for the editor.
- **PluginEditor.cpp**: contains the code for the graphical user interface for the plug-in.

In your own projects, for more complex effects, you might find occasion to add extra source files to the project. But most effects can be implemented in just these four. We will examine each file in turn. Not all lines of each file are included in this text, so please refer to the accompanying materials for the complete source.

These files contain a lot of auto-generated code, but you don't need to do anything about most of this code yet. To start we will focus on the PluginProcessor.cpp file, so open up this file in your IDE. This is where most of the audio-related code is located.

Scroll down until you reach the method called `processBlock()`.

When JUCE auto-generated the project files, a buffer array was created. The `processBlock()` method takes an argument of type `juce::AudioBuffer<float>`. This is a multidimensional array of samples that changes depending on the number of channels and the block size set in your DAW. So if your block size is set to 512 samples and your plugin processes two channels of audio, there will be two arrays of length 512. We will use a `for` loop to change the data in this array by iterating through the array for each channel and changing the value of the samples.

Now edit the code at the end of `processBlock()` so it looks like this:

```
for (int channel = 0; channel < getTotalNumOutputChannels(); ++channel)
{
  auto* channelData = buffer.getWritePointer(channel);
  for (int sample = 0; sample < buffer.getNumSamples(); sample++)
  {
    channelData[sample] = channelData[sample];
  }
}
```

This loops through channels of input data. There could be any number of channels, but the most common situation is dealing with stereo audio, where channel 0 is the left channel and channel 1 is the right channel.

Currently, for every channel, this iterates through the audio buffer but does not change anything. The plug-in should compile and build successfully, but passing an audio signal as input to the plug-in will produce the same signal at the output.

This general structure is the starting point for lots of different audio effect plug-ins.

Now let's use the inner `for` loop to change the value of each sample in the array.

Let's add a floating point gain variable inside `processBlock()` and set its value to something less than 1. We will then multiply every sample by this gain.

```
float gainValue = 0.1f;
for (int channel = 0; channel < getTotalNumOutputChannels(); ++channel)
{
  auto* channelData = buffer.getWritePointer(channel);
  for (int sample = 0; sample < buffer.getNumSamples(); sample++)
  {
    channelData[sample] = gainValue * channelData[sample];
  }
}
```

First, this assigns a variable `channelData` as the write pointer of the audio buffer. It then multiplies the input by a gain equal to 0.1. The input is written back into channel data. The `for` loop then increments the input to the next sample, and repeats.

So this code multiplies every sample value in the audio buffer by a gain of 0.1. When this plugin is placed on a track in your DAW, it will reduce the level of the incoming signal.

Let's try this out now in your DAW. First build the project. This will create a VST3 file in the project directory usually in a location like Builds ->VisualStudioXX->x64->Debug->VST3. Open your DAW of choice and set this folder as a VST location, or move the built file to your VST directory. You should hear that it has applied a gain reduction to any audio used as input to the plug-in.

However, the user has no control over this gain reduction. Let's give the plug-in a user interface.

The `GenericAudioProcessorEditor` is a user interface component that displays the parameters of an AudioProcessor as a basic list of sliders, combo boxes, and switches. It will also automatically lay out the controls on an interface. Here, we will use it to greatly simplify the user interface design and control, allowing us to focus mainly on the audio-related aspects.

For our plugin to use this generic interface, we replace the line

```
return new HelloWorldAudioProcessorEditor (*this);
```

in the PluginProcessor.cpp file with

```
return new juce::GenericAudioProcessorEditor(this);
```

We then open the PluginProcessor.h file, and add this variable to the private section.

```
juce::AudioParameterFloat* gainParam;
```

In the constructor in PluginProcessor.cpp, add a parameter to the audio processor.

```
addParameter(gainParam = new juce::AudioParameterFloat("gain","Gain",0.0
  f,1.0f,0.0f));
```

where

- *gainParam* references the AudioParameterFloat we created in PluginProcessor.h.
- *gain* is the slider's ID.
- *Gain* will be shown next to the slider in the DAW.
- Float values set minimum, maximum, and default values for the slider.

Since we are using the generic interface, it will automatically add a slider to the user interface, which can be used to control this audio parameter. However, the slider doesn't yet do anything. So let's modify the code in the processBlock to be

```
float gainValue = gainParam->get();
```

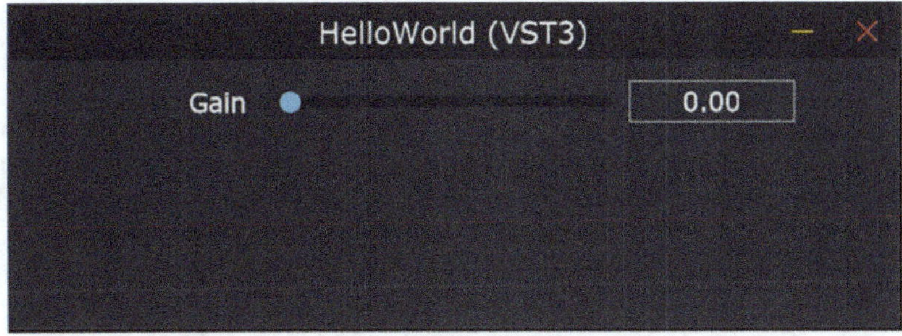

FIGURE 2.6
The interface of the HelloWorld plug-in.

Build and run it again. Now we have user control of the gain slider, as shown in Figure 2.6, so that the user can change the volume of incoming audio data.

Here, we created a basic gain plugin. This isn't very exciting, but it shows the process of developing and testing audio plugins using JUCE. More complicated plugins are made mainly by modifying the process block and adding more UI elements to further control the effect of these modifications.

Summary

This section has examined the code for an example plug-in that implements a simple gain control. Four source files were required, two related to the audio processor and two related to the user interface. The most important task for the audio processor is to implement the *audio callback* function `processBlock()`, which the system calls every time it needs a new block of audio.

Since audio is processed in small blocks, it is necessary to remember the state of the effect between callbacks. So for more complex effects than the simple gain example given here, *instance variables* should be declared in PluginProcessor.h to record information that persists beyond the duration of `processBlock()`.

The other important task was the management of effect parameters. Ordinarily, this would require careful coordination between all four source files; parameters to be declared in PluginProcessor.h alongside instance variables, controls to change their values declared in PluginEditor.h, and methods in both PluginProcessor.cpp and PluginEditor.cpp for getting and setting parameters. However, with the use of the generic user interface, the editor files are not needed at all. We can focus just on the audio-related content in the plugin processor files.

Conclusion

This chapter has shown how to put the principles of building audio effects into practice through the creation of audio plug-ins. The JUCE framework is a convenient environment for creating audio plug-ins since it allows the same code to be compiled into multiple plug-in formats on multiple operating systems. However, the general principles of writing audio callback functions, managing parameters, allocating resources, and creating a user interface are similar across many plug-in formats. The example code included with the book offers templates from which you can start building your own effects, including examples from most chapters and a blank plug-in into which you can add your own code.

Further Reading

Excellent introductions to C++ programming can be found in many sources, including [5–7]. An audio-focused introduction to C++ is included in [8], which also goes into detail on many aspects of audio programming not covered in this text.

There are a large number of texts dedicated to audio programming, each one focusing on different aspects. For working with JUCE, the 'JUCE framework tutorials' videos[1] by the Audio Programmer are especially helpful and popular. And somewhat complementary to aspects of this text are the two texts by Will Pirkle on designing audio effects [9] and virtual instruments [10] and the video series on C++ audio programming on embedded systems [182]. However, machine learning is generally outside the scope of both this text and those resources. For the development of audio plugins that use AI or machine learning, the reader is recommended to consult [11]. Finally, recent useful resources are available for programming with many audio-specific development environments and tools, such as PureData [12] and the Web Audio API [13].

Note

1 https://www.youtube.com/playlist?list=PLLgJJsrdwhPxa6-02-CeHW8ocwSwl2jnu

3

Delay Line Effects

Delay lines are the fundamental building blocks of many of the most important effects. They are rather easy to implement, and only small changes in how they are used allow many different audio effects to be easily constructed. In this chapter, we look at some common effects that are built using delay lines.

Delay

Delay is a simple effect with powerful applications. In the simplest case, adding a single delayed copy of a sound to itself can enliven an instrument's sound in a mix or, at longer delay times, allow a performer to play a duet with him or herself. Many familiar effects, including chorus, flanging, vibrato, and reverb, are also built on delays.

Theory

Basic Delay

The basic delay plays back an audio signal after a specified *delay time*. Depending on the application, the delay time might range from a few milliseconds to several seconds or longer. Figure 3.1 shows a block diagram of the basic delay. It is common to mix the delayed output with the original input, thereby producing two copies of the sound. For this reason, the basic delay is also sometimes known as an *echo* effect (though as we will see, the perception of echo also depends on the delay time).

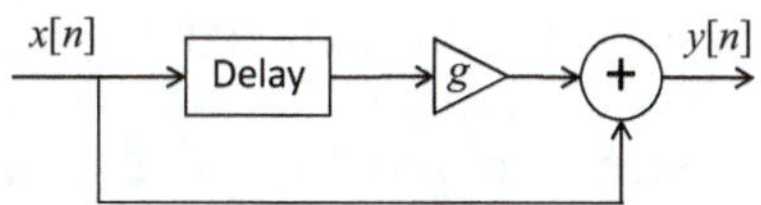

FIGURE 3.1
Diagram of the basic delay unit, or an echo device. The box marked "Delay" is commonly known as a "delay line."

DOI: 10.1201/9781003593942-3

We can express the output audio samples $y[n]$ as a function of the input samples $x[n]$, the delay time N (expressed in samples) and the gain g of the delayed signal:

$$y[n] = x[n] + gx[n-N], \tag{3.1}$$

Delay is a *linear, time-invariant* effect. To show linearity, consider two signals $x_1[n]$ and $x_2[n]$. Let $y_1[n]$ and $y_2[n]$ be the delayed output of each signal individually. If we delay the sum of the signals, we find that the output is the sum of the individual delayed signals:

$$\begin{aligned} y[n] &= (x_1[n]+x_2[n])+g(x_1[n-N]+x_2[n-N]) \\ &= (x_1[n]+gx_1[n-N])+(x_2[n]+gx_2[n-N]) = y_1[n]+y_2[n], \end{aligned} \tag{3.2}$$

Time-invariance implies that shifting the input in time produces an identical shift in the output. Consider taking the delay of $x_d[n]=x[n-M]$ for some number of samples M:

$$y_d[n] = x[n-M] + gx[n-M-N] = y[n-M], \tag{3.3}$$

Using the Z transform, we can also find the frequency response of the basic delay:

$$Y(z) = X(z)+gz^{-N}X(z) \Rightarrow H(z) = \frac{Y(z)}{X(z)} = 1+gz^{-N} = \frac{z^N+g}{z^N}, \tag{3.4}$$

Since the transfer function $H(z)$ has no poles outside the unit circle, the basic delay must be *stable* in all cases, i.e., a bounded input will always produce a bounded output.

Delay with Feedback

The simple "feedforward" delay is limited in application, producing only a single echo. Most audio delay units also have a feedback control (sometimes called regeneration), which sends a scaled copy of the delay output back to the input, as shown in Figure 3.2. Feedback causes the sound to repeat continuously, and assuming a feedback gain less than 1, the echoes will become quieter each time. Though the echoes are theoretically repeated forever, they will eventually become so quiet as to be below the ambient noise in the system and thus be inaudible.

To find the time-domain difference equation for delay with feedback, it helps to consider the signal $d[n]$ at the output of the delay line:

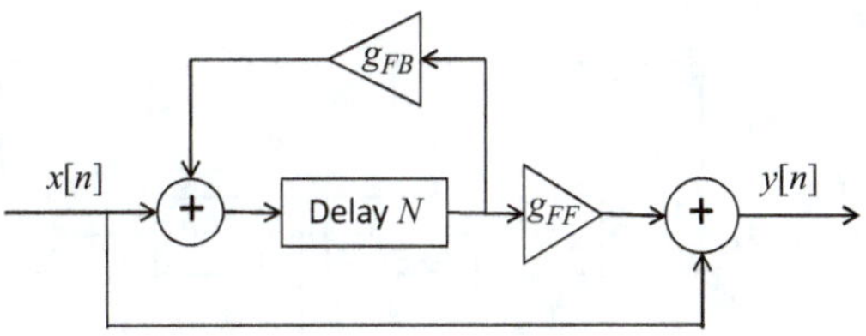

FIGURE 3.2
Diagram of the basic delay unit with feedback.

$$y[n] = x[n] + g_{\text{FF}} d[n] \quad \text{where} \quad d[n] = x[n-N] + g_{\text{FB}} d[n-N]. \tag{3.5}$$

The form of $d[n]$ is similar to the delayed output signal $y[n-N]$, which lets us substitute and write $y[n]$ directly in terms of $x[n]$:

$$\begin{aligned} & y[n-N] = x[n-N] + g_{\text{FF}} d[n-N] \\ & d[n] = \frac{g_{\text{FB}}}{g_{\text{FF}}} y[n-N] + \left(1 - \frac{g_{\text{FB}}}{g_{\text{FF}}}\right) x[n-N] \\ & \Rightarrow y[n] = g_{\text{FB}} y[n-N] + x[n] + (g_{\text{FF}} - g_{\text{FB}}) x[n-N]. \end{aligned} \tag{3.6}$$

Taking the Z transform, we can find the frequency-domain transfer function:

$$H(z) = \frac{Y(z)}{X(z)} = \frac{1 + z^{-N}(g_{\text{FF}} - g_{\text{FB}})}{1 - z^{-N} g_{\text{FB}}} = \frac{z^N + g_{\text{FF}} - g_{\text{FB}}}{z^N - g_{\text{FB}}}. \tag{3.7}$$

Thus, the system has poles at the N complex roots of g_{FB}. Like the basic delay, the delay with feedback is linear and time-invariant. The above transfer function implies the effect will be stable whenever the poles are inside the unit circle, that is when $|g_{\text{FB}}| < 1$. This result aligns with intuition, in that only when the feedback gain is less than 1 will the echoes grow softer over time.

Other Delay Types

Slapback Delay

A *slapback delay* is identical to a basic delay without feedback and a relatively short delay time, typically between 60 and 150 ms. The delayed copy is perceived as a separate sound which appears immediately after the original sound. A longer delay, where there is a noticeable gap between the original and delayed sounds, is often referred to as an *echo*.

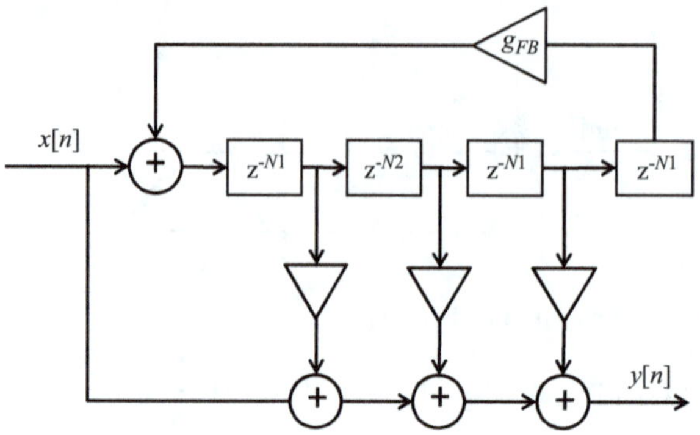

FIGURE 3.3
Flow diagram of a 3-Tap delay. If the last delay value is zero, and only the third tap is used, the system is equivalent to the basic delay.

Multi-Tap Delay

In a standard delay with or without feedback, the time between copies is always the same, since the output signal is taken after the signal reaches the end of the delay line. Multi-tap delay provides more flexibility by taking several additional outputs in the middle of the delay line, where the signal has been delayed only part of the total time. This process is known as *tapping* the delay line, following the analogy of adding taps along a water pipe to get water at various locations. Multi-tap delay is commonly labeled according to the number of taps; for example, a four-tap delay would have four total outputs at various points on the delay line. The delay between each tap is typically not the same, so multi-tap delays allow more complex patterns to be created that can add interesting rhythmic qualities to an instrument. A diagram of multi-tap delay is shown in Figure 3.3.

The multi-tap delay is a more general case of the basic delay design. The multi-tap delay can be further generalized by allowing feedback from the tap outputs to the beginning of the delay line as well. In this case, care must be taken with the feedback gains to avoid creating an unstable system.

Ping-Pong Delay

Ping-pong delay is a multichannel delay-based effect that produces a bouncing sound from one channel to the other, hence the name. It is implemented as a delay with feedback with at least two distinct delay lines (Figure 3.4). Each delay line may be driven by a separate input, or only one input can be used. The output of each delay line, rather than feeding back to itself, attaches to the input of the opposite delay line. In its two-channel configuration, ping-pong delay produces a sound that bounces between left and right channels in a stereo track.

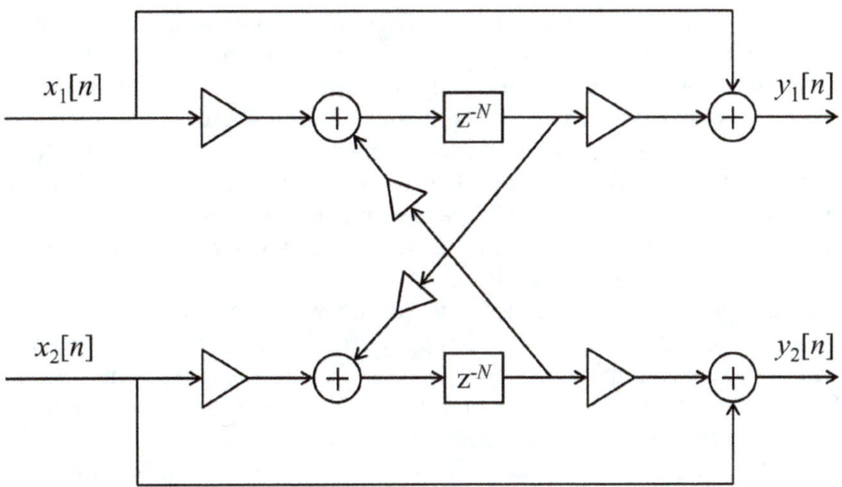

FIGURE 3.4
Flow diagram of a Ping-Pong delay unit.

Implementation

Basic Delay

Before the days of digital audio, delays were one of the more cumbersome effects to implement, relying on physical audio tape or analog "bucket brigade device" integrated circuits to store and retrieve audio signals. In the digital realm, delay is one of the simplest effects. Audio samples are stored in a pre-allocated memory *buffer* as they arrive, while previously stored samples are read from the buffer once the delay time has elapsed. In other words, in a simple delay, each sampling period includes one read operation (retrieving the delayed signal) and one write operation (storing the current signal).

The required *memory footprint* of the buffer is the product of the length of the delay (in samples) and the size of each sample (in bytes). Digital audio processing is commonly handled with 32-bit single-precision floating point samples (float type in C/C++). In that case, a 10-s delay at a sample rate of 44.1 kHz would use 10*44,100*4=1,764,000 bytes (1.7 MB) of memory. This amount of memory is not a significant limitation for any desktop computer, but when implementing a delay using a microcontroller (e.g., to embed into an effects pedal), the memory footprint can easily become the main constraint.

Circular Buffering

The order in which past samples of audio are stored in a memory buffer has a substantial impact on computational performance. Conceptually, it would be simplest to think about samples stored sequentially in memory: the oldest sample is always at the beginning of the buffer, and the most recent sample is always at the end. In this organization, when a new sample arrives, all the samples in the buffer would be shifted back by one, dropping the oldest sample and leaving a gap for the newest sample to be written (Figure 3.5). The problem with this approach is that it is computationally inefficient. For a buffer of length N, writing one new sample first requires moving the other $N-1$ samples to new locations in memory!

Instead, delay-line effects almost always follow an arrangement where each sample is written into the buffer just once and never changes its location in memory. A *write pointer* keeps track of which slot in the buffer is where the next sample will be written, see Figure 3.6. If the memory buffer is implemented as an *array*, then the write pointer is typically an *index* into that array, which takes values between 0 and $N-1$ (where N is the length of the array in samples). Each time a sample is written, the write

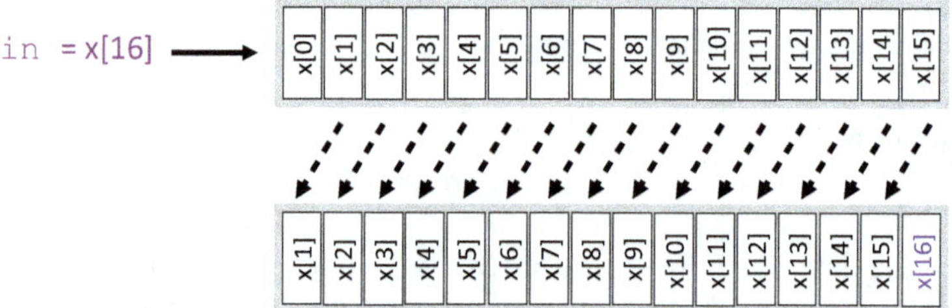

FIGURE 3.5
Delay implemented as a linear buffer. The samples are in order in memory from oldest to newest, but storing each new sample requires moving all the older ones.

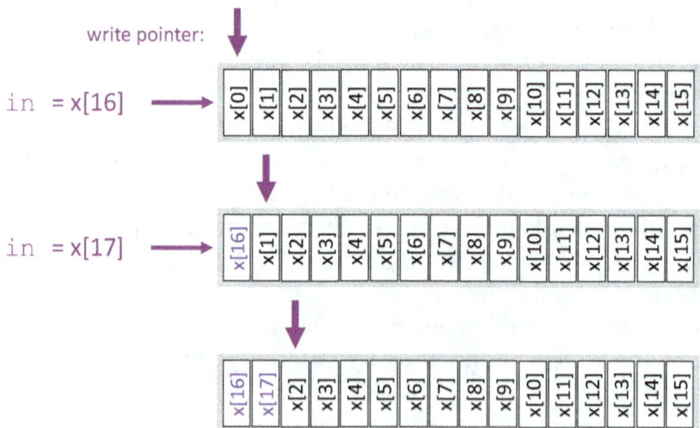

FIGURE 3.6
Delay implemented as a circular buffer. The samples always stay in place, and a write pointer keeps track of where each new sample is stored. When the write pointer reaches the end of the buffer, it wraps around to the beginning.

pointer increments by 1. When the write pointer reaches the end of the buffer, it wraps back around to the beginning (i.e., returns to 0).

This arrangement is known as a *circular buffer*. The beginning and end of the region of memory hold no special meaning: the newest sample in the buffer can always be found one slot before the write pointer, while the write pointer itself points to the *oldest* sample in the buffer (the one that will be overwritten the next time a sample comes in).

Modulo Indexing in Circular Buffers

As we will see later in this chapter, many delay-line effects change the length of their delay over time, whether through adjustment by the user or in response to a low-frequency oscillator (LFO). It is rarely efficient to reallocate memory every time the delay time changes, so a typical approach is to allocate a memory buffer large enough to hold the *longest* possible delay time. In this case, we need some way of looking backwards a fixed distance to retrieve a sample a known length of time in the past. We can do this by counting backwards from the *write pointer*: decreasing indexes correspond to progressively earlier samples in the buffer. Specifically, if W denotes the index of the write pointer, and we want to look k samples before the most recent sample, then we, in principle, should access index $(W-1-k)$ in the array.

The problem with the above strategy is that the write pointer W can take any value between 0 and $N-1$ (where N is the length of the array), and therefore, $(W-1-k)$ might be a negative number. We cannot access negative indices of an array: this would correspond to an invalid location in memory and could make our code crash. But recalling the structure of the circular buffer, we know that when we reach one end of the buffer, we simply wrap around to the opposite end. In other words, to find one sample earlier than index 0, we should access index $N-1$.

The general form for handling these kinds of wrap-arounds in a circular buffer is known as *modulo indexing*: we first compute the desired index we want to read or write, which might be outside the bounds of the buffer. We then apply a modulo (remainder after division) by the *length* of the buffer to ensure that the index is mapped back into the range between 0 and $N-1$. For example, $(N+1)\%N=1$ (where % denotes the modulo operator): an index which would have been past the end of the buffer is wrapped around to the beginning.

However, when implementing a modulo in code, we have to be careful because the modulo of a negative number is still a negative number. Therefore, as a precaution, we should first *add* a multiple of the buffer length. This will have no effect on the final result (since we are taking a modulo by the buffer length), but it will ensure the result is never negative. Therefore, to read k samples before the write pointer, we should read index $(W-1-k+N)\ \%\ N$.

Write and Read Pointers

If we wanted to implement a delay of a fixed length, even if the memory buffer was larger, we could always rely on computing the above formula for every new audio sample. This would give us a reliable delay of k samples regardless of the buffer length N, as long as $k < N$. However, there is a more efficient approach than calculating the formula at every sample: define a *read pointer* that behaves similarly to the *write pointer*, but is used for reading samples out of the buffer rather than writing into the buffer. This is shown in Figure 3.7. Implementing a delay thus involves the following steps:

1. Read a sample from the buffer at the read pointer;
2. Write a sample to the buffer at the write pointer;
3. Advance both the read and write pointers by one sample;
4. If either the read or write pointer has reached the end of the buffer, wrap it around to the beginning.

Here, the delay length D (in samples) is given by the *difference* between the read pointer R and write pointer W:

$$D = (W - R + N)\%N. \tag{3.8}$$

By adjusting the distance between W and R, delay lengths between 1 and $N-1$ samples can be achieved for a buffer size of N.

Variations

Delay with feedback is implemented identically to a simple delay, but rather than storing the raw input signal in the memory buffer, the sum of the input signal and the delayed, scaled output is stored ($x[n] + g_{FB}x[n-N]$ in Figure 3.2). *Ping-pong delay* uses two independent memory buffers, and hence two read operations and two write operations for each sample.

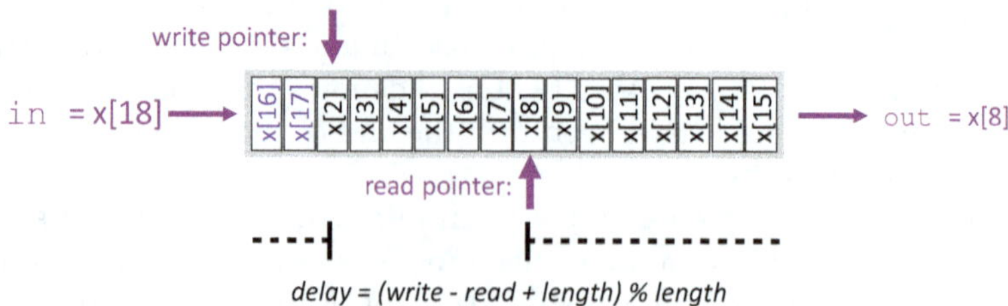

FIGURE 3.7
Delay implemented as a circular buffer with separate read and write pointers. The delay length is given by the difference in location between the pointers using modulo arithmetic.

Multi-tap delay uses a single memory buffer with multiple *read pointers.* At each sample, the current input is written into a slot in the buffer (the *write pointer*), but samples are read back from multiple locations in the buffer (the read pointers). The spacing in memory between the pointers determines the difference in their delay times. This process is further discussed in Chapter 12.

Delay Line Interpolation

Digital delays are implemented using a memory buffer of discrete audio samples. To change the delay time, we change the distance in the buffer between where samples are written and where they are read back. However, many effects, including the flanger and chorus, require a delay that changes over time. To achieve a smooth variation in delay, rounding to the nearest integer number of samples is usually not good enough.

Delay becomes more complex when a non-integer number of samples is required. Consider a simple delay, as in Eq. (3.9), with a length of 0.5 samples,

$$y[n] = x[n] + gx[n-0.5] \Rightarrow y[1] = x[1] + gx[0.5]. \tag{3.9}$$

However $x[n]$ is only defined for integer n, so $x[0.5]$ does not exist. Strictly speaking, it is not correct to think of $x[0.5]$ as "halfway between" $x[0]$ and $x[1]$. But we might ask what the result would be if $x[n]$ were converted from discrete to continuous time, shifted half a sampling period, then re-converted to discrete time. An excellent discussion of the underlying mathematics of this process, which is based on sinc functions $(sin(n)/n)$, can be found in [2]. Interestingly, exact reconstruction of fractional sample values requires knowledge of the entire signal (i.e., $x[n]$ from $n=-\infty$ to ∞), which is clearly impossible in a real-time audio context.

In practice, fractional delays in audio are implemented using interpolation, in which a weighted combination of surrounding samples is used to approximate the fractional sample value. Interpolation involves estimating a value of a continuous function, given discrete points, and can be used to estimate values between points on a delay line. Polynomial interpolation is where the function is estimated to be an Nth order polynomial, $y(t) = c_N t^N + c_{N-1} t^{N-1} + \dots c_1 t^1 + c_0$.

If values out of the delay line closest to the required point are read, this is zeroeth order, or nearest neighbor, interpolation. It is probably the least computationally expensive approach. However, the output now has abrupt jumps between values. The quality of this approach is quite poor. Clicking in the output may be heard as the delay length changes, also known as "zipper noise."

Linear interpolation, or first-order interpolation, is implemented by connecting two known samples by a straight line and then reading the desired value from that line. This is given in the following equation:

$$x(t)=(n+1-t)x[n]+(t-n)x[n+1], \quad n \le t < n+1. \tag{3.10}$$

Linear interpolation is simple to calculate and produces much better results than nearest-neighbor interpolation. However, it is still only a rough approximation to the ideal continuous time case, and it can introduce noise and aliasing into the signal. In many cases, audibly better quality will be obtained with a more computationally complex interpolation method.

One such method is second-order polynomial interpolation. Consider three successive samples, $x[n-1]$, $x[n]$, and $x[n+1]$. We would typically use these samples if we are trying to interpolate a value of x near $x[n]$. Let's define a new function, $y(\tau)=x(t)$, where $\tau=t-n$. Thus $x[n-1]$, $x[n]$, and $x[n+1]$ become $y[-1]$, $y[0]$, and $y[1]$. Assuming that these points are on a second-order polynomial, we have the following three equations;

$$\begin{aligned} y[-1] &= c_2(0-1)^2+c_1(0-1)+c_0 \\ y[0] &= c_2 0^2+c_1 0+c_0 \\ y[1] &= c_2(0+1)^2+c_1(0+1)+c_0 \end{aligned} \tag{3.11}$$

where c_0, c_1, c_2 are the coefficients of some second-order polynomial. This can be solved to give

$$\begin{aligned} c_0 &= y[0] \\ c_1 &= \left(y[1]-y[-1]\right)/2 \\ c_2 &= \left(y[1]-2y[0]+y[-1]\right)/2. \end{aligned} \tag{3.12}$$

So our interpolated values are

$$x(t)=y(\tau)=c_2(t-n)^2+c_1(t-n)+c_0. \tag{3.13}$$

Cubic interpolation requires more computation, but it can give better results with lower added noise resulting from inaccuracies compared to the ideal sinc-function case. There are many forms of cubic interpolation, and a detailed discussion is available from [14]. The simplest case uses the four samples surrounding the interpolated location, $n \le t < n+1$,

$$x(t)\approx c_3(t-n)^3+c_2(t-n)^2+c_1(t-n)+c_0. \tag{3.14}$$

where the coefficients are given by:

$$
\begin{aligned}
c_3 &= -x\left[n-1\right]+x\left[n\right]-x\left[n+1\right]+x\left[n+2\right] \\
c_2 &= x\left[n-1\right]-x\left[n\right]-a_0 \\
c_1 &= x\left[n+1\right]-x\left[n-1\right] \\
c_0 &= x\left[n\right].
\end{aligned}
\tag{3.15}
$$

Code Example

The following C++ code fragment, adapted from the code that accompanies this book, shows the implementation of a basic delay with feedback and without interpolation:

```
// Variables used in this example whose values are set externally:
int numSamples;      // Indicates how many audio samples to process
float *channelData; // Array of audio samples, length numSamples
float *delayData;    // Our own circular delay buffer of audio samples
int delayBufLength; // Length of our delay buffer in samples
int dpr, dpw;        // Read and write pointers into the delay buffer

// User-adjustable effect parameters:
float dryWetMix_;    // Proportion of dry (undelayed) signal to wet
                     // (delayed) signal
float feedback_;     // Feedback level for the delay (0 if no feedback
                     // used)

for (int i = 0; i < numSamples; ++i)
{
    const float in = channelData[i];
    float out = 0.0;

    // In this example, the output is the input plus the contents of the
    // delay buffer (weighted by the mix levels).

    out = (1 - dryWetMix) * in + dryWetMix * delayData[dpr];

    // Store the current information in the delay buffer. delayData[dpr]
    // is the delay sample we just read, i.e. what came out of the
    // buffer.
    // delayData[dpw] is what we write to the buffer, i.e. what goes in

    delayData[dpw] = in + (delayData[dpr] * feedback_);

    if (++dpr >= delayBufLength) dpr = 0;
    if (++dpw >= delayBufLength) dpw = 0;

    // Store the output sample in the buffer, replacing the input
    channelData[i] = out;
}
```

This example assumes that a single channel of input audio data is present in the `channelData` array. In a real-time effect, `channelData` will hold only the most recent block of audio samples, rather than the entire input signal. To implement the delay, we write each input sample into delayData at a position indicated by the write pointer dpw, while reading a delayed sample previously written into the buffer using read pointer dpr. The key computations are as follows:

```
out = (1 - dryWetMix) * in + dryWetMix * delayData[dpr];
```

This applies the basic delay equation, calculating the output in terms of the sum of the input and the output of the delay buffer. Each term is weighted by a parameter (dryWetMix and its inverse). dpr is the read pointer into the delay buffer, which keeps track of the next sample to read out. Next, the content of the delay buffer is updated:

```
delayData[dpw] = in + (delayData[dpr] * feedback_);
```

dpw is the write pointer into the delay buffer. The content of the buffer is updated as the sum of the input and the feedback from the output.

delayData is a circular buffer: after each sample is processed, dpr and dpw are incremented, and when either of them reaches the end of the buffer, it is reset back to position 0. Notice that there is no parameter for the delay length in this code example. It is the difference between the read and write pointers dpr and dpw that determines the delay length. These values must be initialized elsewhere in the effect. If dpr and dpw are used in the audio callback function (see Chapter 2), they should be stored into an instance variable that remembers their values from one call of the function to the next.

If a stereo delay is desired, then delayData should be a multidimensional array. Only one read pointer and one writer pointer are needed, assuming all the channels remain in sync, but in the above code, dpr and dpw would need to be reset for each channel, so that updates to the pointer locations for one channel don't affect the playback of another channel.

Applications

Delays are a very common effect in music production. Even a single basic delay, without feedback, has many uses. A common use is to combine an instrument's sound with a short echo (for example, around 50–100 ms) to create a doubling effect. This can create a wider, more lively sound than the original single version. If the original and delayed copies are panned differently in a stereo mix, short delays can also help make the mix sound "bigger." More complex arrangements, including feedback and multiple taps, can start to simulate the sound of a reverb unit, though reverb (covered in Chapter 10) typically creates more complex patterns than can be simulated with simple delays.

Longer delays become less subtle once the original and delayed copies are easily perceived as two separate acoustic events. One common trick is to synchronize the delay time with the tempo of the music, such that the delayed copies appear in rhythm with the track [15]. If the delay time is especially long, for example, equal to one or more bars of music, a musician can play over him or herself and develop elaborate harmonies and textures.

Sampler and looper pedals are fundamentally based on the delay, with the main difference that instead of continually updating the contents of the delay buffer as new audio comes in, the buffer stores a fixed segment of audio which is played in a loop continuously. Loopers often have additional features to be able to define the start and end of an audio segment, which then can loop continuously – i.e. the range of indexes that the read pointer traverses as it plays. Other features are possible, including the ability to mix additional sounds onto a loop that was previously recorded, the ability to play a loop backwards, or to have multiple simultaneous loops. Some standard delay pedals will include basic looper capability, though often limited to a single loop of a few seconds at most.

Wavetable oscillators, used in some synthesizers, are based on the same principle as the looper: an audio segment is looped continuously. In contrast to looper pedals, in a wavetable, the segment is very short, typically just a single oscillation of a waveform, and the loop is played at different speeds to obtain different pitches.

FRIPPERTRONICS AND CRAFTY GUITARISTS

Robert Fripp (King Crimson, The League of Crafty Guitarists, League of Gentlemen, solo artist…) is known as one of the greatest and most influential guitarists of all time. Evolving out of his work with Brian Eno in 1973, he devised a tape looping technique to layer his guitar sounds in real-time. It used two reel-to-reel tape recorders. The tape traveled from the supply reel of one recorder to the take-up reel of the second one. Then the tape from the second machine is fed back to the first one, and the delay can be changed by adjusting the distance between the two machines. Furthermore, it also provided a recording of the complete overlayed recording, and could be used in live performance. Fripp's girlfriend later named this technique 'Frippertronics,' though we would describe it as a time-varying delay with feedback. Also among Robert Fripp's more unusual contributions, are many of the sounds for the Windows Vista operating system.

Many other famous guitarists are also known for music technology innovations. For instance, Tom Scholz of the band Boston designed a wide range of novel guitar effects devices, including the Rockman amplifier. But one of the most famous guitarists is also one of the people who most influenced music technology, Les Paul. His solid body electric guitar designs were some of the first and most popular, and he is credited with many innovations in multitrack recording.

Vibrato Simulation

Vibrato is defined as a small, quasi-periodic variation in the pitch of a tone. Traditionally, vibrato is not an audio effect but rather a technique used by singers and instrumentalists. On the violin, for example, vibrato is produced by rhythmically rocking the finger back and forth on the fingerboard, slightly changing the length of the string. However, vibrato can be added to any audio signal through the use of *modulated delay lines*.

Vibrato is characterized by its *frequency* (how often the pitch changes) and *width* (total amount of pitch variation). On acoustic instruments, especially wind instruments, vibrato is often accompanied by some degree of amplitude modulation or tremolo (Chapter 6), with the pitch and amplitude of the signal changing in synchrony.

Theory

The vibrato effect works by changing the playback speed of sampled audio. The effect of playback speed is familiar to many listeners: playing a sound faster raises its pitch, and playing it slower lowers the pitch. To add a vibrato, then, the playback speed needs to be periodically varied to be faster and slower than normal.

Implementation of the vibrato effect is based on a *modulated delay line*, a delay line whose delay length changes over time under the control of a *low-frequency oscillator*, as shown in Figure 3.8. Unlike other delay effects, the input signal $x[n]$ is not mixed into the output.

Delay alone does not introduce a pitch shift. Suppose the length $M[n]$ of the delay line does not change. Then at every sampling period n, exactly one sample $x[n]$ goes in and one sample $y[n-M]$ comes out. The sound will be delayed but otherwise identical to the original. But now suppose that the length $M[n]$ decreases each sample by an amount Δm:

$$M[n] = M_{\max} - (\Delta m)n. \tag{3.16}$$

Examining the output of the delay line, we can see that each new input sample n moves the output by $1+\Delta m$ samples:

$$y[n] = x\big[n - M[n]\big] = x\big[(1+\Delta m)n - M_{\max}\big]. \tag{3.17}$$

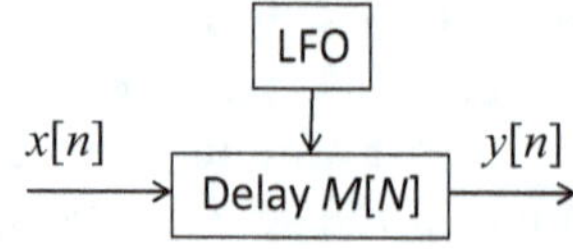

FIGURE 3.8
Modulated delay and pitch shift.

In other words, the playback rate from the buffer is $1+\Delta m$ times the input rate. Correspondingly, the frequencies in the input signal $x[n]$ will all be scaled upwards by a factor of $1+\Delta m$. For similar reasons, if the length of the delay line *increased* with each sample ($\Delta m<0$), the frequencies of the output signal will all be scaled down compared to the original.

Notice that the pitch shift is only sensitive to the change in delay Δm, not its initial value M_{max}. We can generalize to write that pitch shift is dependent on the *derivative* (or, in discrete time, *first difference*) of delay length, with increasing length producing lower pitch:

$$f_{\text{ratio}}[n]=\left(\frac{f_{\text{out}}}{f_{\text{in}}}\right)[n]=1-\left(M[n]-M[n-1]\right). \qquad (3.18)$$

To create a vibrato effect, then, the delay is periodically lengthened and shortened.

Interpolation

Strictly speaking, Eq. (3.17) above is only defined for integer samples of x, i.e., when $(1+\Delta m)n-M_{max}$ is an integer. In general, this is not always the case for a modulated delay line where delay lengths change gradually from sample to sample, so *interpolation* must be used to approximate non-integer values of x. See the earlier section on delay line interpolation for a more detailed discussion of fractional delay and its implementation.

Implementation

Though Eq. (3.18) suggests that arbitrary pitch shifts are possible by choosing the rate Δm at which the delay length varies, real-time audio effects are limited by two practical considerations. First, real-time effects must be *causal*, which implies that the delay length $M[n]$ must be nonnegative. If we want to increase the pitch of a sound, this means that for any finite initial delay M_{max}, $M[n]$ will eventually reach 0 at which point it will no longer be possible to maintain the pitch shift. The second consideration is that computer systems have finite amounts of memory. To decrease the pitch of a sound, $M[n]$ must steadily increase, which requires an ever greater memory buffer to hold the delayed samples. Because the output is being played more slowly than the input, eventually it will be impossible to hold all the intervening audio in memory.

To satisfy these two considerations, the *average change* in delay length over time must be 0. In a vibrato effect, the delay length varies periodically around a fixed central value, producing periodic pitch modulation, but a sustained increase or decrease in pitch is not possible. In Chapter 9, we will examine a technique for real-time pitch shifting using the phase vocoder.

Low-Frequency Oscillator

The delay effect described earlier can be characterized as a linear, time invariant filter. But many of the audio effects that we will encounter are not time invariant. That is, the effect acts like a filter, but now the output as a function of input depends on the time, or in discrete form, the sample number. This is most often accomplished by driving the effect (making some parameters explicitly a function of time) with a *Low-frequency oscillator,* or *LFO*. This is the case for the delay line-based effects described in the following sections.

LFOs do not have a formal definition, but they may be considered to be any periodic signal with a frequency below 20 Hz. They are used to vary delay lines or as a modulating signal in many synthesizers, and they will be used in many of the effects featured later in the book. Like their audio-frequency counterparts, LFOs typically use periodic waveforms like sine, triangle, square, and sawtooth waves. However, any type of waveform is possible, including user-defined waveforms read from a wavetable.

Figure 3.9 depicts three of the most commonly used waveforms. Different LFO waveforms are preferred for different effects. In the vibrato effect, the delay length is typically controlled by a sinusoidal LFO,

$$M[n] = M_{\text{avg}} + W\sin(2\pi nf / f_s), \tag{3.19}$$

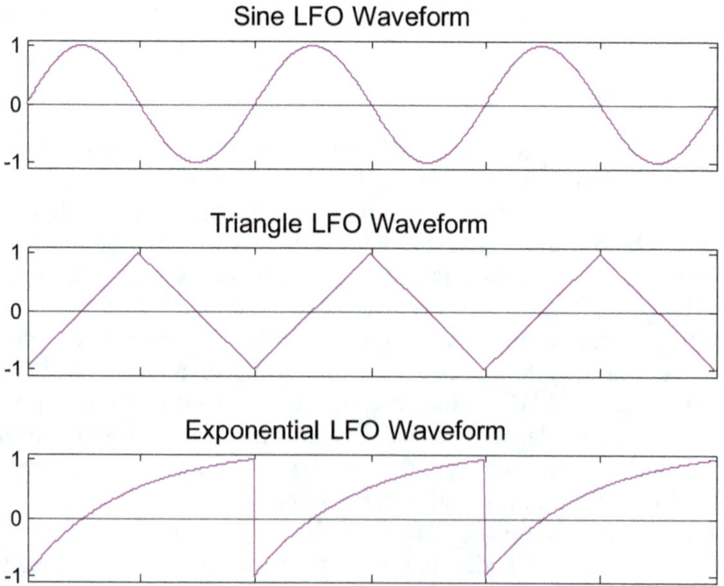

FIGURE 3.9
Three commonly used low-frequency oscillator (LFO) waveforms.

where M_{avg} is the average delay (for real-time effects, it is chosen so that $M[n]$ is always nonnegative), W is the width of the delay modulation, f is the LFO frequency in Hz and f_s is the sample rate. The rate of change from one sample to the next can be approximated by the continuous-time derivative:

$$M[n]-M[n-1]=W\left[\sin\left(\frac{2\pi nf}{f_s}\right)-\sin\left(\frac{2\pi(n-1)f}{f_s}\right)\right]\approx 2\pi fW\cos\left(\frac{2\pi nf}{f_s}\right). \tag{3.20}$$

We can then find the frequency shift for the vibrato effect:

$$f_{\text{ratio}}[n]=1-\left(M[n]-M[n-1]\right)\approx 1-2\pi fW\cos\left(2\pi nf/f_s\right). \tag{3.21}$$

Notice that in the implementation of the LFO, W indicates the amount that the delay length changes, not the amount of pitch shift. Equation (3.21) shows that pitch shift depends on both W and f, such that for the same amount of delay modulation, a faster LFO will produce more pitch shift. As Figure 3.10 demonstrates, this is an expected result, since increasing the frequency of a sine wave while maintaining its amplitude will result in a greater derivative.

Given the desired amount of maximum pitch shift and an LFO frequency, we can calculate the approximate required amount of delay variation:

$$W=\frac{f_{\text{ratio}}[n]-1}{2\pi f}. \tag{3.22}$$

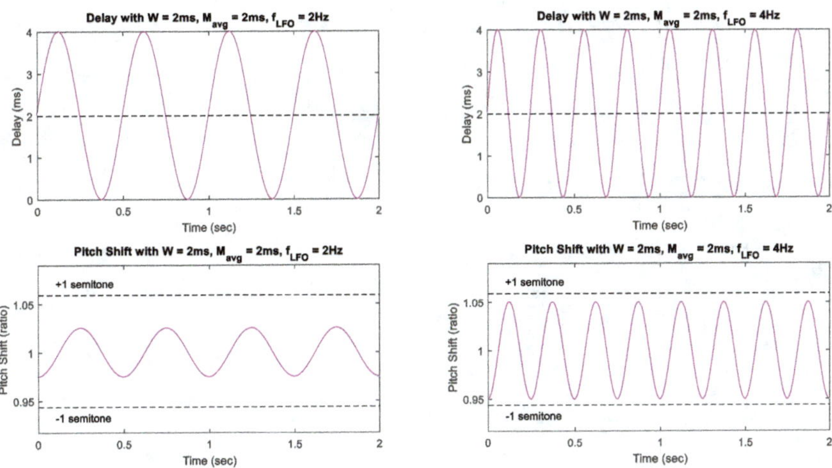

FIGURE 3.10
Vibrato in operation. The LFO waveform on top and the pitch shift on bottom, for LFO frequency 2 Hz (left) and 4 Hz (right).

From this, we can also find the average delay $M_{avg} \geq W$ needed to keep the effect causal. In all but the most extreme cases, M_{avg} will be small enough that no delay will be perceptible at the output of the vibrato effect.

Parameters

The vibrato effect is completely characterized by *LFO frequency, LFO waveform,* and *vibrato (pitch shift) width.* The pitch shift parameter is used to calculate the amount of delay modulation, which is what ultimately produces the vibrato effect. A typical violin vibrato has a frequency on the order of 6 Hz, with frequency variation of around 1% (i.e., approximately 0.99 to 1.01 in frequency ratio). With a sinusoidal LFO, these settings would produce a delay variation W of 0.265 ms in either direction. By way of comparison, two notes a semitone apart differ in frequency by $\sqrt[12]{2} \approx 1.059$ or 5.9%.

Sinusoidal waveforms are best for emulating normal instrumental vibrato, but other waveforms can be used for special effects. For example, a triangle waveform (Figure 3.11a) has only two slopes (rising and falling) and accordingly, the pitch will jump back and forth between two fixed values (Figure 3.11c). A rising sawtooth wave (Figure 3.11b) approximates a pitch-lowering effect since the derivative of delay length is usually positive. However, the periodic discontinuities (Figure 3.11d) in the sawtooth waveform produce artefacts that degrade the quality of the result, so this technique is not normally used for pitch shifting.

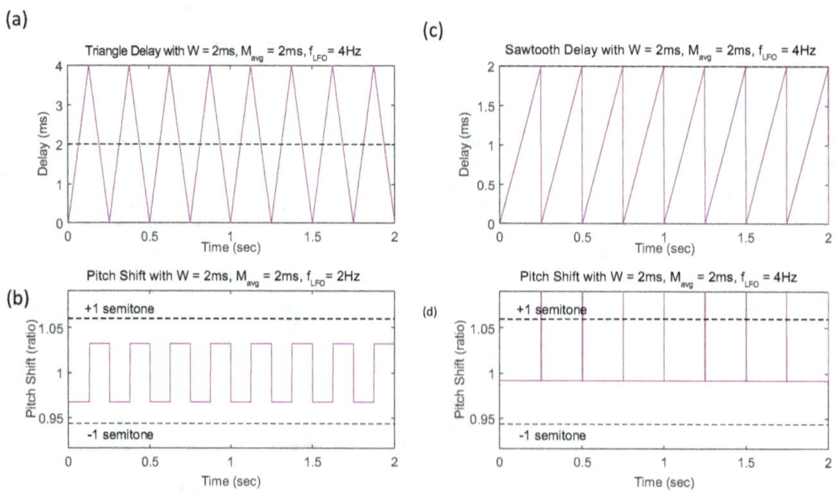

FIGURE 3.11
Triangular (a) and sawtooth LFOs (b), with corresponding pitch shift, (c) and (d).

Code Example

The following C++ code fragment, adapted from the code that accompanies this book, implements a vibrato with sinusoidal LFO and linear interpolation:

```
// Variables used in this example whose values are set externally:
int numSamples;      // Indicates how many audio samples to process
float *channelData;  // Array of audio samples, length numSamples
float *delayData;    // Our own circular delay buffer of audio samples
int delayBufLength;  // Length of our delay buffer in samples
int dpw;             // Write pointer into the delay buffer
float ph;            // Current phase of the LFO, always between 0-1
float inverseSampleRate; // 1/f_s, where f_s is the sampling frequency

// User-adjustable effect parameters:
float frequency_;    // Frequency of the low-frequency oscillator
float sweepWidth_;   // Width of the LFO in samples

for (int i = 0; i < numSamples; ++i)
{
    const float in = channelData[i];
    float interpolatedSample = 0.0;

    // Recalculate read pointer position with respect to the write
    // pointer.
    // A more efficient implementation might increment the read pointer
    // based on
    // the derivative of the LFO without running the whole equation
    // again, but
    // this format makes the operation clearer.
    float currentDelay = sweepWidth_*(0.5f + 0.5f*sinf(2.0 * M_PI * ph));

    // Subtract 3 samples to the delay pointer to make sure we have enough
    // previously written samples to interpolate with
    float dpr = fmodf((float)dpw - (float)(currentDelay * getSampleRate())
                + (float)delayBufLength - 3.0, (float)delayBufLength);

    // Use linear interpolation to read a fractional index into the buffer.
    // Find the fraction by which the read pointer sits between two
    // samples and use this to adjust weights of the samples
    float fraction = dpr - floorf(dpr);
    int previousSample = (int)floorf(dpr);
    int nextSample = (previousSample + 1) % delayBufLength;
    interpolatedSample = fraction*delayData[nextSample]
        + (1.0f-fraction)*delayData[previousSample];

    // Store the current information in the delay buffer.
    delayData[dpw] = in;

    // Increment write pointer at constant rate. Read pointer moves at
    // different
    // rates depending on the settings of the LFO, delay and sweep
    // width.
    if (++dpw >= delayBufLength) dpw = 0;
```

```
    // Store output sample in buffer, replacing input. In vibrato
    // effect, the
    // delaye sample is the only output component (no mixing with the
    // dry signal)
    channelData[i] = interpolatedSample;

    // Update the LFO phase, keeping it in the range 0-1
    ph += frequency_*inverseSampleRate;
    if(ph >= 1.0) ph -= 1.0;
}
```

The code exhibits many similarities to the basic delay in the previous section. One notable difference is that the read pointer `dpr` is now fractional, taking non-integer values. Accordingly, `dpr` cannot be directly used to read the circular buffer `delayData`, since arrays in C++ can only be accessed at integer indices. In this example, the variable `fraction` holds the non-integer component of the read pointer `dpr`; it is used to calculate a weighted average between the two nearest samples in the circular buffer (`previousSample` and `nextSample`). Notice how the index of the sample following `dpr` is calculated:

```
int nextSample = (previousSample + 1) % delayBufLength;
```

The `%` sign is a *modulo operator.* As discussed earlier in this chapter, the use of modulo arithmetic is needed to implement a circular buffer.

For the vibrato effect, no feedback or mixing with the original signal is used, so many of the parameters in the basic delay example are not used here. In the complete code example that accompanies this book, a choice of LFO waveforms and interpolation types is offered.

Applications

Vibrato, when used by vocalists or instrumentalists, can add a sense of warmth and life to a musical line. The width and frequency of vibrato and their evolution over time are important expressive decisions for many performers. Vibrato can also help an instrument or voice stand out from an ensemble. A single musical note will contain energy at discrete, harmonically-related frequencies, but by varying the pitch back and forth, a single note can use more of the frequency spectrum.

Vibrato is sometimes used to cover slight errors in pitch, as it is easier to perceive a steady-pitched sound as being out of tune than one containing vibrato. However, the use of vibrato to cover pitch errors is generally considered poor musical practice.

The vibrato audio effect is not as flexible as a performer's natural vibrato, since the LFO operates at a constant rate and width regardless of the musical material. Also, a simple vibrato implementation does not synchronize with the beginnings and endings of individual notes as a performer would. More advanced implementations do, though, such as can be found in some synthesizers.

Though it is possible to imagine a vibrato effect with a pedal or other control to give the user more flexibility over the LFO, this is rarely seen in practice. Nonetheless, even a fixed-frequency vibrato can add warmth and body to the sound of an instrument, especially when used with reverberation.

Flanging

KEN'S FLANGER

Flanging is an unusual name for an audio effect, and it's certainly not a common word in music or signal processing. The flange refers to a rim or edge, especially on a tape reel. Producers were known to manipulate the flange of a tape reel to achieve nice effects on many early tape recordings. One of the earliest known examples of producing a sound similar to the modern flanger is 'The Big Hurt' by Tony Fisher, recorded in 1959.

But the origin of the name of the audio effect is an unusual one, and has been well documented by Beatles historians Bill Biersach and Mark Lewisohn.

In 1966, The Beatles recorded *Revolver* at Abbey Road. The studio technician, Ken Townsend later said that

> they would relate what sounds they wanted and we then had to go away and come back with a solution… they often liked to double-track their vocals, but it's quite a laborious process and they soon got fed up with it. So, after one particularly trying night-time session doing just that, I was driving home and suddenly had an idea.

What Townsend devised was not the modern flanging, but the closely related chorus effect, or artificial double tracking (ADT). But its implemented using the same approach, slowing down and speeding up a tape machine. The seemingly random variations in speed (and hence also pitch) mimic the effect of a singer trying to harmonize with the original.

John Lennon loved the effect and asked George Martin, the Beatles' producer, to explain it. As Martin recalled,

> I knew he'd never understand it, so I said, 'Now listen, it's very simple. We take the original image and split it through a double vibrocated sploshing flange with double negative feedback ...' He said, 'You're pulling my leg, aren't you?' I replied, 'Well, let's flange it again and see.' From that moment on, whenever he wanted it he'd ask for his voice to be 'flanged,' or call out for 'Ken's flanger.'

Flanging is a delay-based effect originally developed using analog tape machines 50 years ago (see [16] and references therein). It refers to the *flange* or outer rim of the open-reel tape recorders in common use in studios at the time. To create the flanging effect, two tape machines are set up to play the same tape at the same time. Their outputs are mixed together equally, as shown in Figure 3.12.

If the two machines played perfectly in unison, the result would simply be a stronger version of the same signal. Instead, the operator lightly touches the flange of one of the tape machines, slowing it down and thereby lowering the pitch. This action also causes the tape machine to fall slightly behind its counterpart, creating a delay between them. The operator then releases the flange and repeats the process on the other machine, which causes the delay to gradually disappear and then grow in the opposite direction. The process is repeated periodically, alternately pressing each flange.

If too much delay accumulates between the machines, the mixed output will no longer be heard as a single signal but as two distinct copies. For this reason, the delay must be kept well below the threshold of echo perception (see Chapter 10), i.e., only a few milliseconds in each direction, so the result is heard as a single sound rather than two separate sounds.

The flanging effect has been described as a kind of 'whoosh' that passes through the sound. The effect has also been compared to the sound of a

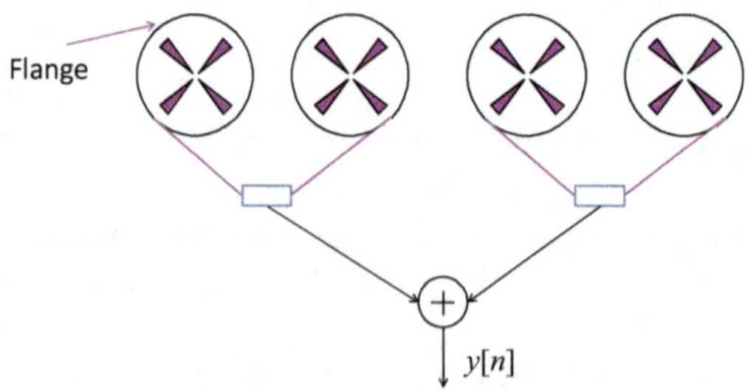

FIGURE 3.12
Two tape machines configured to produce a flanging effect.

jet passing overhead, in that the direct signal and the reflection from the ground arrive at a varying relative delay. And when the delay is modulated rapidly, an audible Doppler shift may be heard [16] (see Chapter 12).

Theory

Principle of Operation

The flanger is based on the principle of constructive and destructive interference. If a sine wave signal is delayed and then added to the original, the sum of the two signals will look quite different, depending on the length of the delay. At one extreme, when the two signals perfectly align in phase, the output signal will be double the magnitude of the input. This is *constructive interference.* At the other extreme, when the delay causes the two signals to be perfectly out of phase, they cancel each other out: an increase in one signal is precisely balanced by a decrease in the other, so they will sum to zero. This is *destructive interference.*

Typical audio signals contain energy at a large number of frequencies. For any given delay value, some frequencies will add destructively and cancel out (*notches* in the frequency response), and others will add constructively (*peaks*). Peaks and notches by themselves do not make a flanger: it is the *motion* of these notches in the frequency spectrum that produces the characteristic flanging sound. As the following sections show, the motion of the peaks and notches is achieved by continuously changing the amount of delay (Figure 3.13).

Basic Flanger

A block diagram of a basic flanger is shown in Figure 3.14. Its operation is closely related to the basic delay discussed previously, with the key difference that the amount of delay varies over time. It is also closely related to the vibrato that was depicted in Figure 3.8. The input-output relation for the flanger can be expressed in the time domain as:

$$y[n] = x[n] + gx\big[n - M[n]\big]. \tag{3.23}$$

The delay length $M[n]$ varies under the control of a separate low-frequency oscillator, discussed in the following sections. This structure is known as a *feedforward comb filter,* since the delayed signals feed forward from the input to the output (with no feedback). To see why this difference equation results in a comb filter, we should consider its Z transform and transfer function:

$$Y(z) = X(z) + gz^{-M[n]}X(z) \Rightarrow H(z) = \frac{Y(z)}{X(z)} = 1 + gz^{-M[n]}. \tag{3.24}$$

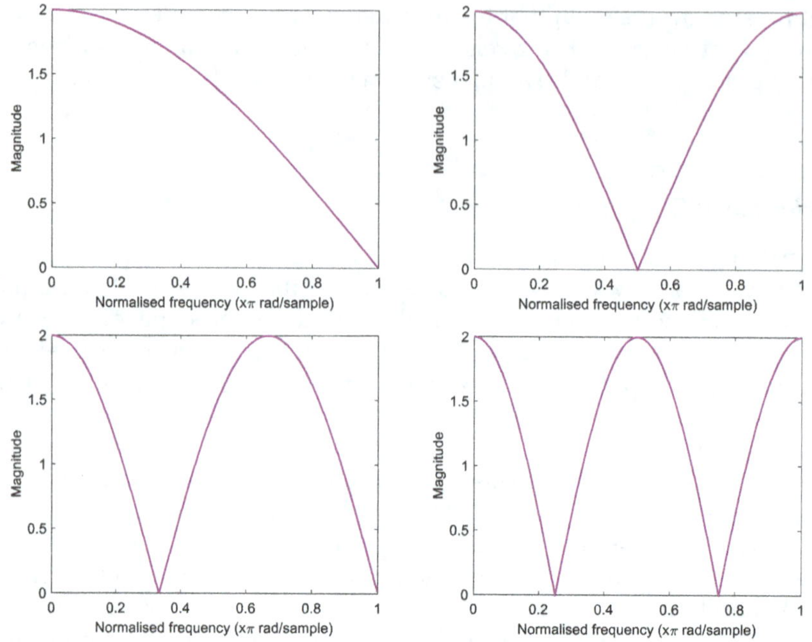

FIGURE 3.13
The magnitude response of a flanger with depth set to 1 and delay times set to 1, 2, 3, and 4 samples.

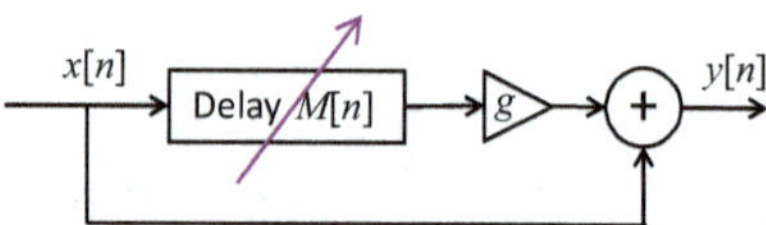

FIGURE 3.14
Block diagram of a basic flanger without feedback. The delay length $M[n]$ changes over time.

To find the frequency response of the flanger for each frequency ω, we substitute $e^{j\omega}$ for z and find the magnitude of the transfer function:

$$\begin{aligned} H(z) &= 1 + ge^{-j\omega M[n]} \\ |Hz)| &= \sqrt{H(z)H^*(z)} = \sqrt{1+2g\cos(\omega M[n]) + g^2}. \end{aligned} \tag{3.25}$$

Notice that the frequency response is periodic: for $g > 0$, we have M peaks in the frequency response, located at the frequencies when the cosine term reaches its maximum value:

$$\omega_p = 2\pi p/M \quad \text{where} \quad p = 0,1,2,\ldots M-1. \tag{3.26}$$

Likewise, for $g > 0$, there are M notches (minima) in the frequency response. These are located where the cosine term reaches its minimum value:

$$\omega_n = (2n+1)\pi/M \quad \text{where} \quad n = 0,1,2,\ldots M-1. \tag{3.27}$$

The pattern of peaks and notches is shown in Figure 3.15. The equations show that a larger delay $M[n]$ produces more notches and a lower frequency for the first notch. As $M[n]$ varies, the notches sweep up and down through the frequency range. The notches are regularly spaced at intervals of f_s/M Hz where f_s is the sampling rate. Their pictorial resemblance in Figure 3.15 to the teeth of a comb is what gives this structure the term *comb filter.* Chapter 5 will examine *phasing*, an effect that produces similar moving notches in the frequency response, which are not regularly spaced.

The depth of the notches depends strongly on the gain g of the delayed signal. When $g = 0$, the frequency response is perfectly flat, as we would expect since $g = 0$ corresponds to no delayed signal. When g is between 0 and 1 (or greater than 1), notches appear in the spectrum, but their depth is finite and depends on the value of g. When $g = 1$, the notches are infinitely deep, as the frequency response exactly equals 0 at the notch frequencies w_n. For this reason, $g = 1$ produces the most pronounced flanging effect.

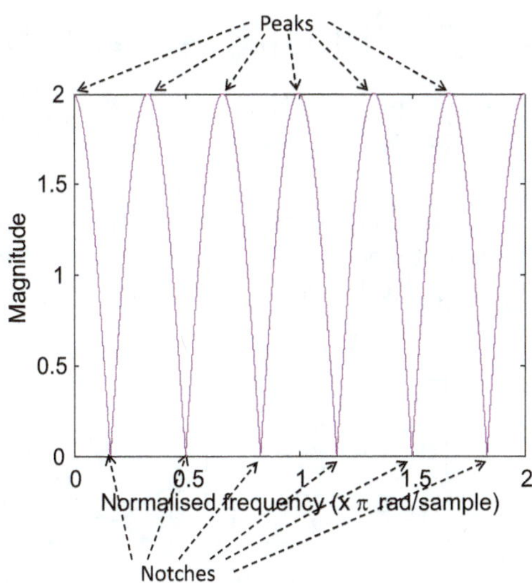

FIGURE 3.15
The frequency response of a simple flanger with a six-sample delay and depth set to 1. The locations of the six peaks and six notches over the whole frequency range from 0 to 2π is shown.

Low-Frequency Oscillator

The characteristic sound of the flanger comes from the *motion* of regularly spaced notches in the frequency response. For this reason, it is critical that the length of the delay $M[n]$ changes over time. Typically, $M[n]$ is varied using an LFO, with a sinusoid being the most common choice for the LFO waveform. Typical delay lengths for the flanger range from 1 to 10 ms, corresponding to notch intervals ranging from 1,000 Hz down to 100 Hz. Further details on LFO parameters are discussed in the Parameters section below.

Flanger with Feedback

Just as feedback could be added to the basic delay, some flangers incorporate a feedback path that routes the scaled output of the delay line back to its input, as shown in Figure 3.16. Feedback on the flanger is also sometimes referred to as 'regeneration.' Like the delay effect with feedback, using feedback in the flanger will result in many successive copies of the input signal spaced several milliseconds apart and gradually decaying over time (Figure 3.17). However, since the delay times in the flanger (typically less than 20 ms) are below the threshold of echo perception (roughly 50–70 ms), these copies are not heard as independent sounds but as coloration or filtering of the input sound.

The difference equation and frequency response for a flanger with feedback can be derived similarly to the delay with feedback:

$$y[n] = g_{FB}y[n-M[n]] + x[n] + (g_{FF} - g_{FB})x[n-M[n]]$$
$$H(z) = \frac{Y(z)}{X(z)} = \frac{1+z^{-M[n]}(g_{FF}-g_{FB})}{1-z^{-M[n]}g_{FB}} = \frac{z^{M[n]}+g_{FF}-g_{FB}}{z^{M[n]}-g_{FB}}. \tag{3.28}$$

We can see that when the feedback gain $g_{FB} = 0$, these terms exactly match the basic flanger without feedback, as expected. In addition to the zeroes of $H(z)$, which are similarly located to the flanger without feedback, the

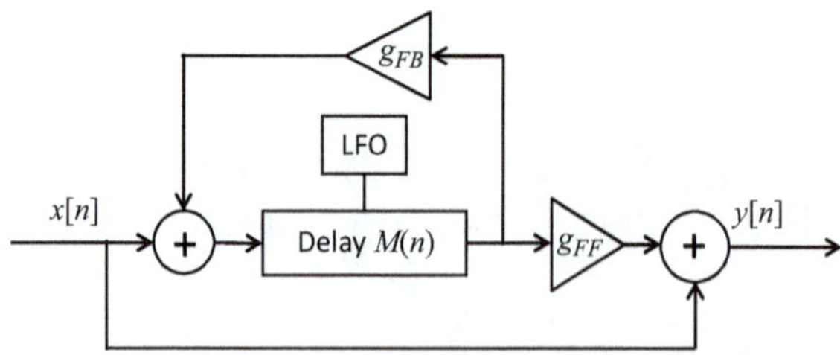

FIGURE 3.16
Flanger with feedback. Delay in all flangers is controlled by a low-frequency oscillator (LFO).

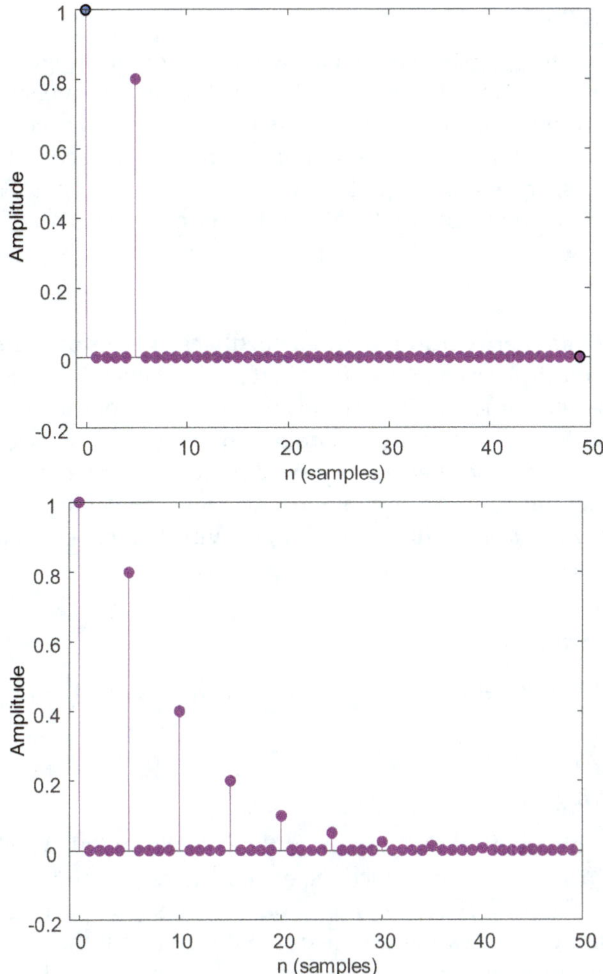

FIGURE 3.17
Impulse responses for a comb filter (i.e., delay effect) with a five-sample delay. On top, gain g=0.8 and no feedback. On bottom, g_{FF}=0.8 and g_{FB}=0.5.

transfer function has poles at the complex roots of g_{FB}. If $g_{FB} < 1$, these will remain inside the unit circle, and the system will be *stable*. This is an intuitive result: feedback gains less than 1 mean that the delayed copies of the sound will gradually decay, where a gain of 1 or more means they will grow (or at least persist with significant amplitude) indefinitely.

The effect of feedback is to make the peaks and notches sharper and more pronounced. Its sound is often described as "intense" or "metallic" and as the feedback gain approaches 1, the pitch f_s/M resulting from the delay line can overwhelm the rest of the sound.

Stereo Flanging

A stereo flanger is constructed out of two monophonic flangers, which are identical in all settings except the *phase* of the low-frequency oscillator. Typically, the two oscillators are in *quadrature phase*, where one leads the other by 90°. In a stereo flanger, the same signal can be used as the input to both channels, or separate signals can be used for each input. The outputs are typically panned fully to the left and right of a stereo mix.

Properties

Because the flanger (with or without feedback) is composed entirely of delays and multiplication, it is a *linear* effect. However, because the properties of the delay line vary over time independently of the input signal, it is *time-variant*, unlike the standard delay: shifting the input signal in time by N samples does not necessarily produce the identical output shifted by N samples, since the delay line length may have changed. The basic flanger is always *stable*, where the flanger with feedback is stable if and only if the feedback gain $g_{FB} < 1$.

Common Parameters

The typical flanger effect contains several controls that the musician can adjust:

Depth (or Mix)

The *depth* controls affect the amount of the delayed signal, which is mixed in with the original. $g = 0$ produces no effect, where $g = 1$ produces the most pronounced flanging effect. Higher depth settings ($g > 1$) produce a louder overall sound due to scaling up the delayed signal, but the flanging effect becomes *less* pronounced: only when the original and delayed copies exactly match in amplitude can perfect cancellation of the notch frequencies occur.

Delay and Sweep Width

The term 'delay' is potentially misleading in the flanger since the length of the delay line varies over time under the control of a low-frequency oscillator. The *delay* control parameter on a flanger affects the minimum amount of delay $M[n]$. The value of the LFO is added to produce larger time-varying values. *Sweep width* controls the total amplitude of the low-frequency oscillator, such that the maximum delay time is given by the sum of the *delay* and *sweep width* controls (Figure 3.18).

As the delay is decreased, the first notch becomes higher in frequency. The delay control thus sets the *highest* frequency that the first notch will

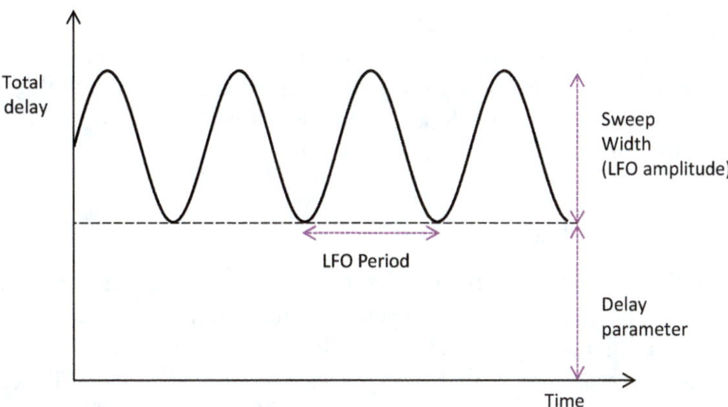

FIGURE 3.18
The maximum delay is the sum of the *sweep width* and *delay* parameters. The delay changes over time according to the *sweep rate*.

reach. If it is set to zero, the notches will disappear entirely when the LFO reaches its minimum value: when the original and delayed signals are exactly aligned, no cancellation will take place at any frequency. Similarly, the sum of the delay and sweep width controls determines the lowest frequency the first notch will reach.

Speed and Waveform

These controls affect the behavior of the LFO controlling the delay length. The *speed* control sets the LFO frequency and typically ranges from 0.1 Hz (10 seconds per cycle) to 10 Hz. *Waveform* is typically chosen from one of several predefined values, including sine, triangle, sawtooth or exponential (triangular in log-frequency). Many flangers do not offer this control and always use a sinusoidal LFO. In this case, the total delay $M[n]$ is given by:

$$M[n] = M_0 + M_W\left[1 + \sin(2\pi f_{\text{LFO}} n / f_s)\right] / 2, \tag{3.29}$$

where M_0 (in samples) is given by the *delay* control, M_W (in samples) is given by the *sweep width* control, f_{LFO} (in Hertz) is given by the *speed* control and f_s (in Hertz) is the sampling frequency. We can see that the value of $M[n]$ varies from M_0 at minimum to $M_0 + M_W$ at maximum, consistent with the expected behavior of these controls.

Feedback (or Regeneration)

The basic flanger has only a feedforward path, in which the delayed signal is added to the original. In a flanger with feedback, the *feedback/regeneration* control sets the gain g_{FB} between the output and the input of the delay line. Possible values are in the range [0, 1), i.e., strictly less than 1, to maintain stability. In practice, values close to 1 are rarely used except for special effects. Even when the system is mathematically stable, large gains at the peaks can result in clipping distortion depending on the level of the input.

Inverted Mode (or Phase)

On some flangers, the feedforward gain g or g_{FF} can be altered in polarity. *Inverted mode* is typically selected with a switch; when activated, g ranges from 0 to –1 instead of 0 to 1. In which case, the peaks and notches in the frequency response will trade places; the lowest peak will occur at $f = f_s/2M$ Hz and the lowest notch at $f = 0$ (DC). Because of the notch at DC, the bass response in the inverted mode is poor, producing a thinner sound unless M is very large (which reduces the frequency of the lowest peak). The different color of the inverted mode flanger can be useful in some musical situations.

Implementation

Buffer Allocation

The flanger, like all delay-based effects, is typically implemented digitally using *circular buffers* (described earlier in this chapter). Memory allocation and de-allocation are highly time-consuming in comparison to basic audio calculations. Since the length of the delay changes with the phase of the LFO, the buffer is pre-allocated to be large enough to accommodate the maximum amount of delay at any point in the LFO cycle, for any settings of the *delay* and *sweep width* parameters.

The actual length of delay at any time is controlled by the distance between the *read pointer* and *write pointer* in the buffer. In a typical implementation, the write pointer will move at a constant speed, advancing one sample in the buffer for each input sample. Moving the read pointer faster than this rate will decrease the amount of delay, while moving it slower will increase the delay.

Interpolation

Since the delay of the flanger changes by small amounts from one sample to the next, it will inevitably take fractional values. As discussed previously in this chapter, mathematically exact *fractional delay* involves calculations requiring knowledge of the complete signal extending to infinity in both directions. This is clearly impractical, so approximations based on linear and cubic interpolation are used, which are suitable for real-time computation. Interpolation is always used when calculating the delayed signal of the flanger.

Code Example

The following C++ code fragment, adapted from the code that accompanies this book, implements a flanger with feedback:

```
// Variables used in this example whose values are set externally:
int numSamples;       // Indicates how many audio samples to process
float *channelData;  // Array of audio samples, length numSamples
float *delayData;     // Our own circular delay buffer of audio samples
int delayBufLength;  // Length of our delay buffer in samples
int dpw;              // Write pointer into the delay buffer
float ph;             // Current phase of the LFO, always between 0-1
float inverseSampleRate; // 1/f_s, where f_s is the sampling frequency

// User-adjustable effect parameters:
float frequency_;    // Frequency of the low-frequency oscillator
float sweepWidth_;   // Width of the LFO in samples
float depth_;        // Amount of delayed signal mixed with original (0-1)
float feedback_;     // Amount of feedback on the delay (>= 0, < 1)

for (int i = 0; i < numSamples; ++i)
{
    const float in = channelData[i];
    float interpolatedSample = 0.0;

    // Recalculate read pointer position with respect to write pointer.
    // More
    // efficient implementation might increment read pointer based on
    // derivative of LFO, but this format makes the operation clearer.

    float currentDelay = sweepWidth_*(0.5f + 0.5f*sinf(2.0 * M_PI * ph));

    // Subtract 3 samples to the delay pointer to make sure we have
    // enough
    // previously written samples to interpolate with
    dpr = fmodf((float)dpw - (float)(currentDelay * getSampleRate())
            + (float)delayBufLength - 3.0, (float)delayBufLength);

    // Linear interpolation to read fractional index into buffer. Find
    // fraction by
    // which read pointer sits between 2 samples, used to adjust weights
    // of samples
    float fraction = dpr - floorf(dpr);
    int previousSample = (int)floorf(dpr);
    int nextSample = (previousSample + 1) % delayBufLength;
    interpolatedSample = fraction*delayData[nextSample]
        + (1.0f-fraction)*delayData[previousSample];

    // Store current information in buffer. With feedback, what we read
    // is included
```

```
    // in what gets stored in buffer, otherwise it's just simple delay
    // line.
    delayData[dpw] = in + (interpolatedSample * feedback_);

    // Increment write pointer at constant rate. Read pointer moves at
    // different
    // rates depending on  settings of the LFO, delay and sweep width.
    if (++dpw >= delayBufLength) dpw = 0;

    // Store the output sample in the buffer, replacing the input
    channelData[i] = in + depth_ * interpolatedSample;

    // Update the LFO phase, keeping it in the range 0-1
    ph += frequency_*inverseSampleRate;
    if(ph >= 1.0) ph -= 1.0;
}
```

This code example is nearly identical to the code for vibrato. The main differences appear at the end of the example, where feedback is used on the delay buffer, and the original (dry) signal is mixed with the output:

```
delayData[dpw] = in + (interpolatedSample * feedback_);
// [...]
channelData[i] = in + depth_ * interpolatedSample;
```

For this reason, the flanger also has feedback and depth parameters where the vibrato example did not. This example code for the flanger could be used with slight modifications and different parameter values (mainly longer delay time) to implement a chorus. Complete flanger and chorus examples accompany this book, including variable LFO waveform and stereo options.

Applications

The flanger originated as a way of conveniently simulating the double-tracking effect on vocals, but its application goes well beyond the voice. Flanging is commonly used as a guitar effect (where it can be implemented with analog or digital electronics) and is often applied to drums and other instruments. The frequency of the LFO can be aligned to the tempo of the music for beat-synchronous effects.

Resonant Pitches

Recall that audio signals of a single pitch are typically composed of *harmonically-related* sinusoids, i.e., integer multiples of a fundamental frequency. Because the peaks and notches in the flanger frequency response are always uniformly spaced, they can impose a discernible resonant pitch on the audio signal. The effect is similar to being inside a resonant tube whose length changes over time according to the amount of delay. The resonance effect is particularly strong when feedback is used and when the *depth* control is at its maximum.

Avoiding Disappearing Instruments

The notches in a flanger are spaced at regular intervals in frequency, much like the harmonics of a musical instrument, where the signal consists of regular multiples of a fundamental frequency. If a flanger is applied to an instrument sound and the notches happen to line up precisely with the instrument's harmonics, it is possible for the instrument to disappear entirely! In practice, the effect will never be perfect, and the instrument will not be completed eliminated, but strange amplitude modulation effects could take place as the notches sweep up and down. This problem does not occur when flanging is applied to more noise-like signals, such as drums. Flanging can also be used on an entire mix, where the frequency content is likely to be complex enough to avoid these modulation effects.

The flanger and the chorus are nearly identical in implementation, both being based on *modulated delay lines*. The primary difference is that the chorus uses longer delay times (30 ms is a typical value) to accentuate the perception of multiple instruments playing together. Flanging and chorus are used in similar situations, though because of the greater delay between copies of the sound and corresponding perception of multiple instruments, chorus is somewhat less likely to be used on complex audio sources such as an entire mix.

Chorus

In music, the chorus effect occurs when several individual sounds with similar pitch and timbre play in unison. This phenomenon occurs naturally with a group of singers or violinists, who will always exhibit slight variations in pitch and timing, even when playing in unison. These slight variations are crucial to producing the lush or shimmering sound we are accustomed to hearing from large choirs or string sections. The chorus audio effect simulates these timing and pitch variations, making a single instrument source sound as if there were several instruments playing together.

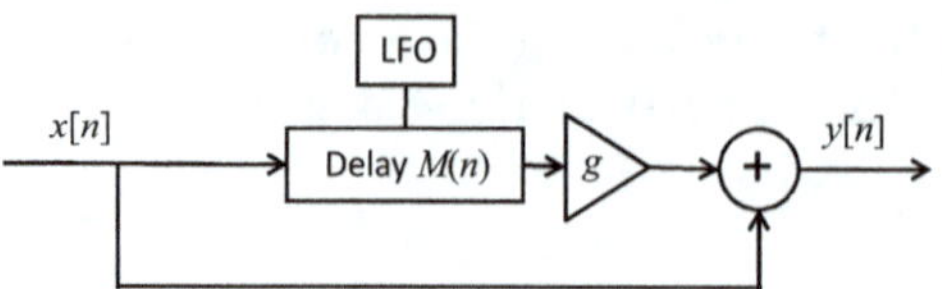

FIGURE 3.19
The flow diagram for the chorus effect including its LFO dependence. The delay changes with time.

Theory

Basic Chorus

Figure 3.19 shows a block diagram of a basic chorus, in which a delayed copy of the input signal is mixed with the original. Like the *flanger* and *vibrato* effect, the delay length varies with time (*modulated delay line*). The input-output relation can be written as:

$$y[n] = x[n] + gx\left[n - M[n]\right]. \tag{3.30}$$

Notice that this formula is identical to the basic flanger presented in the previous section. In general, the chorus effect is nearly identical to the flanger, using the same structure with different parameters. The main difference is the *delay length*, which in a chorus is usually between 20 and 30 ms, in contrast to delays between 1 and 10 ms in the flanger. We saw previously that the flanger produces patterns of *constructive* and *destructive interference*, resulting in a frequency response characterized by evenly-spaced peaks and notches:

$$\left|H\left(e^{j\omega}\right)\right| = \sqrt{1 + 2g\cos(\omega M[n]) + g^2}. \tag{3.31}$$

With the peaks (points of maximum frequency response) located at:

$$\omega_p = 2\pi p/M \quad \text{where} \quad p = 0,1,2,\ldots M-1. \tag{3.32}$$

And the notches (minimum frequency response) at:

$$\omega_n = \pi(2n+1)/M \quad \text{where} \quad n = 0,1,2,\ldots M-1. \tag{3.33}$$

Thus, the *comb filtering* produced by the flanger also occurs in the chorus. However, the longer delay $M[n]$ substantially alters its perceived effect. At a sample rate of 48 kHz, a 30 ms delay corresponds to M=1,440 samples. There will therefore be 1,440 peaks and 1,440 notches in the frequency response, each located at intervals of f_s/M (recalling that $\omega = 2\pi$ corresponds to the sampling frequency f_s). Thus a peak will occur every 33.3 Hz, with notches likewise spaced every 33.3 Hz. These are close enough together that the characteristic sweeping timbre of the flanger is no longer perceptible. In particular, any sense that the comb filter has a definite *pitch* (owing to its regularly spaced peaks) will be lost at such close spacing. Nonetheless, though the sound is different than the flanger, this comb filtering is an important part of the sonic signature of the chorus effect.

Considered a different way, a delay on the order of 20–30 ms begins to approach the threshold where two separate sonic events can be perceived, though a clear perception of an echo requires a longer delay still (100 ms or more). So the chorus can also be heard as two separate copies of the same sound, whose exact timing relationship changes over time as $M[n]$ changes. Both understandings are mathematically correct; the only difference lies in human audio perception.

Low-Frequency Oscillator

In the chorus, as in the flanger, the delay length $M[n]$ varies under the control of a *low-frequency oscillator* (*LFO*). Several waveforms can be used for the LFO, with sinusoids being the most common. In comparison to the flanger, slower LFO *sweep rates* (3 Hz or less) but higher LFO *sweep widths* (5 ms or more) are typically used. Like all modulated delay effects, *interpolation* is used to calculate the output of the delay line whenever $M[n]$ is not an integer.

Pitch-Shifting in the Chorus

The wider sweep width (delay variation) in the chorus has an important consequence on the pitch of the delayed sound. As was shown for the vibrato effect, changing the length of a delay line introduces a *pitch shift* into its output, where lengthening the delay scales all frequencies in a signal down, and reducing the delay scales them up. Recall the formula for pitch shift as a function of LFO frequency f, sweep width W, and sample rate f_s:

$$f_{\text{ratio}}[n] \approx 1 - 2\pi f W \cos(2\pi n f / f_s). \tag{3.34}$$

Given that the cosine function ranges from –1 to 1, we can find the maximum pitch shift for any given set of parameters as:

$$f_{\text{ratio,max}} = 1 + 2\pi f W. \tag{3.35}$$

For $f = 1$ Hz and sweep width $W = 10$ ms, this results in a pitch ratio of 1.063 (6.3% variation), slightly more than a semitone (5.9%) in either direction. This is a noticeable amount of tuning variation between the original and delayed copies of the signal, which can simulate and even exaggerate the natural variation in pitch between musicians playing in unison.

Note that in comparison to the vibrato effect, the chorus mixes the original and delayed copies, so a single pitch shift is not heard.

Multi-Voice Chorus

The basic chorus can be considered as a *single voice chorus* in that it adds a single delayed copy to the original signal. A *multi-voice chorus*, by contrast, involves several delayed copies of the input signal mixed together, with each delayed copy moving independently. Figure 3.20 shows a diagram for an arrangement with two delayed copies (*dual-voice*). Each individual voice can be analyzed identically to the basic chorus described in the preceding sections, but the sum total of all voices will produce a more complex, richer tone, suggesting multiple instruments played in unison. The Implementation section below discusses control strategies for the voices in a multi-voice chorus.

Stereo Chorus

Stereo chorus is a variation on multi-voice chorus, where each delayed copy of the signal is panned to a different location in the stereo field. When two voices are used, the delayed signals are typically panned completely to the left and right, with the original signal in the center. In this case, the two voices are usually run in *quadrature phase*: each LFO has the same *sweep rate* and same *sweep width*, but they differ in phase by 90°. When more than two voices are used, they may be spread evenly across the stereo field or split into two groups, with one group panned hard left and the other group panned hard right.

Properties

The chorus is implemented identically to a flanger without feedback, so it shares the same properties. Notably, since delay and mixing are *linear* operations, the chorus is a linear effect (for any number of voices).

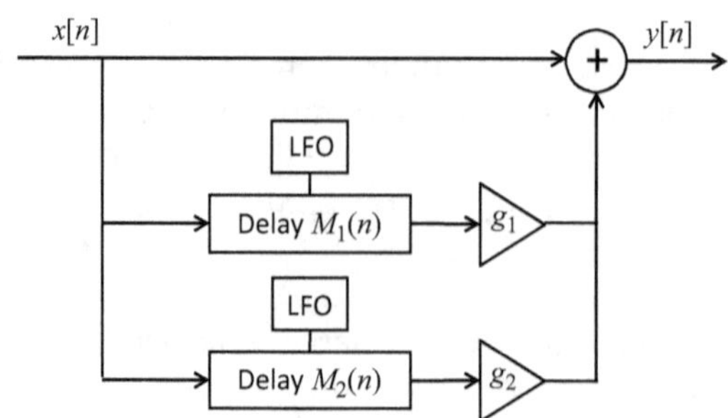

FIGURE 3.20
A multi-voice chorus diagram.

The delay $M[n]$ changes over time under the control of the LFO, so the chorus is a *time-variant* effect: an input signal applied at one time may produce a different result than the same signal applied at a different time if the LFO phase differs. Since the chorus never involves feedback, it is always *stable*, with a bounded input producing a bounded output.

Common Parameters

Chorus effects have several user-adjustable controls:

Depth (or Mix)

Like the flanger, the *depth* controls affect the amount of delayed signal(s) which are mixed in with the original. $g=0$ produces no effect, where $g=1$ produces the most pronounced chorus effect. Higher depth settings ($g>1$) make the delayed copies louder than the original, a setting rarely found in practice as it produces a weaker chorus effect than $g=1$. Some simple chorus effects may not have this control (always setting g to 1). Confusingly, the term *depth* is also sometimes used to refer to *sweep width*, or the amount of variation in the delay. When examining an existing chorus effect, it is thus important to find out what the depth control means.

Delay and Sweep Width

The *delay* parameter on the chorus controls the minimum amount of delay $M[n]$. Typical values are on the order of 20 to 30 ms, and this setting represents one of the primary differences between flanger and chorus. The *sweep width* (which is sometimes called *sweep depth*, but should not be confused with *depth/mix*) controls the amount of additional delay added by the LFO. In other words, *sweep width* controls the amplitude of the LFO, and the maximum delay is given by the sum of *delay* and *sweep width*. The relationship between the delay and sweep width parameters was depicted in Figure 3.18. Typical values for sweep width range from 1–2 ms to 10 ms or more. A larger sweep width will result in more pitch, creating a warbling effect, whereas changing the delay parameter will not affect the pitch modulation.

Speed and Waveform

As in the flanger, the *speed* (or *sweep rate*) sets the number of cycles per second of the LFO controlling the delay time. In addition to producing a more quickly oscillating chorus effect, higher speed will produce more pronounced pitch modulation for the same *sweep width*, since the delay line length will be changing more quickly over time. Typical values in the chorus are slower than in the flanger, ranging from roughly 0.1 to 3 Hz.

The *waveform* control selects one of several predefined LFO waveforms, including sine, triangle, sawtooth, or exponential (triangular in log-frequency). In addition to controlling how the voices move in time, each waveform will affect the type of pitch modulation. For example, the derivative of a sine wave is a cosine, which is always changing, so the pitch is always changing as well. A triangular waveform, though, only has two discrete slopes, so the pitch will jump back and forth between two fixed values. Many chorus effects do not offer a *waveform* parameter and always use a sine LFO. Like the flanger, this results in a total delay $M[n]$ given by:

$$M[n] = M_0 + M_W\left[1 + \sin(2\pi n f_{LFO}/f_s)\right]/2. \quad (3.36)$$

where M_0 (in samples) is given by the *delay* control, M_W (in samples) is given by the *sweep width* control, f_{LFO} (in Hertz) is given by the *speed* control and f_s (in Hertz) is the sampling frequency. $M[n]$ thus varies from M_0 to $M_0 + M_W$.

Number of Voices

As discussed in the previous section, a *multi-voice* chorus uses more than one delayed copy of the input sound, simulating the effect of more than two instruments being played in unison. Many chorus units give the option of choosing the number of voices. In a simple implementation, each voice could be controlled by the same LFO, but with a different phase. The delay time will be different for each voice since they are at different points in the waveform, but they will remain synchronized to one another over time. More complex implementations can use different LFO waveforms and speeds for each voice.

Other Variations

When multiple instruments play in unison, the variations between them are likely to be more random than periodic. Instead of using a fixed-speed LFO, the delay time between voices could be changed in a more irregular, quasi-random fashion (keeping in mind that abrupt changes in delay will produce audible pitch artifacts). Another variation is to modulate the amplitude of each voice to model the fact that musicians playing in unison will not all have the same relative loudness.

Summary: Flanger and Chorus Compared

The chorus and flanger effects are nearly identical in structure, both being based on modulated delay lines. The main differences are in the parameter settings: the chorus uses a longer *delay time* (30 ms is a typical value) than the flanger and often a larger *sweep width*. These together produce more sense of separation between the original and delayed copies of the signal and more pitch modulation. The chorus tends to use a lower *speed* or *sweep rate* than the flanger, though there is a significant area of overlap. One structural difference is that the flanger can use *feedback* to produce a more intense effect, whereas this is almost never found in the chorus. On the other hand, the chorus can use more than one delayed copy of the sound (*multi-voice* chorus) where the flanger only uses one copy (except in a stereo flanger). When chorus and flanger effects are implemented in stereo, the same procedure is used in both cases, panning one delayed copy to the left and one to the right, though a multi-voice chorus with more than two delayed copies allows further variations on this procedure.

Flanging and chorus are used in similar situations, though because of the greater delay between copies of the sound and corresponding perception of multiple instruments, chorus is somewhat less likely to be used on complex audio sources such as an entire mix.

Further Reading

Use and key concepts related to delay are nicely covered in [17]. Delay is a very mature audio effect, with recent research focusing more on perception and preference than on new implementations [18]. Many of the related concepts, such as fractional delay filters, are also well-established [19]. On the other hand, LFO-driven delay-based effects might be considered under-researched, though there has been significant recent work on neural modeling of such effects [20]. Another active area is estimation of time delays, which is useful for estimating sound source locations, and often important in comb filter prevention [21], source separation [22], source localization [23], and dereverberation [24,25] algorithms.

Problems

1. Suppose we want to delay a signal by 2.5 samples. Briefly explain two different methods of calculating the new signal and their relative advantages and disadvantages.
2. Consider a signal $x[0]=0.8, x[1]=0.4, x[2]=0.1, x[3]=-0.15, x[4]=-0.4$. Use zero, first-, and second-order interpolation to estimate the value $x[1.7]$.

3.
    i. Draw a block diagram for delay with feedback. Label the input $x[n]$, output $y[n]$ and any other commonly used parameters.
    ii. Under what conditions is the system stable? Why?
4. A delay without feedback produces notches at 300 and 900 Hz. List at least three other frequencies where notches will also be present, and explain why. For a sample rate of 48 kHz, how many samples of delay could produce this comb filter?
5. There are two microphones on a guitar, one close microphone placed only 5 cm away, picking up the direct sound, and another placed 1.5 m away from the guitar, picking up the room sound. How much delay should be added to the close microphone to align the two signals?
6.
    i. Draw a block diagram of a basic flanger. Now draw a block diagram of a flanger incorporating feedback or regeneration.
    ii. Derive the frequency response of a flanger without feedback.
    iii. Based on your block diagram in part (i), write the difference equation(s) for a flanger **with** feedback, relating the output $y[n]$ to the input $x[n]$. Define any other variables you use in your equation(s).
    iv. Describe whether or not the flanger is *linear, time-invariant,* and *stable.* Explain why, and whether the answer depends on parameter settings or whether feedback is used.
7. Define the following parameters for a flanger: *depth, delay, sweep width,* and *sweep rate.* Describe the effect of varying their settings.
8. Why do we not hear an echo when flanging is applied?
9. How does chorus differ from flanger in terms of the LFO settings and use of feedback?
10. Suppose we implement the flanger with a circular buffer. We may hear artefacts as the delay length changes. Explain why this occurs and suggest a solution to overcome it.

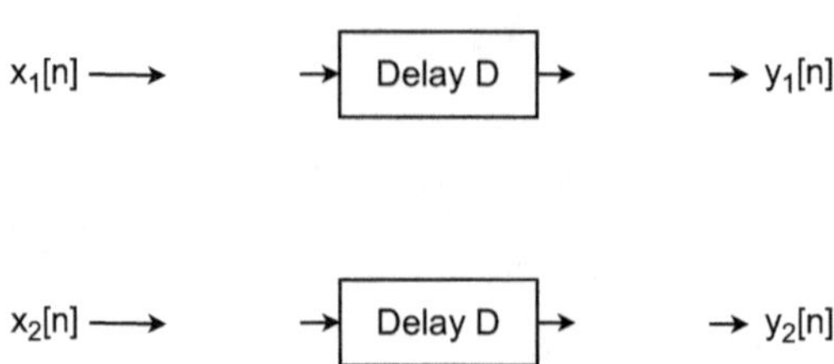

FIGURE 3.21
Template for a ping-pong delay.

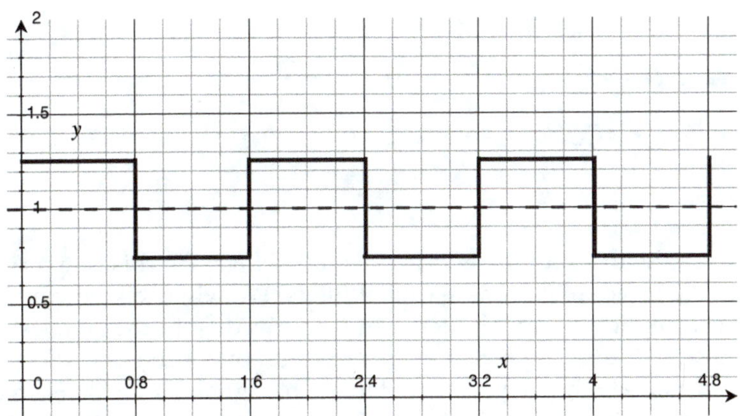

FIGURE 3.22
Relative pitch at the output of a vibrato unit.

11.
  i. Starting with the template in Figure 3.21, draw a block diagram for a ping-pong delay with mono input and stereo output. Label the inputs $x_1[n]$ and $x_2[n]$, the outputs $y_1[n]$ and $y_2[n]$, and any other commonly used parameters.
  ii. Give the transfer function and difference equation for ping-pong delay. That is, find $Y_1(z)$ and $Y_2(z)$ in terms of $X_1(z)$ and $X_2(z)$.
  iii. Suppose the ping pong delay was extended to four channel-output, with the input sound bounced sequentially around them. Draw a block diagram for this arrangement.
  iv. With delays of length D, how large does the total buffer need to be (in samples) for the two and four-channel ping-pong delays?

12.
  i. For a vibrato effect, suppose that the delay time is given by $d(t)=M+W\sin(2pft)$, where M is the average delay, W is the sweep width and f is the LFO frequency. If $M=5$ ms and $f=6\,\text{Hz}$, find the value of W needed to give a maximum pitch shift of 1.03 (roughly half a semitone). You can leave the result as a fraction.
  ii. The plot in Figure 3.22 shows the relative pitch at the output of a vibrato unit. What LFO waveform was used to produce this result, and why?

4

Filter Design

Filters are the basis for many types of audio effects. In this chapter, we'll look at the basics of designing digital filters. We will show how many of the most important digital filters can be derived from the simplest possible designs. Several types of filters are commonly used in audio effects for manipulating the frequency content of the signal. These filters, in their ideal form, are depicted in Figure 4.1. They can be summarized as follows.

Low pass, high pass, and band pass filters: the first class of filters aims to totally eliminate certain frequency ranges from the signal. The *low pass filter* passes low frequencies below some *crossover frequency*[1] ω_c while eliminating all frequencies above the crossover. The *high pass filter* is a mirror image of this. It passes all frequencies above the crossover frequency ω_c and eliminates all frequencies below it. *Band pass filters* pass a range of frequencies. They are defined by the *center frequency* ω_c and the *bandwidth B,* which specify the location and width of the *passband,* respectively. Frequencies above and below this passband are blocked. The inverse of the band pass filter is the *band stop filter,* which blocks frequencies in its *stop band* while passing frequencies above and below.

Shelving, peaking, and notch filters: this class of filters does not aim to entirely eliminate any frequencies, but rather aims to adjust the relative gain of specific portions of the frequency spectrum. *Shelving filters* come in two forms: *low shelving* and *high shelving* filters. They are analogous to low pass and high pass filters, but instead of eliminating high or low frequency, they provide a boost or cut on the shelf region and leave all other frequency content unaffected. *Peaking* and *notch filters* are analogous to band pass and band stop filters in that they affect only a narrow range of the spectrum. They apply a boost or cut to a specific frequency band while leaving the area around it unchanged. Like band pass and band stop filters, peaking and notch filters are often specified by a *center frequency* and *bandwidth.* The center frequency defines the location of maximum gain in a peaking filter, or minimum gain in a notch filter. The bandwidth defines the size of the region around the center frequency. The gain is also typically a parameter in peaking and notch filter design. These will be explained in detail later in this chapter.

DOI: 10.1201/9781003593942-4

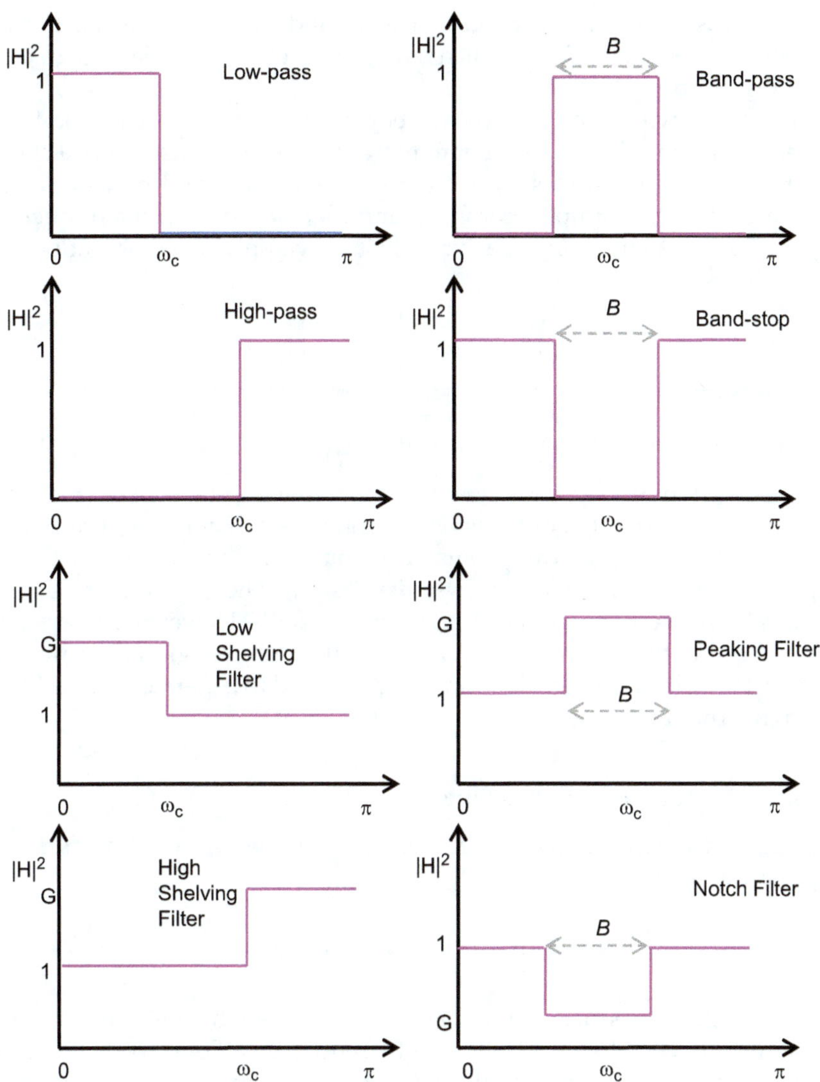

FIGURE 4.1
Ideal filters.

In Chapter 5, we will see how these filters are used in audio effects like equalizers. But let us first study their general behavior. In practice, we cannot achieve the perfectly sharp crossover between pass band and stop band shown in Figure 4.1, and in fact a perfectly sharp crossover would not be aesthetically desirable. Rather, in audio contexts, we usually desire a smooth transition between regions. As we will see, the *order* of a filter (a rough measure of its complexity) also determines how sharp the crossover is between pass band and stop band. Audio effects most commonly use low-order filters, which have the benefits of simpler design and less susceptibility to error.

In the following sections, we will begin with a very simple prototype filter and gradually generalize it to more complex designs. We will show that all the common classes of filter can be constructed by a series of operations on this simple prototype, and that we can alter the frequency spectrum of a signal almost arbitrarily just by applying combinations of such filters.

Filter Construction and Transformation

In this section, we examine how a simple filter can be constructed and how it can be extended or transformed into other designs. We keep most of the discussion regarding transformations quite general, so that they apply to high-order as well as low-order designs. There is a fair amount of maths in this section, and we don't skip the details. However, readers can skip ahead to the next section on Popular IIR Filter Design if they are just interested in how these transformations are used to generate various filter designs from Figure 4.1.

Simple, Prototype Low Pass Filter

Consider averaging every two consecutive samples. In the time domain, the output is given as;

$$y[n]=\left(x[n]+x[n-1]\right)/2. \tag{4.1}$$

This equation produces a low pass filter. To see why, let's consider its response at very low and very high frequencies. For a very low frequency input, the signal hardly changes from sample to sample. In fact, if the input signal has frequency 0, $x[n] = A\cos(2\pi\ 0\ n/f_s)$, then it is constant. This is known as DC input, since direct current electrical signals have this quality. For DC input, the output of this filter is identical to the input.

Now suppose we have a very high frequency signal at half the sampling frequency. So $x[n] = A\cos(2\pi(f_s/2)n/f_s) = A\cos(\pi n)$. Thus, the signal switches sign from sample to sample, and the output of this filter is zero.

Let's look at this filter in the frequency domain. Recall from Chapter 1 that we can express this in the Z domain as $Y(z)=\left(X(z)+z^{-1}X(z)\right)/2$. Its transfer function, written in positive powers of z, is

$$H(z)=\frac{z+1}{2z}, \tag{4.2}$$

and the square magnitude of this transfer function is

$$|H(z)|^2=\frac{1+z}{2z}\frac{1+z^{-1}}{2z^{-1}}=\frac{2+z+z^{-1}}{4}=\frac{1+\cos\omega}{2}. \tag{4.3}$$

Hence, this acts as a low pass filter, allowing low frequencies to pass through to the output, but removing high-frequency content. You can easily see that for $f=0$, $H(z=1)=1$ and for $f=f_s/2$, $H(z=-1)=0$. But what happens halfway, at $f=f_s/4$? Here, $\omega=2\pi f/f_s=\pi/2$, we see that $|H(z=j)|=1/\sqrt{2}$, or –3 dB. We call this the crossover frequency of our low pass filter.

Defining the Gain at Crossover Frequency

There is no single definition of the crossover frequency. Generally, the crossover frequency is where the frequency response makes the transition between two values. In some cases, the crossover frequency is defined as that frequency at which the power level of the signal decreases by 3 dB from its maximum value, known as the –3 dB point. This is problematic because some filters, like a small shelving filter, may never decrease as much as 3 dB.

One of the most effective definitions, and the main one that we will use in this chapter, is that the crossover frequency f_c is that frequency at which the square magnitude of the filter's transfer function is half way between its two extremes.

That is, consider a filter with transfer function $H(f)$, or equivalently $H(z)$ where $z=e^{j\omega}=e^{j2\pi f/f_s}$. Then the gain at the crossover frequency, $G_c=|H(f_c)|$, may be defined using the arithmetic mean of the extremes of the square magnitude response,

$$G_c^2=\left(G_{\max}^2+G_{\min}^2\right)/2. \tag{4.4}$$

This is the simplest definition, and one we will use in examples that show how to generate simple low-order filters.

This definition works well for low pass and high pass filters, and the shelving filters, but the band pass, band stop, peaking, and notch filters have additional parameters that relate to bandwidth. We previously specified that the center frequency of such filters is where the filter reaches its maximum or minimum value, but what about bandwidth? Well, the gain at the bandwidth may be defined similarly to the gain at the crossover frequency. But now there are two crossover frequencies, one on either side of the center frequency.

In Chapter 5, we will consider another definition of the gain at the crossover frequency that is often used for shelving, peaking and notch filters, and has particular benefits in terms of ensuring symmetric designs when using decibel scales.

Changing the Gain at the Crossover Frequency

Suppose that at some frequency ω_G we have

$$|H\left(z=e^{j\omega_G}\right)|^2=G^2. \tag{4.5}$$

Now consider a transformation, given by

$$F(z)=\frac{z-\alpha}{1-\alpha z},\ \alpha=\frac{\tan(\omega_G/2)-1}{\tan(\omega_G/2)+1}. \tag{4.6}$$

This has a few important properties. Every value inside the unit circle is mapped to a value inside the unit circle; those outside are mapped to values outside the unit circle, and those on the unit circle stay on the unit circle. So we can view this transfer function as preserving stability. Furthermore, it maps $z=1$ to 1 and $z=-1$ to -1.

But because of how α was defined, it has another important property,

$$F\left(z=e^{j\pi/2}\right)=e^{j\omega_G}. \tag{4.7}$$

That is, F maps this frequency ω_G to $\pi/2$. Hence, the square magnitude of $H(F(z))$ at $\omega = \pi/2$ is the same as the square magnitude of $H(z)$ at $\omega = \omega_G$.

In particular, for our first-order prototype, it can be shown from Eqs. (4.3) and (4.5) that,

$$\tan(\omega_G/2)=(1/G^2-1)^{1/2}. \tag{4.8}$$

So we know exactly the transformation to apply in order to change the gain at the crossover frequency to a given value G.

This also provides an easy way to change our definition of the crossover frequency. Decide what we want the crossover gain to be, then use (4.8) to find the frequency with this gain, and then (4.6) to derive the filter that uses this new gain at the crossover frequency.

Shifting the Crossover Frequency

So far, our crossover frequency has been set to $\pi/2$. We would like to be able to set the crossover frequency to any value between 0 and π. This

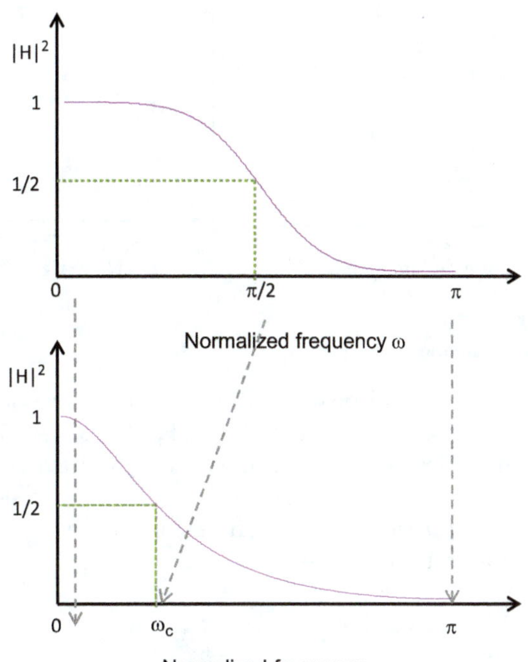

FIGURE 4.2
Shifting the crossover frequency of a low pass filter.

would allow us to change the location that provides the transition between where frequencies are passed and blocked.

Consider another transformation, given by

$$F(z) = \frac{z-\beta}{1-\beta z}, \beta = \frac{1-\tan(\omega_c/2)}{1+\tan(\omega_c/2)}, \tag{4.9}$$

This transfer function will map some new crossover frequency ω_c to the crossover frequency of our prototype filter, $\pi/2$, as shown in Figure 4.2.

$$F\left(z = e^{j\omega_c}\right) = e^{j\pi/2}. \tag{4.10}$$

Creating a Low Shelving Filter

We can construct a low shelving filter by transforming our prototype filter, such that the square magnitude response is transformed from H^2 to $(G^2 - 1)H^2 + 1$, depicted in Figure 4.3. This transformation changes the extreme square magnitudes 0 and 1 of a low pass design to 1 and G^2.

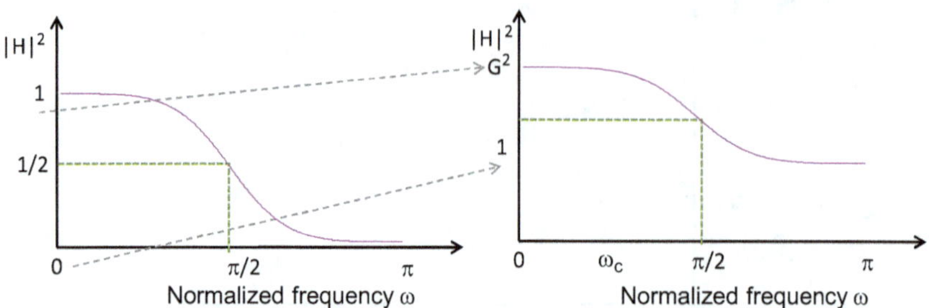

FIGURE 4.3
Shelving filter transformation.

Recall that the poles will push up the magnitude response for nearby frequencies, and the zeros will pull it down. For the low shelving filter, we want to keep the poles on the imaginary axis, giving a sharp crossover frequency. But now we shift the zeros so that, for each first-order section $H(z = 1) = g = G^{1/N}$ and $H(z = -1) = 1$. That is, if the first-order section of a prototype low pass filter is written as

$$H_{LP} = k\frac{z-q}{z-p}. \tag{4.11}$$

Then the first-order section of the low shelving filter becomes,

$$H_{LS} = \frac{1}{2}\frac{\left[1+p+g(1-p)\right]z-\left[1+p-g(1-p)\right]}{z-p}. \tag{4.12}$$

So $H(z = 1) = g$, $H(z = -1) = 1$.

Inverting the Magnitude Response

For a high pass filter, we want $|H(z = 1)|^2 = 0$, $|H(z = -1)|^2 = 1$ and $|H(z = \omega_c)|^2 = G^2$. So we apply the following simple transformation,

$$F_{HP}(z) = -z. \tag{4.13}$$

This transformation is shown in Figure 4.4.

Low pass to Band Pass Transformation

Now we want to create a band pass filter from our simple filter with center frequency $\pi/2$. We would like to transform the frequency range 0 to π to the frequency range $-\pi$ to π. If this transformation is applied to the input before a low pass filter is applied, then our low pass filter becomes a band

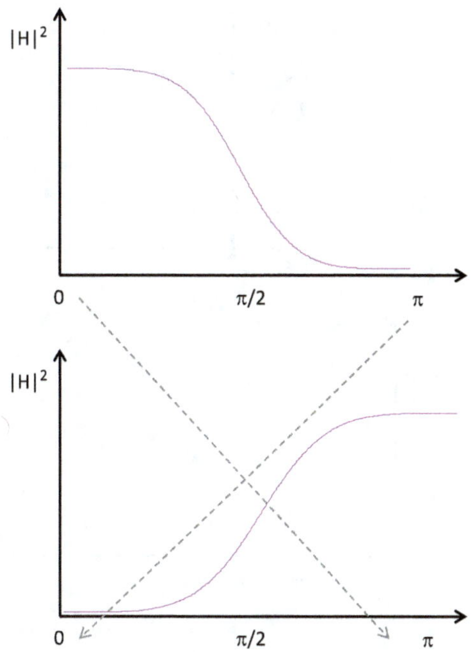

FIGURE 4.4
Reversing the z domain to turn a low pass into a high pass filter.

pass filter, as shown in Figure 4.5. To do this, consider a transfer function $F(z)$ with the following constraints;

$$
\begin{aligned}
&F(z) = -\frac{z^2 + \alpha_1 z + \alpha_2}{\alpha_2 z^2 + \alpha_1 z + 1} \\
&F\left(z = e^{j0}\right) = e^{-j\pi} = -1 \\
&F\left(z = e^{j\omega_l}\right) = e^{-j\pi/2} = -j \\
&F\left(z = e^{j\omega_c}\right) = e^{j0} = 1 \\
&F\left(z = e^{j\omega_u}\right) = e^{j\pi/2} = j \\
&F\left(z = e^{j\pi}\right) = e^{j\pi} = -1.
\end{aligned}
\tag{4.14}
$$

F will move the lower and upper crossover frequencies to $\pm\pi/2$, where our prototype low pass filter has its crossover frequency, and it will move the center frequency to 0, where our prototype low pass filter has gain equal to 1.

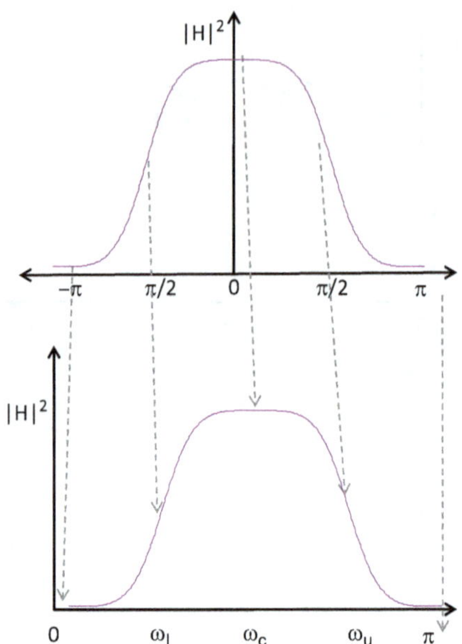

FIGURE 4.5
Transforming a low pass filter into a band pass filter.

We can solve Eq. (4.14) to arrive at

$$H_{\mathrm{BP}}(z) = H(F(Z)) = k\frac{z^2 - z(1+q)\cos\omega_c + q}{z^2 - z(1+p)\cos\omega_c + p}. \tag{4.15}$$

This transfer function has second-order polynomials in the numerator and the denominator. So our first-order section has now become a second-order section.

High-Order IIR Filter Design

In most audio production applications, low-order filters are used. The smooth transitions are often advantageous in that the effect of filtering is subtle and not perceived as an artefact. However, they lack the ability for fine grain control. To address this, higher order IIR filters can be employed. They provide steep transitions at crossover frequencies for shelving filters, or at the upper and lower crossover frequencies for peaking and notch filters.

Orfanidis [26] suggested techniques for the design of high-order minimum phase filters based on a variety of classic filters: Butterworth, Chebyshev I, Chebyshev II, and elliptic polynomials. However, all these high-order designs begin with analog prototypes. But it is possible to

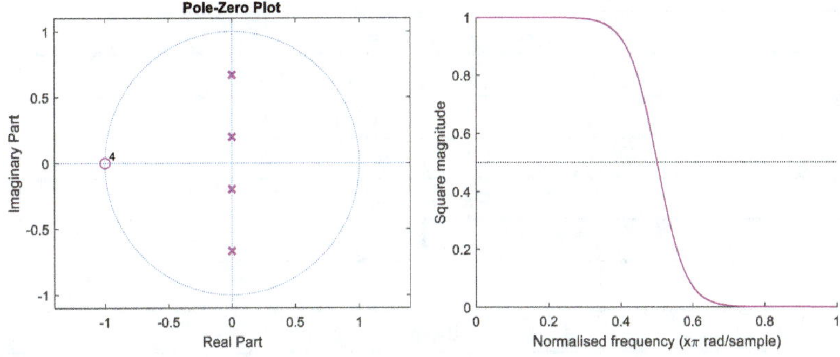

FIGURE 4.6
Pole zero plot (top) and square magnitude response for a fourth-order prototype low pass filter.

design high-order filters entirely in the digital domain, as described in [27,28] and summarized herein.

Let's return to our simple low pass filter. It has one zero at $z = -1$ and one pole at $z = 0$. Replace the prototype first-order filter with an Nth order prototype filter,

$$H_P(z) = \prod_{n=1}^{N} \frac{1}{2\cos\gamma_n} \frac{z+1}{z - j\tan\gamma_n}, \quad \gamma_n = \frac{(2n-1-N)\pi}{4N}, \quad n = 1,2\ldots N. \tag{4.16}$$

A fourth-order prototype filter is depicted in Figure 4.6. These filters have the same crossover frequency and the same behavior at DC and at Nyquist (half the sampling frequency, $\omega = \pi$) as the first-order filter. But now the poles result in a very sharp transition at $\omega = \pi/2$. In fact, the square magnitude is now

$$|H_P(z)|^2 = \frac{1}{1 + \tan^{2N}(\omega/2)}. \tag{4.17}$$

We now have a high-order filter, composed of first-order sections. Note that these first-order sections have complex valued coefficients. But they come in complex conjugate pairs, so they can also be written as second-order sections with real valued coefficients.

Popular IIR Filter Design

It is possible to design finite impulse response (FIR) filters, where the output is dependent only on current and previous inputs, and not on previous outputs, to create filters whose ideal forms were depicted in Figure 4.1.

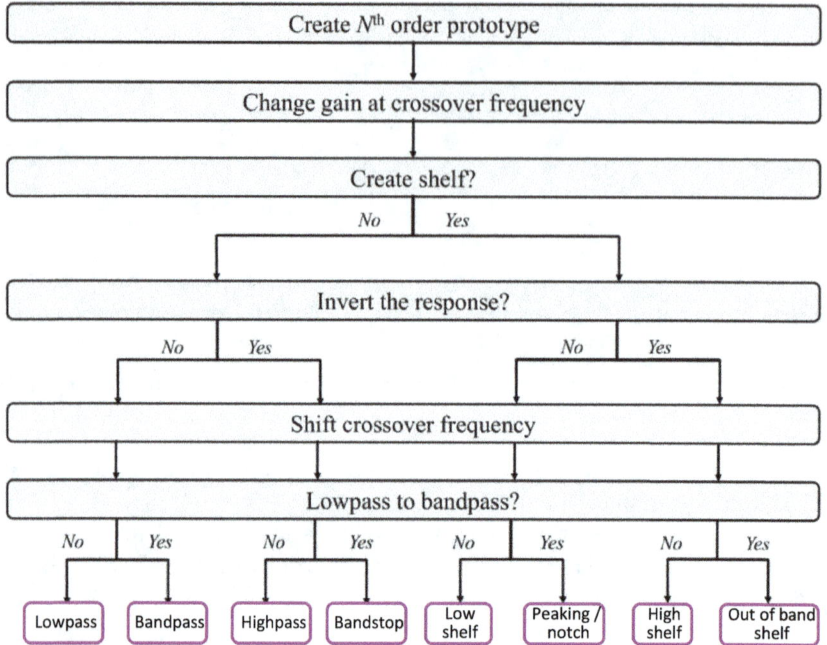

FIGURE 4.7
Flow diagram of a design technique for high-order IIR filters. Starting with a simple Nth order prototype filter, a series of transformations are made on each first-order section to transform this into any of the main higher order filter types.

However, FIR designs either give poor approximations to the ideal forms or must be very high order and hence introduce significant delay and computational issues. Thus, preferred designs are often IIR (infinite impulse response) filters, where the output is fed back to the input. This section will use the transformations from the previous section to construct such filters.

The steps to follow in these transformations are depicted in Figure 4.7. We first construct a high-order prototype filter of some order and with prescribed gain at $\omega = \pi/2$. Then a sequence of transformations can be applied to each first-order section, in order to turn this filter into any of the filters given in Figure 4.1. For each filter, we will describe the steps and give an example of the lowest order where crossover frequency or bandwidth is defined using Eq. (4.4).

Low Pass

As mentioned, a low pass filter has a transfer function H_{LP} with magnitude 1 at frequency 0 and magnitude 0 at frequency $\omega = \pi$ $(f = f_s/2)$. That is, the magnitude of the lowest frequencies is unaffected, and the highest frequencies are eliminated. At some crossover frequency ω_c, it has magnitude G_c, and this represents the transition where frequencies below ω_c are considered passed and above this are rejected.

In order to generate a high-order low pass filter, we first generate our high-order prototype filter. Then by applying these transformations to each first-order section in our high-order prototype filter, we can generate a high-order low pass filter.

We use Eq. (4.6) to transform each first-order section in the Nth order prototype filter such that each first-order section has magnitude $G_c^{1/N}$, rather than $(1/2)^{1/(2N)}$, at $\omega = \pi/2$ (that is, the whole filter has square magnitude G_c^2, rather than 1/2 at $\omega = \pi/2$). Then we apply Eq. (4.9) to shift the crossover frequency so that each first-order section has magnitude $G_c^{1/N}$ at ω_c, rather than at $\pi/2$.

Consider a simple first-order case where we define the gain at the crossover frequency such that the square magnitude is the average of the two extremes, $G_c^2 = (0^2 + 1^2)/2 = 1/2$.

We start with the simple prototype low pass filter whose transfer function is given by (4.2). Since this definition of gain at the crossover is the same definition used in the prototype filter, there is no need to change the gain at the crossover frequency.

Now we shift the crossover frequency from $\pi/2$ to ω_c,

$$H_{LP}(z) = \frac{(1-\beta)}{2}\frac{z+1}{z-\beta} = \frac{(z+1)\tan(\omega_c/2)}{z(\tan(\omega_c/2)+1)+\tan(\omega_c/2)-1}. \tag{4.18}$$

Where β was defined as in (4.9).

Figure 4.8 shows a pole zero plot and square magnitude response of a first-order and a fourth-order low pass filter with crossover frequency of $\pi/4$. In both cases, the zeros are at –1. Notice that, for the fourth-order design, the poles are placed on the arc of a circle so as to pull down the magnitude sharply at ω_c.

High Pass

A high pass filter has a transfer function H_{HP} with magnitude 0 at frequency $\omega = 0$, magnitude 1 at frequency $\omega = \pi$, and magnitude G_c at some crossover frequency ω_c.

To generate a high pass filter with crossover frequency ω_c, we begin with the high-order prototype filter. Then we transform each first-order section, such that it has magnitude $G_c^{1/N}$ at $\omega = \pi/2$. Then we invert the magnitude response of each of these sections, and then shift the crossover frequency of each section from $\pi/2$ to ω_c.

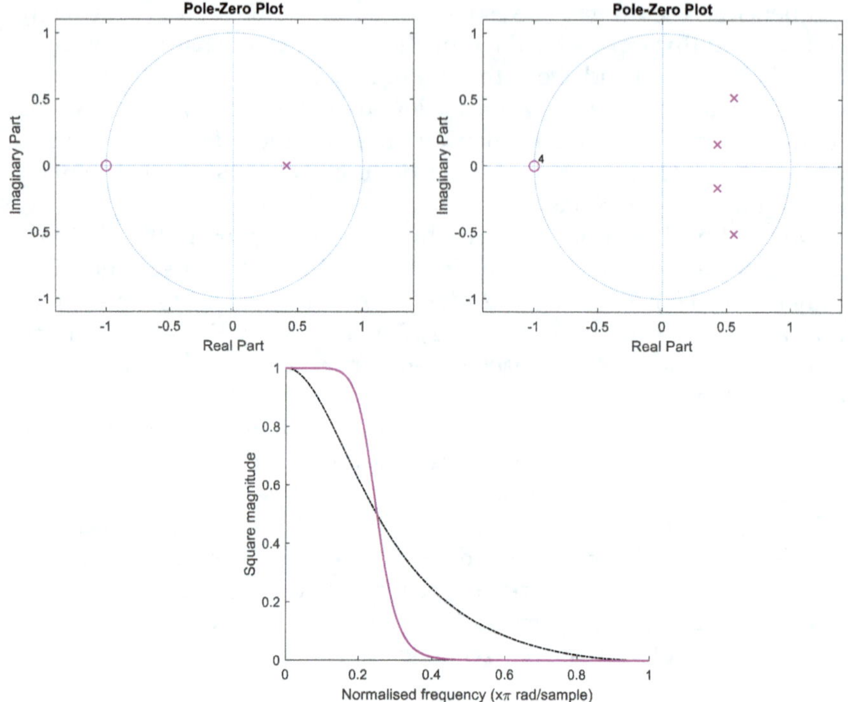

FIGURE 4.8
Pole zero plot for a first-order (top left) and fourth-order (top right) low pass filter with center frequency $\omega_c = \pi/4$. On the bottom, square magnitude response for the first-order (black) and fourth-order (purple) filters.

Let us again consider the first-order case where $G^2 = 1/2$. There is no need to change the gain at the crossover frequency, so we begin with Eq. (4.2). Then we invert the magnitude response using (4.13), giving

$$H_{\mathrm{HP},\omega_c=\pi/2}(z) = \frac{1}{2}\frac{z-1}{z}. \tag{4.19}$$

And now we shift the crossover frequency,

$$H_{\mathrm{HP}}(z) = \frac{1}{2}\frac{(1+\beta)(z-1)}{z-\beta}) = \frac{z-1}{(\tan(\omega_c/2)+1)z+\tan(\omega_c/2)-1}, \tag{4.20}$$

Figure 4.9 shows a pole zero plot and square magnitude response of a first-order and a fourth-order low pass filter with crossover frequency of $\pi/4$. These are identical to the low pass filters, except now the zeros have been moved from –1 to +1, giving maximal attenuation at the low frequencies.

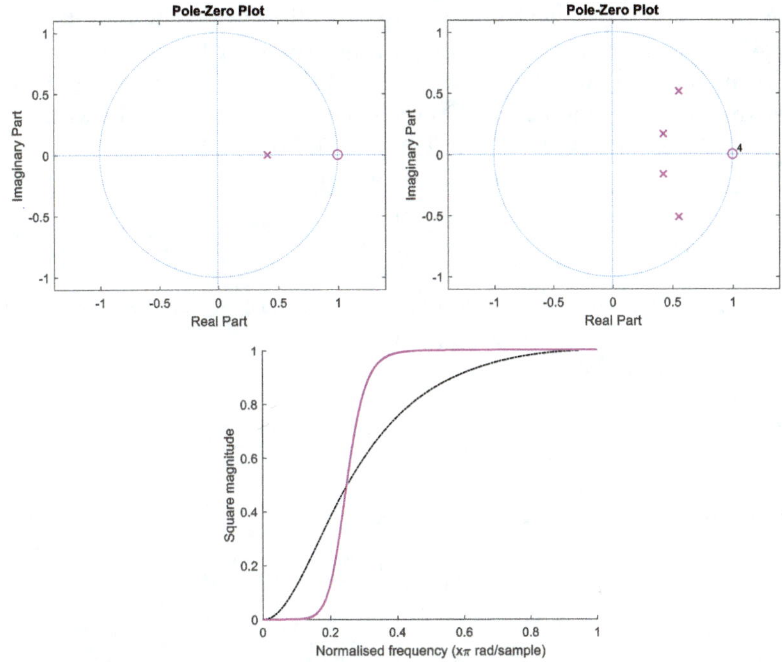

FIGURE 4.9
Pole zero plot for a first-order (top left) and fourth-order (top right) high pass filter with center frequency $\omega_c = \pi/4$. On the bottom, square magnitude response for the first-order (black) and fourth-order (purple) filters.

Low Shelf

A low shelving filter has a transfer function H_{LS} with magnitude G at frequency $\omega = 0$ (representing the low shelf), magnitude 1 at frequency $\omega = \pi$, and magnitude G_c at some crossover frequency ω_c.

As before, we start with our high-order prototype filter. We will change the gain at the crossover frequency to some value g. We will transform the range of square magnitudes of the prototype filter to the range of square magnitudes of the shelving filter. That is, we want the extreme square magnitudes, 0 and 1, of the low pass filter, 0 and 1, to map to the extreme square magnitudes of the low pass filter, 1 and G^2, with g^2 mapping to $G_c{}^2$. Thus,

$$\frac{g^2 - 0}{1 - 0} = \frac{G_c{}^2 - 1}{G^2 - 1} \rightarrow g^2 = \frac{G_c{}^2 - 1}{G^2 - 1}. \tag{4.21}$$

So for each first-order section of an Nth order prototype filter, we apply Eqs. (4.6) and (4.21) to change the gain at the crossover frequency to this new value $g^{1/N}$. Then we transform each section to a shelving filter using Eq. (4.12) and then use Eq. (4.9) to shift the crossover frequency to ω_c.

Consider again our example first-order filter with the gain at the crossover frequency defined such that the square magnitude is the average of the two extremes $G_c^2 = (G^2 + 1^2)/2$. Then Eq. (4.21) simplifies to

$$g^2 = \frac{(G^2+1)/2-1}{G^2-1} = 1/2, \tag{4.22}$$

and as with the low pass filter, this choice of crossover frequency is the same as that used in the prototype filter.

Now we create the shelf,

$$H_{\text{LS},\omega_c=\pi/2}(z) = \frac{1}{2}\frac{z[1+G]+G-1}{z}. \tag{4.23}$$

And finally, we shift the crossover frequency,

$$H_{\text{LS}}(z) = \frac{1}{2}\frac{z(1+G+(1-G)\beta)-[1-G+(1+G)\beta]}{z-\beta}, \text{ where } \beta$$

$$= \frac{1-\tan(\omega_c/2)}{1+\tan(\omega_c/2)}, \tag{4.24}$$

which reduces to

$$H_{\text{LS}}(z) = \frac{(1+G\tan(\omega_c/2))z+G\tan(\omega_c/2)-1}{(1+\tan(\omega_c/2))z+\tan(\omega_c/2)-1}. \tag{4.25}$$

In Figure 4.10, we've given a pole zero plot and square magnitude response of a first-order and a fourth-order low shelving filter with crossover frequency of $\pi/4$. By placing the poles slightly closer to 1 than the zeros, it has the effect of pushing up the low frequencies. If the poles and zeros were further apart, then the shelf would be higher (a larger value of G). If the zeros were placed closer to 1 than the poles, then the shelf would be at a value $G<1$.

High Shelf

A high shelving filter has a transfer function H_{HS} with magnitude 1 at frequency $\omega = 0$, magnitude G at frequency $\omega = \pi$ (representing the high shelf), and magnitude G_c at some crossover frequency ω_c.

As before, we will change the square magnitude at the crossover frequency of each first-order section of an Nth order prototype filter, from

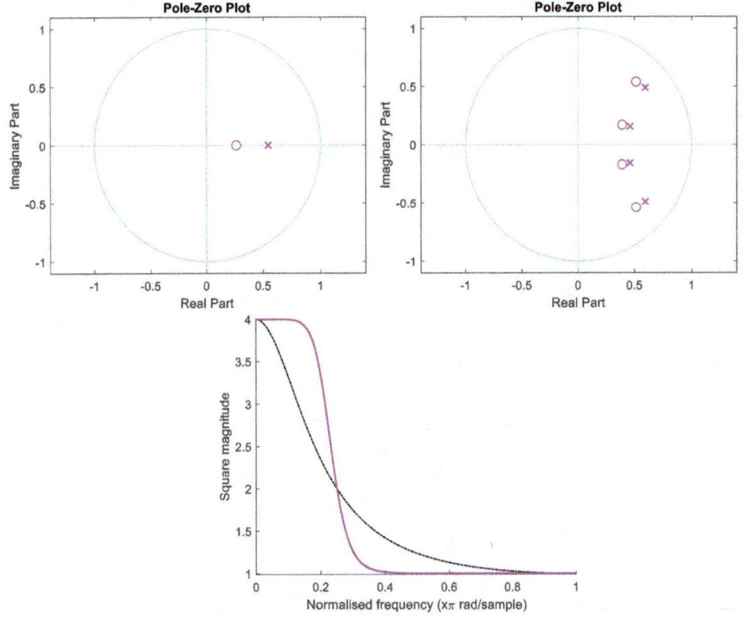

FIGURE 4.10
Pole zero plot for a first-order (top left) and fourth-order (top right) low shelving filter with center frequency $\omega_c = \pi/4$ and gain at center frequency $G = 2$. On the bottom, square magnitude response for the first-order (black) and fourth-order (purple) filters.

$(1/2)^{1/N}$ to $g^{2/N}$, where g^2 is defined as in (4.21). Then we transform each section to a shelving filter, then invert the magnitude response and then shift the crossover frequency to ω_c.

For our first-order filter, the procedure is the same as with the low shelf, giving Eq. (4.23) after we transform this to a shelving filter. Inverting the magnitude response then gives,

$$H(z) = \frac{1}{2}\frac{z[1+G]-G+1}{z}. \tag{4.26}$$

And after shifting the crossover frequency, we arrive at,

$$H_{HS}(z) = \frac{(G+\tan(\omega_c/2))z+\tan(\omega_c/2)-G}{(1+\tan(\omega_c/2))z+\tan(\omega_c/2)-1}. \tag{4.27}$$

First- and fourth-order high shelving filters are depicted in Figure 4.11. Compared with the low shelving filter, the pole and zero positions have been switched. Note that the pole zero plots are also equivalent to a low shelving filter with $G < 1$, only the constant term would be different.

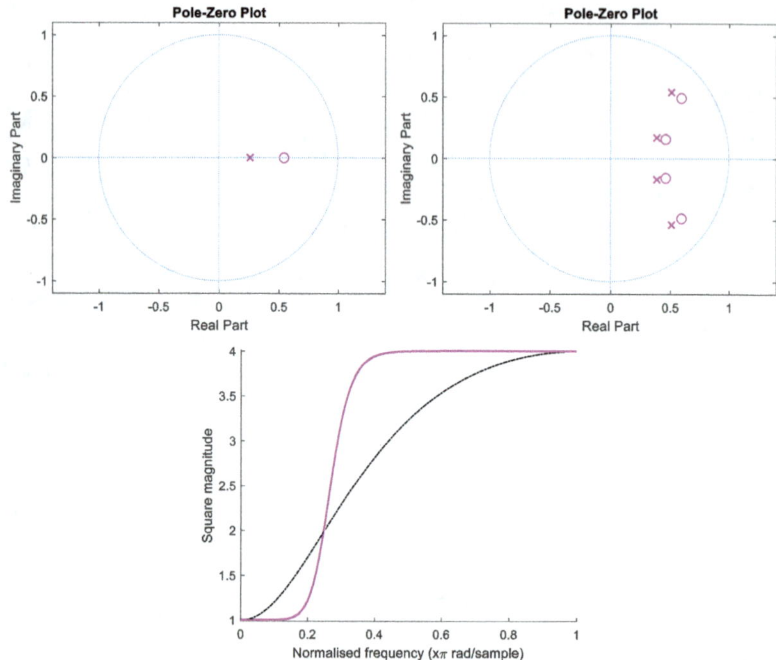

FIGURE 4.11
Pole zero plot for a first-order (top left) and fourth-order (top right) high shelving filter with $\omega_c = \pi/4$ and $G = 2$. On the bottom, square magnitude response for the first-order (black) and fourth-order (purple) filters.

Band Pass Filters

For a band pass filter, $H_{BP}(\omega = 0) = 0$, $H_{BP}(\omega = \pi/2)=0$, $H_{BP}(\omega = \omega_c) = 1$, and for the upper and lower crossover frequencies, $|H_{BP}(\omega=\omega_l)| = |H_{BP}(\omega=\omega_u)| = G_c$. Bandwidth is defined as $B = \omega_u - \omega_l$. So this filter is designed to only pass a range of frequencies around the crossover frequency and suppress all other content.

To design a high-order band pass filter, we consider each first-order section of our prototype filter. We change the gain at the crossover frequency to G_c using Eq. (4.6), then shift the crossover frequency to the bandwidth with Eq. (4.9), where ω_c in Eq. (4.9) is replaced with B. Finally, we transform this to a bandpass filter using Eq. (4.15).

For our first-order filter with the gain at the crossover frequencies defined such that the square magnitude is the average of the two extremes $G_c^2 = (0^2 + 1^2)/2 = 1/2$, again there is no need to change the gain at the crossover frequency of the prototype low pass filter.

So we now shift the center frequency of the prototype filter to B.

$$H(z) = \frac{(z+1)\tan(B/2)}{z(\tan(B/2)+1)+\tan(B/2)-1}. \tag{4.28}$$

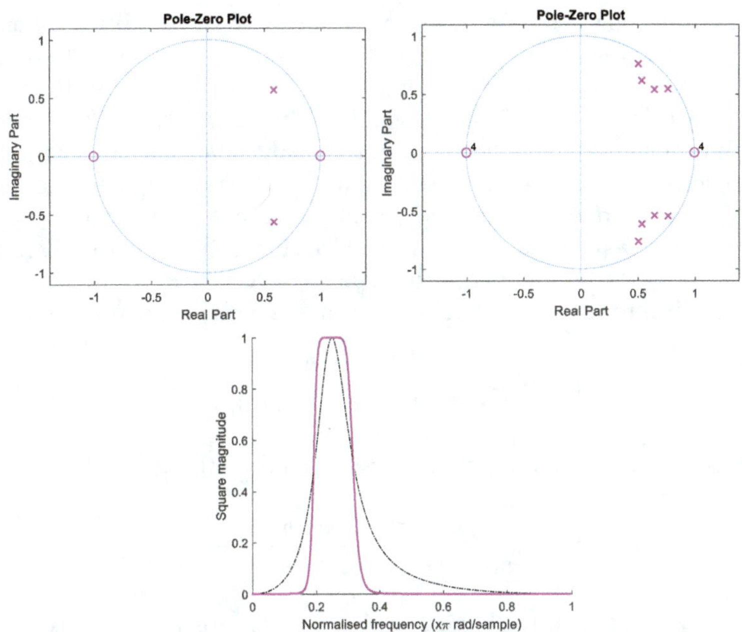

FIGURE 4.12
Pole zero plot for a second-order (top left) and eighth-order (top right) band pass filter with $\omega_c = \pi/4$ and $B = \pi/8$. On the bottom, square magnitude response for the first-order (black) and fourth-order (purple) filters.

Then transform to a band pass filter with bandwidth B and center frequency ω_c, which gives

$$H_{BP}(z) = \frac{\tan(B/2)\left(z^2 - 1\right)}{\left(\tan(B/2) + 1\right)z^2 - 2\cos\omega_c z + 1 - \tan(B/2)}. \tag{4.29}$$

Band pass filters are shown in Figure 4.12. Note that these are twice the order of the previous examples (second and eighth, as opposed to first and fourth) since the band pass transformation doubles the order of the filter. Now there are zeros at both 1 and –1, but poles surround the center frequency, allowing for a sharp transition from attenuating to passing frequency content. If the bandwidth is increased, then the distance of the poles from the center frequency would also increase.

Band Stop Filters

For a band stop filter, $H_{BP}(\omega = 0)=1$, $H_{BP}(\omega = \pi/2)=1$, $H_{BP}(\omega = \omega_c)=0$, and bandwidth is $B = \omega_u - \omega_l$ where $|H_{BP}(\omega = \omega_l)| = |H_{BP}(\omega = \omega_u)| = G_c$. So this filter is designed to suppress a range of frequencies around the center frequency and pass all other content.

To design a band stop filter, we again start with the prototype filter. For each first-order section, we change the gain at the crossover frequency. Then invert the magnitude response, then shift the crossover frequency, then transform this to a band pass filter. This is similar to the design of a high-order band pass filter, except that by inverting the magnitude response using Eq. (1.14), the resultant filter has band stop behavior.

For our first-order filter with the gain at the upper and lower crossover frequencies given as $G_c^2 = 1/2$, there is no need to change the gain at the crossover frequency. So, starting with the prototype, we invert the magnitude response and then shift the crossover frequency, as in the high pass filter, giving

$$H_{\mathrm{HP}}(z) = \frac{z-1}{(\tan(B/2)+1)z + \tan(B/2) - 1}. \tag{4.30}$$

Then transform to a band pass filter with bandwidth B.

$$H_{\mathrm{BS}}(z) = \frac{z^2 - 2z\cos\omega_c + 1}{(\tan(B/2)+1)z^2 - 2\cos\omega_c + 1 - \tan(B/2)}. \tag{4.31}$$

The pole zero plots of the band stop filters in Figure 4.13 are very similar to the band pass filters in Figure 4.12. But now the zeros have been moved

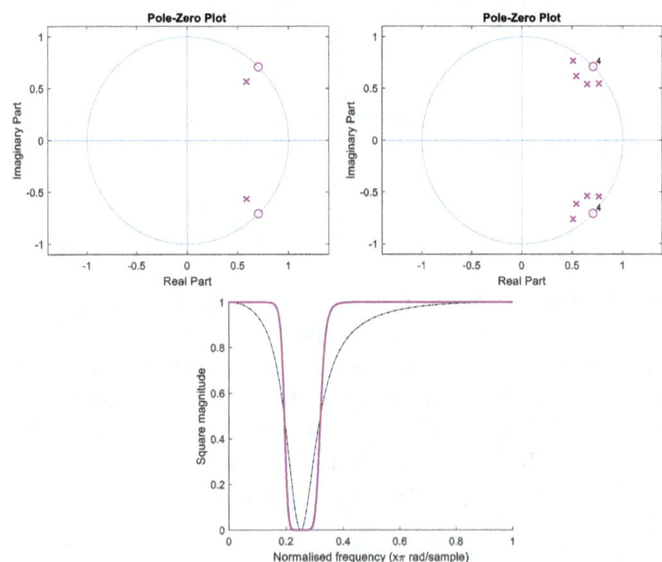

FIGURE 4.13
Pole zero plot for a second-order (top left) and eighth-order (top right) band stop filter with $\omega_c = \pi/4$ and $B = \pi/8$. On the bottom, square magnitude response for the first-order (black) and fourth-order (purple) filters.

from +1 and –1 to $\cos\omega_c + j\sin\omega_c$ and $\cos\omega_c - j\sin\omega_c$. This has the effect of the poles and zeros cancelling out far from the center frequency (leaving the magnitudes unaffected), and having the zeros dominate at the center frequency (attenuating the magnitude to zero).

Peaking and Notch Filters

For a peaking or notch filter, $H_{PN}(\omega = 0) = 1$, $H_{PN}(\omega = \pi/2) = 1$, $H_{PN}(\omega = \omega_c) = G$, and for the upper and lower crossover frequencies, $|H_{PN}(\omega = \omega_l)| = |H_{PN}(\omega = \omega_u)| = G_c$ [27]. Bandwidth is defined as $B = \omega_u - \omega_l$. When G is greater than 1, this filter provides a boost around ω_c and is known as a peaking filter. When G is less than 1, this filter attenuates the frequency content near ω_c and is known as a notch filter. Clearly, if $G = 0$, the filter completely removes frequency content near ω_c and is a band stop filter.

To design a high-order peaking or notch filter, we consider each first-order section of our prototype filter. We use Eqs. (4.6) and (4.21) to change the gain at the crossover frequency. Next we transform this to a shelving filter using Eq. (4.12). Then we shift the crossover frequency to the bandwidth with Eq. (4.9), where ω_c in Eq. (4.9) is replaced with B. Finally, we transform this to a band pass filter using Eq. (4.15).

For our first-order filter with the magnitude of the transfer function at bandwidth defined as we defined the magnitude at the crossover frequency for the shelving filter, we can follow the same steps as were taken with the shelving filter to give Eq. (4.25) except now the crossover frequency is replaced by the bandwidth B.

$$H_{LS}(z) = \frac{\left(1+G\tan(V/2)\right)z + G\tan(B/2) - 1}{\left(1+\tan(B/2)\right)z + \tan(B/2) - 1}. \tag{4.32}$$

Then transform to a band pass filter with bandwidth B.

$$H_{PN}(z) = \frac{\left(G\tan(B/2)+1\right)z^2 - 2z\cos\omega_c + 1 - G\tan(B/2)}{\left(\tan(B/2)+1\right)z^2 - 2z\cos\omega_c + 1 - \tan(B/2)}. \tag{4.33}$$

The pole zero and square magnitude plots for first- and fourth-order designs are shown in Figure 4.14.

Instead of bandwidth B, peak filters are often parameterized using quality factor Q. These two terms are related by $Q = \omega_c/B$.

To summarize, when we use the simple definition of bandwidth or crossover frequency, the standard filters mentioned above all have relatively straightforward forms for first-order designs (where bandwidth is not specified), and second-order designs (for band pass, band stop and peaking/notch filters). These are given in Table 4.1.

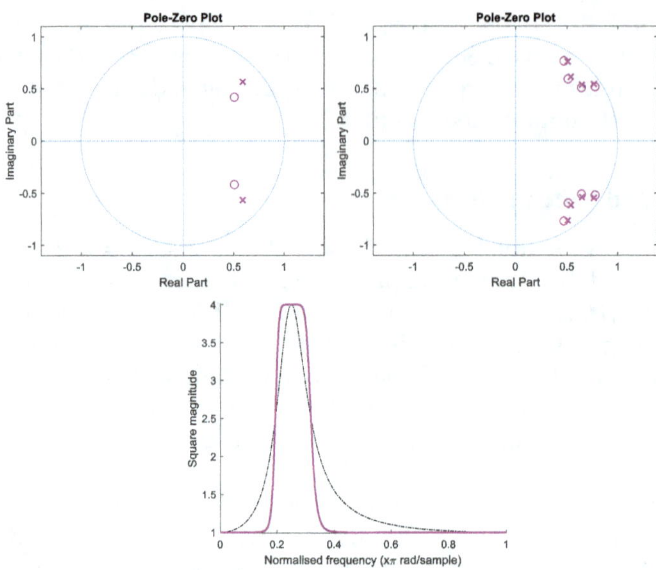

FIGURE 4.14
Pole zero plot for a second-order (top left) and eighth-order (top right) peaking filter with $\omega_c = \pi/4$, $B = \pi/8$ and $G = 2$. On the bottom, square magnitude response for the first-order (black) and fourth-order (purple) filters.

TABLE 4.1

Transfer Functions of Common First- and Second-Order Filters, and Their Equivalent Forms Based on Simpler Filters

	Transfer Function	Equivalence
First-order low pass (H_{LP})	$\frac{\tan(\omega_c/2)+\tan(\omega_c/2)z^{-1}}{1+\tan(\omega_c/2)-\left[1-\tan(\omega_c/2)\right]z^{-1}}$	
First-order high pass (H_{HP})	$\frac{1-z^{-1}}{1+\tan(\omega_c/2)-\left[1-\tan(\omega_c/2)\right]z^{-1}}$	$1-H_{LP}(z)$
First-order low shelf (H_{LS})	$\frac{1+g\tan(\omega_c/2)-\left[1-G\tan(\omega_c/2)\right]z^{-1}}{1+\tan(\omega_c/2)-\left[1-\tan(\omega_c/2)\right]z^{-1}}$	$GH_{LP}(z)$ $+H_{HP}(z)$
First-order high shelf (H_{HS})	$\frac{\tan(\omega_c/2)+G+\left[\tan(\omega_c/2)-G\right]z^{-1}}{1+\tan(\omega_c/2)-\left[1-\tan(\omega_c/2)\right]z^{-1}}$	$H_{LP}(z)$ $+GH_{HP}(z)$
Second-order band pass (H_{BP})	$\frac{\tan(B/2)-\tan(B/2)z^{-2}}{1+\tan(B/2)-2\cos\omega_c z^{-1}+\left[1-\tan(B/2)\right]z^{-2}}$	-
Second-order band stop (H_{BS})	$\frac{1-2\cos\omega_c z^{-1}+z^{-2}}{1+\tan(B/2)-2\cos\omega_c z^{-1}+\left[1-\tan(B/2)\right]z^{-2}}$	$1-H_{BP}(z)$
Peaking or notch filter (H_{PN})	$\frac{1+G\tan(B/2)-2\cos\omega_c z^{-1}+\left[1-G\tan(B/2)\right]z^{-2}}{1+\tan(B/2)-2\cos\omega_c z^{-1}+\left[1-\tan(B/2)\right]z^{-2}}$	$GH_{BP}(z)+$ $H_{BS}(z)$

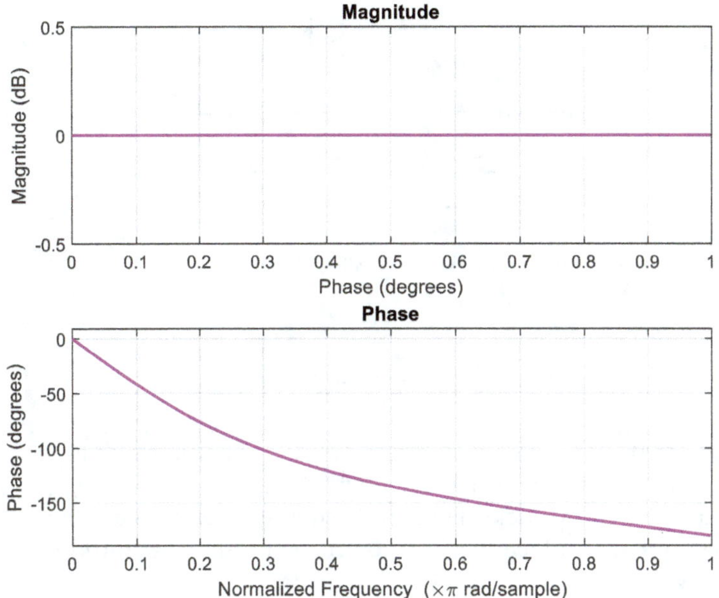

FIGURE 4.15
The magnitude and frequency response for a particular allpass filter (first-order, crossover frequency of $\omega_c = \pi/4$). All frequencies have unity gain but a different phase shift.

The Allpass Filter

In addition to the filter types discussed so far, there is one more type that commonly appears in audio effects. This is the *allpass filter*, and it passes all frequencies with no amplification or attenuation. That is, the magnitude of the gain is 1 at all frequencies. But if the gain doesn't change, what does the filter do? The answer lies in the *phase* of the signal. The allpass filter introduces a frequency-dependent phase shift, which is useful for many effects, especially the *phaser* (Chapter 5). Also, a simple delay is a type of allpass filter, since it changes the phase but not the magnitude of each frequency component. Figure 4.15 is a plot of the magnitude and phase response of another possible allpass filter.

The allpass filter may be given as

$$H(z) = \pm \frac{a_N z^N + a_{N-1} z^{N-1} \ldots + a_1 z + 1}{z^N + a_1 z^{N-1} \ldots + a_{N-1} z + a_N}. \tag{4.34}$$

We can easily check that this has the allpass property, $|H(z)|^2 = 1$.

If we look at any first-order section from this high-order filter, it can be written in the form,

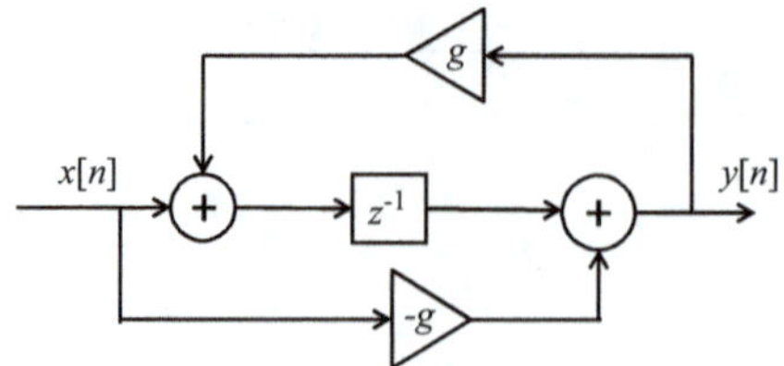

FIGURE 4.16
The block diagram for the first-order digital allpass filter.

$$H_n(z) = \pm \frac{zc^* - 1}{z - c}. \tag{4.35}$$

and we see that for a pole at c, there is a corresponding zero at $1/c^*$.

A digital allpass filter can be created rather simply with a single sample delay and two gain blocks. Figure 4.16 shows the block diagram; note that the feedforward and feedback gains are identical in magnitude but opposite in sign.

This is only a first-order allpass filter, as in the following equation;

$$H(z) = -\frac{zg - 1}{z - g}. \tag{4.36}$$

It's simply a version of Eq. (4.35) with real valued coefficients.

More complicated, higher order versions with multiple delay blocks must be constructed in order to create an arbitrary phase response. See Chapter 5 for how this is accomplished with the phaser.

Implementation

How do we go from filter formulas such as those in Table 4.1 to working code that implements the filter? The first step is to calculate the numerical value of each *coefficient* for the desired parameters. The filters in this chapter are specified in terms of discrete-time frequency $\omega_c = 2\pi f_c / f_s$ which could range from 0 to π, based on a desired continuous-time crossover frequency f_c, and a sample rate f_s, both in Hertz.

The calculation of coefficients is often more computationally expensive than applying them to the signal, so filter implementations will often only recalculate the coefficients when the filter parameters change. If the filter parameters change continuously, as in several effects in the next chapter, the coefficients might be recalculated at a fraction of the audio sample rate (for example, once every 16 samples).

From here, we should have an equation of the following form (for a second-order filter; for a first-order filter, $a_2 = b_2 = 0$):

$$H(z) = \frac{b_0 + b_1 z^{-1} + b_2 z^{-2}}{a_0 + a_1 z^{-1} + a_2 z^{-2}}. \tag{4.37}$$

Before we can proceed further with implementation, our filter must have $a_0 = 1$. If that is not yet the case, then normalize all the coefficients by a_0 so this condition is satisfied.

The next step is to rewrite the equation in the time domain (discussed further in Chapter 1), yielding the following formula:

$$y[n] + a_1 y[n-1] + a_2 y[n-2] = b0y[n] + b1y[n-1] + b2y[n-2]. \tag{4.38}$$

Rearranging, we get the familiar formula to calculate one sample of $y[n]$ based on previous values of x and y:

$$y[n] = b0y[n] + b1y[n-1] + b2y[n-2] - a_1 y[n-1] - a_2 y[n-2]. \tag{4.39}$$

We have already calculated the coefficients a_i and b_i, so the only other thing we need to know is where to find the previous samples $x[n-i]$ and $y[n-i]$. In general, the only way we can reliably access these previous samples is to save them ourselves when they come in, by using *global variables* or *instance variables* which remember their value across calls to the audio callback function. Consider the following code snippet:

```
int numSamples;          // Indicates how many audio samples to process
float *channelData;      // Array of audio samples, length numSamples
float coefficients[6]; // Previously calculated filter coefficients
float x1, x2, y1, y2;   // Previous values of the input and output
for (int i = 0; i < numSamples; ++i)
{
    const float in = channelData[i];

    float out = coefficients[0] * in /* b0 */
    + coefficients[1] * x1 /* b1 */
    + coefficients[2] * x2 /* b2 */
    - coefficients[4] * y1 /* a1 */
    - coefficients[5] * y2; /* a2 */

    x2 = x1; // Update the previous values of x and y
    x1 = in;
    y2 = y1;
    y1 = out;

    channelData[i] = out;
}
```

Here, the `for()` loop would be part of the audio callback function, while `coefficients`, `x1`, `x2`, `y1`, and `y2` are all global or instance variables, declared outside of the audio callback. `x1` corresponds to `x[n-1]` (and similarly for the other variables). The reason this works is because of the four lines toward the end of the example, where the current inputs and outputs are stored in `x1` and `y1`, respectively, and the values of `x1` and `y1` are moved to `x2` and `y2`. At the next iteration of the `for()` loop (even if it doesn't take place until the next audio callback), `x1` and `y1` will now hold data that is one sample old.

The order of the lines of code updating the previous samples is very important. It is necessary to run the first line (`x2 = x1`) before the second one (`x1 = in`). Otherwise, the previous value of `x1` will be lost and `x1` and `x2` will both end up with the same value.

When implementing code to apply filter coefficients from other sources, pay close attention to the sign of the a coefficients. Different sources will sometimes have different conventions for whether the a coefficients should be added or subtracted to calculate the output.

Furthermore, for a high-order filter (most commonly found with FIR filters), a *circular buffer* will be a more efficient way of keeping track of previous inputs than updating individual variables. Circular buffers are discussed further in Chapter 3.

Applications of Filter Fundamentals

Exponential Moving Average Filter

Let's now take a look at one simple, but very important filter. The exponential moving average filter, also known as a smoothing filter or one-pole moving average filter, is given by

$$y[n]=\alpha y[n-1]+(1-\alpha)x[n]. \tag{4.40}$$

where α is the amount of decay between adjacent samples. For instance, if α is 0.75, the value of each sample in the output signal is three-quarters of the previous output and one-quarter of the new input. The higher the value of x, the slower the decay. This filter is often used to smooth the effect of processing a signal, or to derive a smoothly changing estimate of signal level. As we will see in Chapter 7, it features prominently in dynamics processing.

The transfer function is $H(z)=(1-\alpha)z/(z-\alpha)$, which gives one zero at zero, one pole at α and gain of $(1-\alpha)$.

The impulse response of this filter is:

$$y[n]=\alpha^n(1-\alpha) \tag{4.41}$$

Now consider the step function, $x[n]=\begin{cases} 1 & n \geq 0 \\ 0 & n<0 \end{cases}$. The step response of this filter is:

$$y[n]=1-\alpha^n \text{ for } x[n]=1, n \geq 1. \tag{4.42}$$

The time constant τ is defined as the time it takes for this system to reach $1 - 1/e$ of its final value, i.e., $y[\tau f_s]=1-1/e$. Thus, from (4.42), we have

$$\alpha=e^{-1/(\tau f_s)}, \tau=-1/(f_s \ln \alpha). \tag{4.43}$$

This is an important filter, since it can be used to very simply smooth a signal. Generally, one gives the time constant τ, finds α from Eq. (4.43), and implements the filter in the time domain using (4.40). Note that this particular filter is primarily used for smoothing a signal in the time domain, and not explicitly for modifying frequency content. It simply provides a smooth output that can still quickly respond to sudden changes in the input.

Loudspeaker Crossovers

Low pass, high pass and band pass filters are commonly used in *crossover networks* (or just *crossovers*) for loudspeakers. Larger hi-fi speakers almost always use more than one speaker driver to cover the entire audio frequency range. *Woofers* are large drivers (10 to 30 cm diameter) used for the lower frequencies, and *tweeters* are smaller cone or dome-shaped drivers (2 to 5 cm diameter) used for the higher frequencies. Some speakers also include a third *midrange* driver to cover the frequencies between the woofer and tweeter. Others use a single free-standing *subwoofer* to cover the lowest bass frequencies for the entire audio system; since the human ear cannot localize low bass frequencies, a single large subwoofer can be used to cover the low frequency content for a stereo or surround-sound system, which allows the remaining speakers to be smaller and less expensive.

For a loudspeaker to work properly, the same signal cannot be sent to every driver. Woofers reproduce high frequencies poorly and with significant distortion, and sending bass frequencies to a tweeter can cause mechanical damage. The role of the crossover is to divide the audio signal into two or three frequency ranges, which are sent to each driver. In a speaker with a woofer and tweeter (*two-way*), the crossover consists of a low pass filter for the woofer and a high pass filter for the tweeter (Figure 4.17a). The two filters have the same crossover frequency (typically

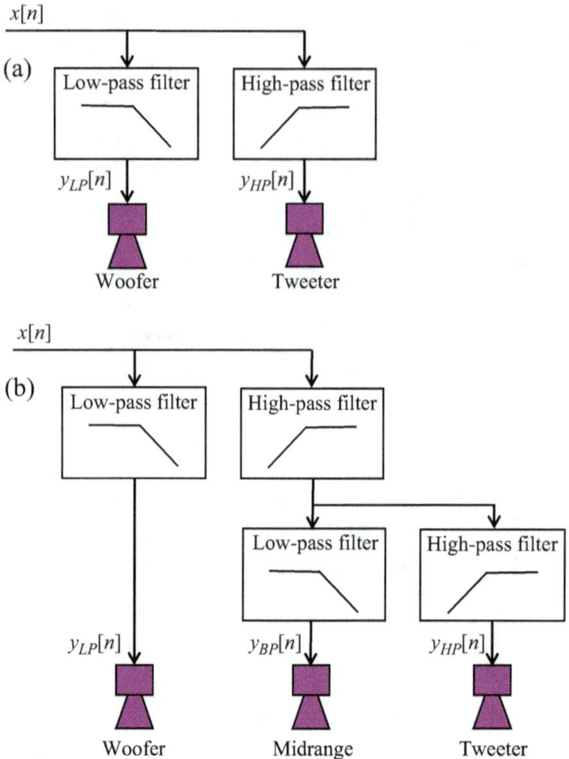

FIGURE 4.17
A loudspeaker crossover separates the incoming audio signal into several different frequency bands so that each loudspeaker is provided with content within a preferred frequency range. Two-way (a) and three-way (b) loudspeaker crossovers are depicted.

around 1–3 kHz), which means that any given input frequency should go either to the woofer or to the tweeter, but not both. Of course, practical filters do not have perfectly sharp crossovers, so there will always be a gradual transition between the two drivers. The gradual transition is a desirable property in practice, and very few crossovers use filters higher than fourth order. Second-order filters are most commonly used in crossovers, and the occasional first-order design can also be found.

When woofer, midrange, and tweeter are used (a *three-way* speaker), there are two crossover points to consider: the transition between the woofer and the midrange and between the midrange and tweeter. In practice, a three-way crossover looks like two two-way crossovers in series (Figure 4.17b). A low pass filter is used for the woofer and a high pass filter for the rest of the signal. The output of the high pass filter is then split again into midrange and tweeter components. An equivalent way to understand the operation is that the woofer has a low pass filter, the midrange a band pass filter, and the tweeter a high pass filter. The woofer-midrange crossover frequency is typically in the range of 200 Hz–1 kHz, and the midrange-tweeter crossover in the range 2–4 kHz.

In the early days of loudspeakers, crossovers were built using high-power capacitors and inductors that could operate directly on the output of an audio power amplifier. This is still a common practice, but increasingly, self-powered speakers such as studio monitors will perform the crossover filtering on the line-level audio signal and use a separate power amplifier for each driver; this process is known as *bi-amplification* for a two-way speaker or *tri-amplification* for a three-way speaker. Crossovers at line level can generally have lower distortion and tighter component tolerances, and they can be more sophisticated in correcting slight imperfections in the frequency responses of the drivers.

Good crossover design can be as important as the quality of the drivers in determining the sound of a finished loudspeaker. Many subtle design variations are possible, which are discussed in detail in [29]. Crossovers are easily implemented digitally using the basic filters presented in this chapter, the main practical constraint being that there must be at least one digital-to-analog converter channel and one amplifier for every driver (e.g., six channels for a stereo three-way speaker).

Further Reading

There are many classic texts on filter design. Major contributors, already mentioned, include Orfanidis [3], Moorer [30], and Massenburg [31]. The particular approach herein, though, is somewhat unusual in that the filters are designed directly in the digital domain, without resorting to analogue prototypes. Variations on this (with more detail) are given in [27,28,32]. Also, filter design is almost ubiquitous wherever signal processing is performed, so filter design tools are an integral part of most software tools and packages that perform signal processing. MATLAB, in particular, comes with a vast array of tools for designing almost any filter, using almost any established approach to filter design.

There are always new tasks and problems in signal processing, so research continues in the design of custom filters that address those challenges. A new approach to filter design, though, relies on training a neural network with millions of random filters, in order to find the best set of coefficients to match some design requirements [33]. New techniques have also been described to make IIR filters differentiable [34], so that they can be embedded into deep learning systems.

Problems

1. Design a second-order peaking filter for a 48 kHz signal with center frequency 6 kHz and gain of 6 dB, and bandwidth of 1 kHz. Assume bandwidth is defined as when the square magnitude is the average of the two extremes, $G_B^2 = (1 + G^2)/2$, where G is the linear gain.
2. Consider the second-order filter having transfer function $H(z) = \frac{r^2z^2 - 2rz\cos\theta + 1}{z^2 - 2rz\cos\theta + r^2}$. Show that it is an all-pass filter, that is, show that $|H(z)| = 1$. Find the zeros of the filter as a function of the poles.
3. Consider again the exponential moving average filter. Give the square magnitude response. Also, find the crossover frequency, if it is defined as where the square magnitude is halfway between the two extremes, $|H(z = 1)|^2$ and $|H(z = -1)|^2$.
4. Consider the moving average window $y[n] = \frac{1}{W}\sum_{i=0}^{W-1} x[n-i]$.

   What is the transfer function, impulse response and step response of this filter? Define the time constant as was done for the exponential moving average, the time it takes for this system to reach $1 - 1/e$ of its final value, i.e., $y[\tau f_s] = 1 - 1/e$.

   If the time constant of the moving average window is set equal to the time constant of the exponential moving average, what is the relationship between W and α?
5. An out-of-band shelving filter has unity gain at center frequency ω_c, bandwidth B, and magnitude G at frequency 0 and at $f_s/2$. It may be constructed by converting the prototype to a shelving filter, reversing the filter, shifting the cutoff frequency to the bandwidth, and then transforming this to a bandpass filter. A second-order design could also be constructed by manipulating other filter designs. Give the transfer function for a second-order out-of-band shelving filter.

Note

1 The crossover frequency is sometimes called the corner, cut-off, or transition frequency, but we will use the term crossover throughout.

5

Filter Effects

In a sense, any audio effect could be considered a *filter* in that every effect performs mathematical operations on the input signal that produce some change in the content. This chapter, however, is concerned with effects based on the canonical types of filters defined in Chapter 4; *low pass, high pass, band pass, peaking/notch, shelving,* and *allpass* filters. The effects that fall into this category are *equalizers,* a broad class of effects that adjust the frequency balance of the input audio signal; *wah-wah,* a musical effect typically used on guitar, which is based on a peaking filter; and the *phaser,* an effect based on allpass filters that produces a sweeping, spacious sound similar to the flanger (Chapter 3).

Equalization

Equalization (*EQ* for short) is one of the most common audio effects. It is the process of adjusting the relative strength of different frequency bands within a signal. The name "equalization" comes from the desire to obtain a flat (equal) frequency response from an audio system by compensating for non-ideal equipment or room acoustics. Peaks or troughs in the frequency response of a system are often described as 'coloration' of the sound, and equalization can be used to remove this coloration [17,35,36].

Equalization covers a broad class of effects, ranging from simple tone controls to sophisticated graphic and parametric equalizers. All EQ effects are based on *filters,* and most equalizers consist of multiple subfilters, each of which affects a single frequency band. In the following, we will go through the steps of designing an equalizer out of such subfilters and discuss the properties these subfilters should have.

Theory

Equalizers range in complexity from simple two-knob tone controls to complex multi-band graphic and parametric equalizers, but at a fundamental level, all designs are based on the same collection of filter structures. Equalizer design, therefore, depends heavily on choosing the correct

DOI: 10.1201/9781003593942-5

parameters for each filter, keeping in mind which parameters should be fixed by the design and which should be user-adjustable.

Recall from Chapter 4 that shelving filters come in *low* and *high* shelf configurations. A low shelving filter has adjustable gain at low frequencies and unity gain at high frequencies, and the reverse is true for the high shelving filter.

However, when filters are used for equalization, there is usually a fundamental difference in design as compared to the designs introduced in Chapter 4. Previously, the crossover frequencies for shelving filters were defined based on the arithmetic mean of the extremes of the square magnitude of the frequency response. But when shaping the spectrum of an audio signal, we are interested in decibel magnitudes. On that scale, we want to ensure that certain symmetries hold. To do this, we define crossover frequencies based on the *geometric mean* of the extremes of the square magnitude response,

$$G_c^{\,2} = \sqrt{G_{\max}^2 \cdot G_{\min}^2}\,. \tag{5.1}$$

This geometric mean is equivalent to the arithmetic mean of the gains on a decibel scale.

Using the geometric mean, the gain at the crossover frequency, G_c, is defined to be $\sqrt{G}$, where G is the gain of the shelf. We can now use the procedure described in Chapter 4 to design the shelving filters. The resulting transfer function for the low shelving filter is as follows:

$$H_{\text{LS}}(z) = \frac{[G\tan(\omega_c/2)+\sqrt{G}]z + G\tan(\omega_c/2)-\sqrt{G}}{[\tan(\omega_c/2)+\sqrt{G}]z + \tan(\omega_c/2)-\sqrt{G}}. \tag{5.2}$$

And for the high shelving filter:

$$H_{\text{HS}}(z) = \frac{[\sqrt{G}\tan(\omega_c/2)+G]z + \sqrt{G}\tan(\omega_c/2)-G}{[\sqrt{G}\tan(\omega_c/2)+1]z + \sqrt{G}\tan(\omega_c/2)-1}. \tag{5.3}$$

Note also that this definition is not usable with low pass, high pass, band pass, or band stop filters. There, one extreme of the square magnitude response has a gain of 0, which results in the crossover frequency being located at that same extreme of the square magnitude response.

Due to this choice, symmetric magnitude frequency responses, above and below the unit gain level, will be produced for gains G and $1/G$ (i.e., equal boost or cut in dB), as is usually wanted. Figure 5.1 illustrates the magnitude responses for low and high shelving filters with different crossover frequencies and gains. It can be observed that a boost and a cut are symmetric on a decibel scale. Also, when G=1, the pole and the zero

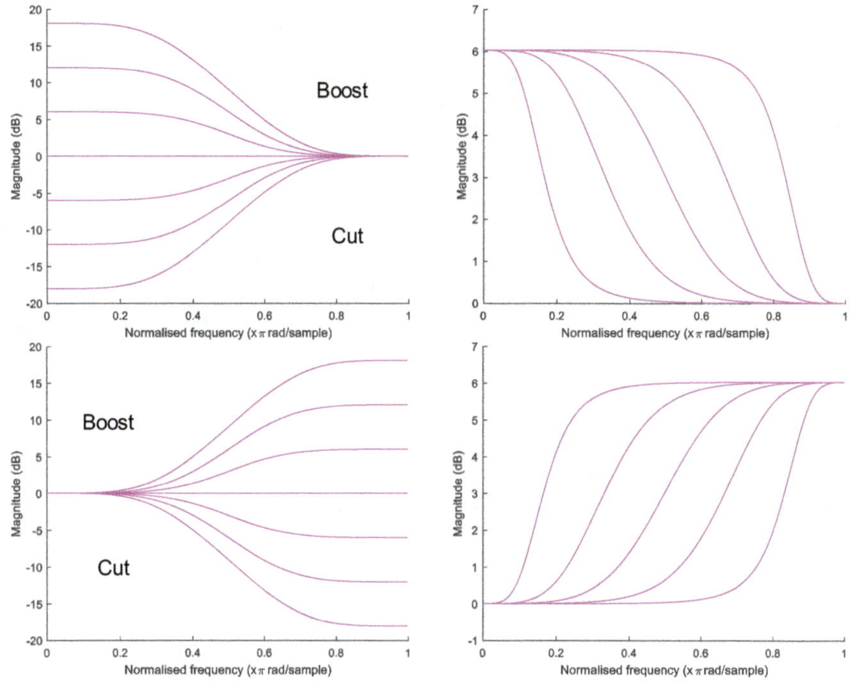

FIGURE 5.1
Frequency response of a low shelving filter (left) and a high shelving filter (right), for varying the gain (top) or varying the center frequency (bottom).

of the transfer function, Eq. (5.2) for the low shelf or (5.3) for the high shelf, coincide, and the magnitude response becomes unity.

Comparing (5.2) with (5.3) shows that these transfer functions are related as $H_{HS}(z)=G/H_{LS}(z)$. This implies that a cascade of a low shelf and a high shelf with the same crossover frequency ω_c and gain G will produce a constant gain G across all frequencies.

For the peaking or notch filter, we will again use the geometric mean, so now the gain at bandwidth is defined to be $\sqrt{G}$. Following the same steps as in Chapter 4 but with this new definition of gain at bandwith, we can derive the transfer function as follows;

$$H_{PN}(z)=\frac{\left[\sqrt{G}+G\tan(B/2)\right]z-2\sqrt{G}\cos\omega_c z+\sqrt{G}-G\tan(B/2)}{\left[\sqrt{G}+\tan(B/2)\right]z-2\sqrt{G}\cos\omega_c z+\sqrt{G}-\tan(B/2)}. \quad (5.4)$$

Figure 5.2 presents examples of peak and notch filters with various gains, crossover frequencies, and bandwidths. Figure 5.2a shows that these filters are symmetric on the dB scale. It is also seen that when the center frequency is high, the magnitude response itself becomes asymmetric so

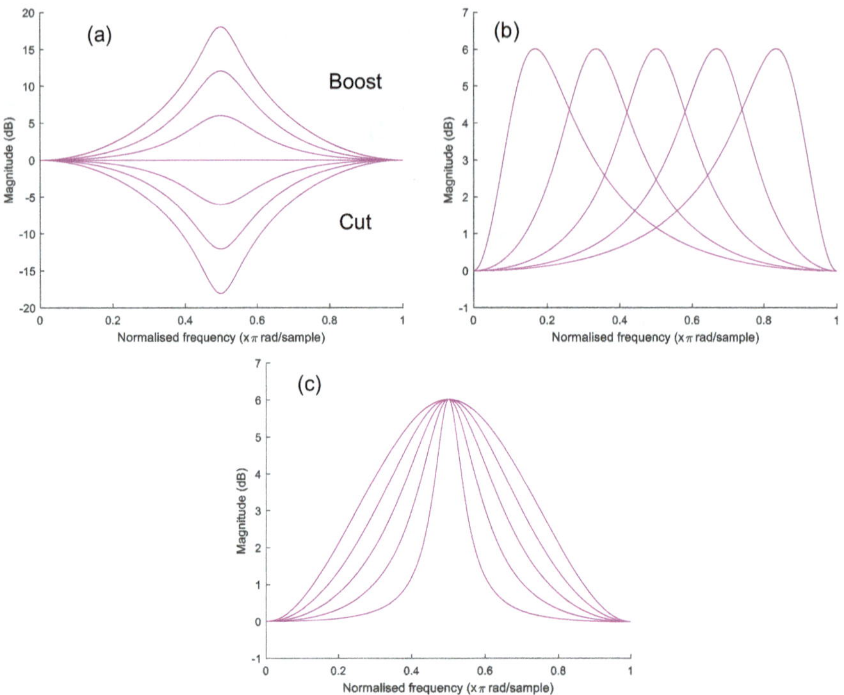

FIGURE 5.2
Magnitude responses for a peaking or notch filter when adjusting (a) gain, (b) center frequency, and (c) bandwidth.

that the upper *skirt* (i.e., the magnitude response of a peaking filter on either side of the center frequency) is steeper than the lower one. This feature is caused by the requirement that the digital equalizing filters have a unity gain at the Nyquist limit.

Two-Knob Tone Controls

Most stereo systems feature *tone controls,* which provide a simple and quick way to adjust the sound to suit the listener's taste and compensate for the frequency response of the room. Tone controls are the simplest and possibly most common equalization system. A basic version consists of two knobs, typically labeled 'bass' and 'treble'. These knobs are used to control the *gain* of low and high frequencies, respectively, through the use of shelving filters. Figure 5.3a shows a block diagram of a typical implementation.

In tone controls, a low shelving filter is used in the bass control and a high shelving filter is used in the treble control. In each case, the maximum gain of the shelf, G, is adjustable. Where $G>1$, the bass (low pass case)

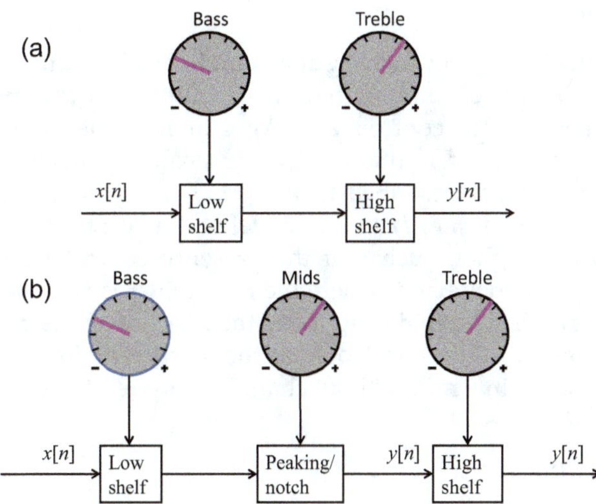

FIGURE 5.3
(a) Bass and treble tone controls implemented as a low shelving filter and high shelving filter placed in series. (b) Three tone controls, including a peaking/notch filter to adjust the midrange.

or treble (high pass case) will be boosted; where $G<1$, it will be cut. First-order shelving filters are typically used, which produce a gradual 6 dB/octave transition between low and high frequencies, as shown previously in Figure 5.1.

Typical tone controls have an adjustment range of ±12 dB, corresponding to approximately $0.25<G<4$ for each filter. The *crossover frequency* ω_c is usually fixed. The values for bass and treble vary by manufacturer, but the bass control usually uses a lower ω_c than the treble control. Crossover frequencies for the bass control might range from 100 to 300 Hz, and the treble control from 3 to 10 kHz.

Three-Knob Tone Controls

In two-knob tone controls, the *midrange* frequencies (between bass and treble) are usually left unchanged. On some units, in addition to control of the bass and treble, there may be a "midrange" or "mid" control. This control is usually implemented as a peaking or notch filter, as in Figure 5.2.

The knob on the midrange control affects the gain G at center frequency, which generally takes the same range of values as the bass and treble controls (e.g., $0.25<G<4$). The center frequency ω_c is generally fixed to be midway between the bass and treble controls, and the bandwidth B is chosen so that the midrange control affects the frequencies that are left unadjusted by the bass and treble controls.

Presence Control

Some systems, including many guitar amplifiers and mixing consoles, will have a 'presence' knob or button in addition to the tone controls mentioned above. This controls a peaking filter in the upper midrange frequencies. The transfer function for the peaking filter is identical to the midrange control described in the previous section, but the center frequency and the *quality factor* or *Q* (defined as center frequency over bandwidth) are higher, such that the presence control affects roughly the 2–6 kHz frequency band, where the human ear is very sensitive. The presence control is intended to simulate the effect of an instrument being physically present in the same room as the listener, and it can help bring out an instrument in a mix without changing its overall level.

Loudness Control

Many stereo amplifiers feature a "loudness" control, either as a knob or an on-off button. In this context, *loudness* has a different meaning than *volume* or *amplitude* (which refer to the total sound pressure level, or SPL, produced by the audio system). The sensitivity of human hearing is heavily dependent on frequency: for the same SPL, midrange frequencies will be perceived to be louder than very low or very high frequencies.

Figure 5.4 shows a plot of *equal loudness contours* for human hearing. Each curve indicates the SPL required to maintain the same perceived loudness (equivalent to 90 Phon, 80 Phon… 10 Phon, where 1 phon=1 dBSPL at a frequency of 1 kHz) across all frequencies. Notice that the curves become flatter as the overall SPL increases. This means that, especially at low listening levels, bass and treble need to be boosted with respect to the midrange to be perceived as equally loud.

The *loudness* control is therefore designed to boost the low and high frequencies of the input signal. Two *shelving* filters (one for bass, one for treble) can be used for this purpose. When the loudness control is implemented as a continuous knob, its setting controls the *gain* of the low- and high-frequency shelves, with the midrange gain fixed at 1. When it is implemented as a button, it switches between a flat frequency response (no boost) and an inverse of the A-Weighting loudness curve, which is a rough approximation of the equal loudness contour near 40 phon. Thus, it is used to boost the frequency content that we perceive to have additional attenuation at low listening levels.

Graphic Equalizers

The graphic equalizer is a tool for precisely adjusting the gain of multiple frequency regions. In contrast to the simple 2- or 3-knob tone control, a graphic equalizer can provide up to 30 controls for manipulating the

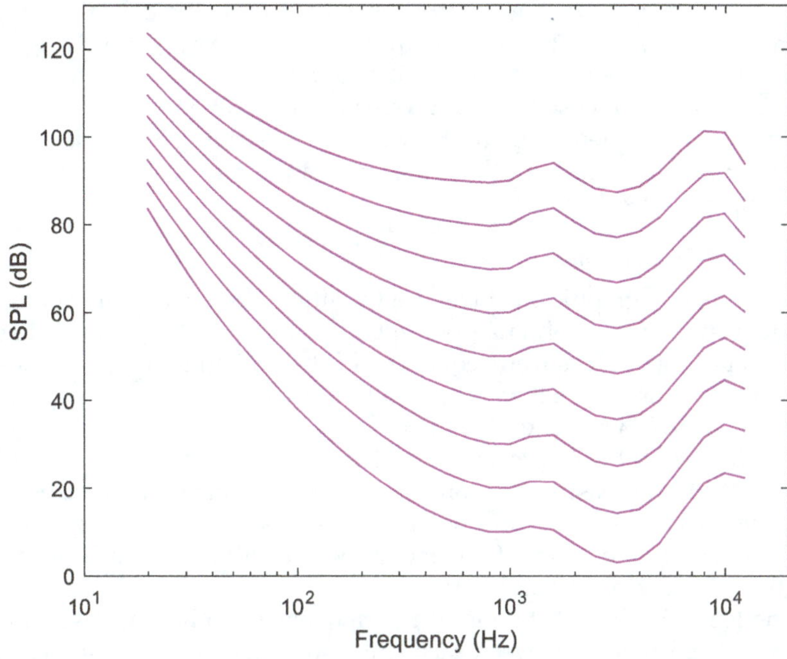

FIGURE 5.4
Equal loudness contours by phon (90 phon- top curve, 10 phon- bottom curve), as given in the ISO 226 standard. Each curve represents the Sound Pressure Level (SPL) required for which a listener will perceive a constant loudness when presented with pure steady tones across the frequency range.

frequency response. Structurally, a graphic EQ is simply a set of filters, each with a fixed center frequency. The only user control is the amount of boost or cut in each frequency band, which is often controlled with vertical sliders. The term "graphic" refers to the fact that the position of the sliders' knobs can be understood as a graph of the equalizer's magnitude response versus frequency. In other words, the position of the sliders resembles the frequency response itself, which makes the graphic equalizer intuitive to use despite the number of controls.

It should be noted that a wide variety of techniques exist for designing filters that achieve an arbitrary frequency response [28], including modern machine learning approaches that offer unprecedented accuracy [33]. However, the graphic equalizer still offers many advantages; it is intuitive and easily modified in real-time by the user.

The basic unit of the graphic equalizer is the *band*. A band is a region in frequency defined by a *center frequency* ω_c and a *bandwidth* B or *quality factor* Q. Recall that these three terms are related by $Q=\omega_c/B$.

The gain of each band is controllable by the user. Typical gains range from –12 to +12 dB ($0.25<G<4$), with 0 dB ($G=1$) meaning no change or "flat". The same peaking or notch filter found in the midrange tone control, Eq. (5.4), can be used to create a single band of a graphic equalizer.

The center frequency ω_c and bandwidth B are not user-adjustable. To choose the right values for these parameters, we must consider the relationship between the bands.

Bands in a Graphic Equalizer

The bands in a graphic equalizer are usually distributed logarithmically in frequency to match human perception. Let us denote the normalized lower and upper crossover frequencies of the ith band with $\omega_{l,i}$ and $\omega_{u,i}$, respectively. As before, bandwidth is the difference between the upper and lower crossover frequencies, $B_i = \omega_{u,i} - \omega_{l,i}$.

The frequency bands are adjacent [37], so the upper crossover of band i will be the lower crossover of band $i+1$, $\omega_{u,i}=\omega_{l,i+1}$. That is, input audio frequencies below this crossover will be primarily affected by the gain control for band i, where input frequencies above it will be primarily affected by the gain control for band $i+1$.

The logarithmic distribution of the frequency bands can be specified using a fixed ratio R between each band, so $\omega_{l,i+1}=R\cdot\omega_{l,i}$, $\omega_{u,i+1}=R\cdot\omega_{u,i}$, or $B_{i+1}=R\cdot B_i$. We also consider the geometric mean of the two crossover frequencies, $\omega_{M,i} \equiv \sqrt{\omega_{l,i}\omega_{u,i}}$, where you can see that we have the same relationship for the distribution of these values, $\omega_{M,i+1}=R\cdot\omega_{M,i}$.

Two common designs are *octave* and 1/3-*octave* graphic equalizers. An octave is a musical interval defined by a doubling in frequency, so octave graphic equalizers will have the ratio $R=2$ between each band. In a 1/3-octave design, each octave contains three bands, which implies $R^3=2$ or R~1.26. So starting at 100 Hz, an octave spacing would have geometric mean frequencies at 200, 400, 800 Hz... and a 1/3-octave spacing would have filters centered at 126, 159, 200 Hz, etc.

The number of bands is determined by their spacing and the requirement to cover the entire audible spectrum. Octave graphic equalizers usually have 10 bands, ranging from 31 Hz at the lowest to 16 kHz at the highest. 1/3-octave designs have 30 bands ranging from 25 Hz to 20 kHz. These frequencies, shown in Table 5.1, are standardized by the ISO (International Standards Organization) [38].

The bandwidth can be easily related to the geometric mean of the upper and lower crossover frequencies,

$$
\begin{aligned}
&\omega_{M,i} \equiv \sqrt{\omega_{l,i}\omega_{u,i}} = \sqrt{R}\cdot\omega_{l,i} = \omega_{u,i}/\sqrt{R} \\
&\rightarrow B_i = \omega_{u,i} - \omega_{l,i} = \left(\sqrt{R} - 1/\sqrt{R}\right)\omega_{M,i}
\end{aligned}
\quad (5.5)
$$

TABLE 5.1

The ISO Standard for Octave and 1/3 Octave Frequency Bands

Octave Bands			1/3 Octave Bands		
Lower Frequency f_l (Hz)	**Geometric Mean Frequency f_M (Hz)**	**Upper Frequency f_u (Hz)**	**Lower Frequency f_l (Hz)**	**Geometric Mean Frequency f_M (Hz)**	**Upper Frequency f_u (Hz)**
22	31.5	44	22.4	25	28.2
			28.2	31.5	35.5
			35.5	40	44.7
			44.7	50	56.2
44	63	88	56.2	63	70.8
			70.8	80	89.1
			89.1	100	112
88	125	177	112	125	141
			141	160	178
			178	200	224
177	250	355	224	250	282
			282	315	355
			355	400	447
355	500	710	447	500	562
			562	630	708
			708	800	891
710	1,000	1,420	891	1,000	1,122
			1,122	1,250	1,413
			1,413	1,600	1,778
1,420	2,000	2,840	1,778	2,000	2,239
			2,239	2,500	2,818
			2,818	3,150	3,548
2,840	4,000	5,680	3,548	4,000	4,467
			4,467	5,000	5,623
			5,623	6,300	7,079
5,680	8,000	11,360	7,079	8,000	8,913
			8,913	10,000	11,220
			11,220	12,500	14,130
11,360	16,000	22,720	14,130	16,000	17,780
			17,780	20,000	22,390

Note that the geometric mean of the crossover frequencies of a filter, $\omega_{M,i}$ is not usually the true center frequency where the filter reaches its maximum or minimum value, $\omega_{c,i}$. From Chapter 4 and a bit of trigonometry, we can find a relationship between the upper and lower crossover frequencies and the center frequency of a band pass, band stop, peaking, or notch filter,

$$\tan^2(\omega_{c,i}/2) = \tan(\omega_{u,i}/2)\tan(\omega_{l,i}/2). \tag{5.6}$$

However, the geometric mean is usually quite close to the center frequency. Thus, the bandwidth scales roughly proportionally with the center frequency. Higher bands will have a larger bandwidth than lower ones. Since $Q=\omega_c/B$, this is another way of saying that the Q factor is nearly constant for each band in a graphic equalizer. From (5.5), we can estimate Q as

$$\begin{aligned} Q &= \omega_{c,i}/B_i \sim \omega_{M,i}/B_i \\ &= 1/\left(\sqrt{R} - 1/\sqrt{R}\right) = \frac{\sqrt{R}}{R-1}. \end{aligned} \tag{5.7}$$

So for an octave (10-band) equalizer, $Q \sim \sqrt{2} \sim 1.41$ since $R=2$. For a third-octave (30-band) equalizer, we find $Q \sim \omega_{c,i}/B_i \sim \dfrac{\sqrt[6]{2}}{\sqrt[3]{2}-1} \sim 4.32$ since $R = \sqrt[3]{2} \sim 1.26$.

Ideally, the subfilter for the ith band has a magnitude of the desired gain G_i inside the band and gain at bandwidth $\sqrt{G}$, so that at the crossover frequency, when $G_{i=}G_{i+1=}G$, we have

$$\left| H_{i+1}\left(e^{j\omega_{L,i+1}}\right) H_i\left(e^{j\omega_{U,i}}\right) \right| = G. \tag{5.8}$$

The individual filters are high-order low-shelving filters, as derived previously.

We base our design on the lower crossover frequencies, so that we express other parameters as

$$\begin{aligned} \omega_{u,i} &= R\omega_{l,i} \\ B_i &= \omega_{u,i} - \omega_{l,i} = (R-1)\omega_{L,i} \\ \omega_{C,i} &= 2\tan^{-1}\sqrt{\tan(\omega_{u,i}/2)\tan(\omega_{l,i}/2)} \end{aligned} \tag{5.9}$$

So the design of a graphic equalizer is as follows;

1. Choose the distribution of filters, i.e., set R for octaves, one-third octaves,...
2. Choose the first lower crossover frequency.
3. From this, find the bandwidth and center frequency.
4. For a given gain, generate the peaking/notch filter.
5. Find the next lower crossover frequency.
6. Repeat steps 3–5 until the whole frequency range is covered.

This procedure could equally have been performed with band pass filters rather than peaking/notch filters. However, then the filters would have been arranged in parallel, not in series.

Parametric Equalizers

The parametric equalizer is perhaps the most powerful and flexible of the equalizer types. Where each band of a graphic equalizer is adjustable only in gain, each band in a parametric equalizer has three adjustments: gain, center frequency and *Q* (or bandwidth). Most parametric equalizer bands are implemented with the same peaking/notch filter we have previously seen, Eq. (5.4). Notice that the three controls relate directly to the parameters G, $\omega_{c,}$ and B (where $B=\omega_c/Q$):

Since each band is more complex on a parametric equalizer compared to a graphic equalizer, a complete parametric equalizer unit will typically have fewer bands (an EQ section on a mixing console might have 1 or 2 bands; a dedicated rack-mount unit, 4 or 6 bands).

In addition to the gain, center frequency, and *Q* controls, some bands may have a switch to act as a *shelving* filter rather than the typical peaking/notch filter. In *shelving* mode (Figure 5.1), gain *G* is applied to all frequencies below the center frequency (*low shelving filter,* sometimes used in the lowest-frequency band) or to all frequencies above the center frequency (*high shelving filter,* sometimes used in the highest-frequency band). These shelving filters are similar to those found in the basic tone control, but where the tone control usually uses first-order shelving filters, the parametric equalizer uses either first or second-order filters.

WHO INVENTED THE PARAMETRIC EQUALISER?

Early filters included bass and treble controls without adjustable center frequency, bandwidth, and cut or boost. So sound engineers could only make broad, overall changes to a signal. When graphic equalizers arrived, engineers were still limited to the constraints imposed by the number and location of bands.

In the 1960s, Harold Seidel of Western Electric and Bell Telephone devised a tunable parametric filter. Daniel Flickinger then introduced an important tunable equalizer in early 1971. His circuit allowed arbitrary selection of frequency and cut/boost level in three overlapping bands over the entire audio spectrum.

In 1966, Burgess Macneal and George Massenburg began work on a new recording console for International Telecomm Incorporated. During the building of the console, Macneal and Massenburg, who was still a teenager, conceptualized an idea for a sweep-tunable EQ that would avoid inductors and switches. Soon after, Bob Meushaw, a friend of Massenburg, built a three-band, frequency-adjustable, fixed-Q equalizer.

When asked who invented the parametric equalizer, Massenburg stated

> four people could possibly lay claim to the modern concept: Bob Meushaw, Burgess Macneal, Daniel Flickinger, and myself...Our (Bob's, Burgess' and my) sweep-tunable EQ was borne, more or less, out of an idea that Burgess and I had around 1966 or 1967 for an EQ ... By 1969 I was spending all of my time designing circuitry sufficient to get to an elegant user interface: we perceived this as three controls adjusting, independently, the parameters for each of three bands for a recording console.... I remember agonizing over the topology for the EQ for months, and asking everyone I knew for help. I wrote and delivered the AES paper on Parametrics at the Los Angeles show in 1972 [31] ... it's the first mention of 'Parametric' associated with sweep-tunable EQ... what I'm proudest of is less in designing devices alone, and more in exploring the ever-expanding applications and uses of gear, and then applying that knowledge to designs.

George Massenburg went on to win many awards for both his technical contributions to recording technology and his critically acclaimed recordings. And today, the parametric equalizer is pervasive in audio signal processing and a fundamental tool in digital filtering techniques as well.

Summary

Every equalization effect, from the simplest tone control to the most complex parametric EQ, is based on the same set of filter types. The main differences between them are the number of individual subfilters and the controls presented to the user. First- and second-order filters are most commonly used in all equalizers, as these provide sufficient flexibility while minimizing complexity and artifacts sometimes found in higher-order filters. The next section discusses the practical implementation of the required filters for each type of equalizer.

Implementation

General Notes

All types of equalizers discussed in this chapter were originally analog effects, and their digital implementation follows the analog designs as closely as possible. *IIR* (*infinite impulse response*) filters are used throughout. Partly this is to emulate analog models, but processing delay is also a concern whenever any effect is used in live performance. While the maximum allowable total delay in an audio system may be a matter of discussion, it is safe to require each individual device to have the lowest possible processing delay, to allow cascading of several devices. This suggests using minimum-phase IIR filters instead of linear phase FIR filters, which typically exhibit higher latency for the same performance.

While some simple equalization effects, such as the *presence* control, can be implemented with a single filter section, most effects require two or more independent filters (referred to here as *subfilters*) to produce the overall result. In the case of the graphic equalizer, 30 or more subfilters might be used, and the way they are connected to one another becomes extremely important. Two topologies are commonly used: *parallel* connections, where audio is processed independently through several subfilters and the results added together, and *cascade* (or *series*) *connections*, where the output of one subfilter becomes the input to the next one. The optimal choice for each type of equalizer is discussed in the following sections.

Tone Control Architecture

The two-knob tone control is implemented with a *low shelving* filter for the bass and a *high shelving* filter for the treble. First-order filters are used to create smooth (6dB/octave) transitions between affected and unaffected frequency regions. When a third midrange knob is used, it is often created from a second-order *peaking/notch* filter. *Cascade* connections are most commonly used for these controls, as shown in Figure 5.3b.

Calculating the output of a first-order IIR filter requires not only the current input sample $x[n]$, but also the previous input $x[n-1]$ and the previous output $y[n-1]$. The second-order *peaking/notch* filter also requires values for $x[n-2]$ and $y[n-2]$. Therefore, a complete three-knob tone control will require 11 stored previous values.

Calculating Filter Coefficients

The coefficients for each filter in the tone control depend on the settings of the three knobs. As seen in Eqs. (5.2–5.4), calculating the coefficients requires a complex series of operations including trigonometric functions and floating-point division, all of which are more computationally expensive than multiplication. It is therefore advisable to only recalculate the filter coefficients when the knob positions change. The calculated coefficients can then be stored and recalled each time they are needed to process new audio samples.

Presence and Loudness Architecture

The *presence* control can be implemented with a single *peaking/notch* filter with a center frequency around 4 kHz and a Q of approximately 1. This corresponds to a bandwidth of 4 kHz, thus covering the 2–6 kHz frequency range. The presence knob adjusts the *gain* of this band; as with all peaking/notch filters, the gain outside of the band is 1.

The *loudness* control is architecturally nearly identical to the two-knob tone control. A *low shelving* and *high shelving* filter connected in *cascade* mode (Figure 5.3a). Their crossover frequencies are fixed by design (100 Hz in the bass and 8 kHz in the treble are typical values), just as the crossover frequencies are fixed in the tone control. However, a single knob or button controls the passband gain of both filters. The midrange gain remains at 1.

Graphic Equalizer Architecture

The graphic equalizer can be considered a generalization of the basic tone control with more bands; however, it is often implemented differently. The bass and treble controls in a basic tone control are connected in *cascade*, with each control affecting part of the frequency range and leaving the rest unaffected. A cascade of *peaking/notch* filters can be used in the graphic equalizer, but it involves the series connection of up to 31 subfilters, which can have unexpected, harmful effects.

The discussion in this chapter has focused mostly on the *magnitude response* of each filter, but *phase distortion* needs to be considered as well. Every IIR filter will have a nonlinear phase response with respect to frequency, and when filters are connected in cascade, the phase response of each section will add. In some cases, the combined phase shifts can produce audible artifacts that color the sound. FIR filters, which are less commonly used in equalization, can have linear phase, avoiding this particular form of distortion. However, FIR filters typically have a longer *group delay* for the same filter performance, so cascading many FIR filters in series might produce an audible delay between input and output.

Instead of using a cascade of peaking/notch filters, graphic equalizers often use a collection of band pass filters arranged in *parallel*, as shown in Figure 5.5. The audio input is split and sent to the input of every band pass filter. Each filter allows only a small frequency band to pass through. The center frequencies and bandwidths are configured so that if all the outputs were added together, the original signal would be reconstructed. The controls on the graphic equalizer are then implemented by changing the gain of each band pass filter output before the signals are summed together.

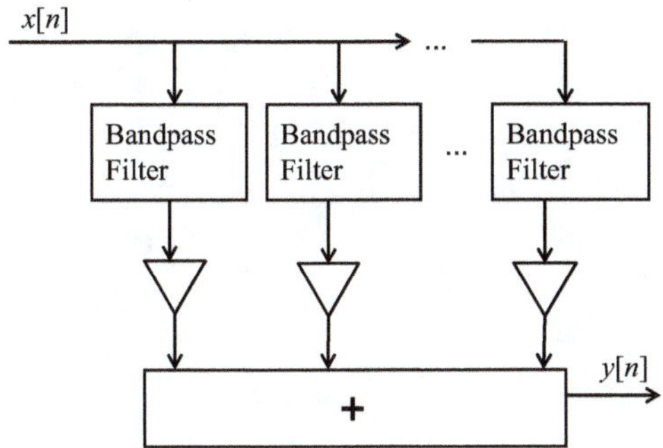

FIGURE 5.5
A diagram of a graphic equalizer, implemented with band pass filters placed in parallel, with N bands of control.

Parallel connection of band pass filters avoids the accumulating phase errors (and, potentially, *quantization noise*) found in the cascade. It also has a secondary benefit: the coefficients of each band pass filter depend only on their *center frequency* and Q and do not change with the setting of the gain sliders. Coefficients can therefore be calculated once when the effect is initialized, given information on the *sample rate*. Changing gain only involves changing a single number that multiplies the output of the band pass filter.

Parametric Equalizer Architecture

A single section of a parametric equalizer is created from a second-order *peaking/notch* filter (or in certain cases, a second-order *shelving* filter). When multiple sections are used, they are always connected in *cascade* so that the effects of each subfilter are cumulative. Even though the parametric equalizer is among the most complex for the user to control, its implementation is no more complex than a simple tone control. Also like the tone control, filter coefficients should be recalculated every time the user changes any knob, but for reasons of efficiency, should *not* be recalculated every audio sample.

Code Example

Implementing a parametric equalizer includes two main tasks: first, the coefficients of the IIR filter must be calculated whenever the user changes the center frequency, gain, or bandwidth; and second, the filter must be applied to each audio sample. This code shows the calculation of coefficients. For efficiency, it is run only when the user updates the controls:

```
void ParametricEQFilter::makeParametric(const double discreteFrequency,
    const double Q,
    const double gainFactor) noexcept
{
  /* Limit the bandwidth so we don't get a nonsense result from tan(B/2)
  */
  const double bandwidth = jmin(discreteFrequency / Q, M_PI * 0.99);
  const double two_cos_wc = -2.0*cos(discreteFrequency);
  const double tan_half_bw = tan(bandwidth / 2.0);
  const double g_tan_half_bw = gainFactor * tan_half_bw;

  /* setCoefficients() takes arguments: b0, b1, b2, a0, a1, a2
   * It will normalise the filter according to the value of a0
   * to allow standard time-domain implementations
   */

    setCoefficients (1.0 + g_tan_half_bw, /* b0 */
                     two_cos_wc, /* b1 */
                     1.0 - g_tan_half_bw, /* b2 */
                     1.0 + tan_half_bw, /* a0 */
                     two_cos_wc, /* a1 */
                     1.0 - tan_half_bw /* a2 */);
}
```

This code is based on the Juce `IIRFilter` class, which implements a generic two-pole, two-zero IIR filter. The `ParametricEQFilter` object is a subclass of IIRFilter, which implements the coefficients for a parametric equalizer. `setCoefficients()` saves the values of the six coefficients so they can later be used to calculate output samples. `jmin()` is a simple macro provided by Juce to return the minimum of two numbers.

The following code applies the filter to a block of audio samples for a single channel:

```
int numSamples;         // Indicates how many audio samples to process
float *channelData;     // Array of audio samples, length numSamples
float coefficients[6]; // Previously calculated filter coefficients
float x1, x2, y1, y2;  // Previous values of the input and output

for (int i = 0; i < numSamples; ++i)
{
    const float in = channelData[i];

    float out = coefficients[0] * in /* b0 */
    + coefficients[1] * x1 /* b1 */
    + coefficients[2] * x2 /* b2 */
    - coefficients[4] * y1 /* a1 */
    - coefficients[5] * y2; /* a2 */

    x2 = x1;
    x1 = in;
    y2 = y1;
    y1 = out;

    channelData[i] = out;
}
```

Here, `coefficients` is an array containing the values `b0`, `b1`, `b2`, `a0`, `a1`, and `a2`, as previously calculated and stored using `setCoefficients()`. Notice that `a0` is not used in this calculation. When the coefficients are set, they are normalized so `a0 = 1`, eliminating the need for it in further calculations. The variables $x1$, $x2$, $y1$, and $y2$ hold the last two inputs and outputs, respectively. For example, if `in` represents $x[n]$, then `x1` represents $x[n-1]$ and `x2` represents $x[n-2]$.

Applications

Graphic Equalizer Application

Graphic equalization is more commonly found in live performance and recording studios than in most home stereo systems. One common use of graphic equalization is to "tune" a room, adjusting the equalizer to roughly compensate for resonances in the room or imperfections in the frequency response of the speakers [39]. The goal is to achieve a desired frequency response, flattening out extremes, reducing coloration in the sound, and achieving greater sonic consistency among performance venues. However, graphic equalizers are occasionally found in consumer stereo systems and even in digital music player software, where they can be used as a more flexible form of tone control for adjusting the sound to taste.

Parametric Equalizer Application

Parametric equalizers allow the operator to add peaks or notches at arbitrary locations in the audio spectrum. Adding a peak can be useful to help an instrument be heard in a complex mix (see also the *presence control* earlier in the chapter), or to deliberately add coloration to an instrument's sound by boosting or reducing a particular frequency range [40]. Notches can be used to attenuate unwanted sounds, including removing power line hum (50 or 60 Hz and sometimes their harmonics) and reducing feedback. To remove artifacts without affecting the rest of the sound, a narrow bandwidth would be used. To deliberately add coloration to an instrument's sound by reducing a particular frequency range, a wider bandwidth might be used.

Wah-Wah

The sound of the *wah-wah* effect resembles its name: wah-wah is a filter-based effect that imparts a speech-like quality to the input sound, similar to a voice saying the syllable "wah". Wah-wah is most commonly known as a guitar effect which was popularized by Jimi Hendrix, Eric Clapton, and others in the late 1960s. However, its origins go back to the early days of jazz, when trumpet and trombone players achieved a similar sound using mutes.

The wah-wah audio effect uses a *band pass* or *peaking* filter whose center frequency is changed by a foot pedal. In some pedals, the mix between the original and filtered signal can be controlled by a separate knob, as shown in Figure 5.6 [41].

Theory

Basis in Speech

In speech, *formants* are the peaks in the frequency spectrum when a human voice utters a sound and are due to resonances in the vocal tract. For many vowel sounds, at least three formants can be easily identified. Humans listen for and assign meaning to the relative spacing of the first three formants of the human vocal tract.

Formants are distinct from the *fundamental frequency* (or pitch) of the voice, which is the frequency at which the vocal folds vibrate. We hear and notice the fundamental frequency, but in most languages, its exact location is not important when assigning meaning to vocal sounds. We also notice relative shifts of the fundamental frequency, but these are often associated with emotional states, such as when someone's voice goes up in pitch when under stress. But the relative positioning of formants, especially the first two formants, represents important information in how we interpret vowel sounds.

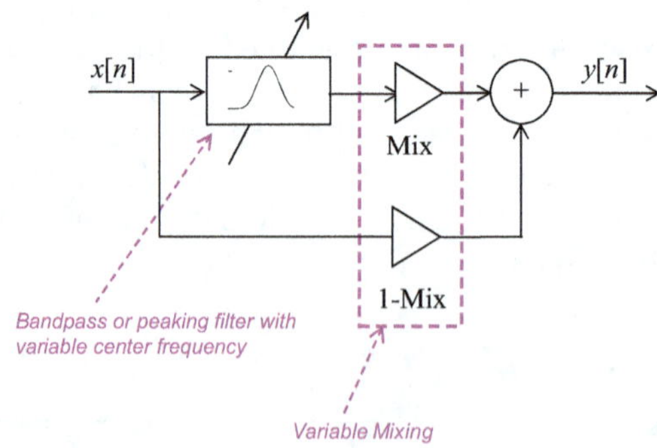

FIGURE 5.6
General wah-wah effect block diagram.

THE SERENDIPITOUS INVENTION OF THE WAH-WAH PEDAL

The first wah-wah pedal is attributed to Brad Plunkett in 1966, who worked at Warwick Electronics Inc., which owned Thomas Organ Company. Warwick Electronics acquired the Vox name due to the brand name's popularity and association with The Beatles. Their subsidiary, Thomas Organ Company, needed a modified design for the Vox amplifier, which had a midrange boost, so that it would be less expensive to manufacture.

In a 2005 interview in Universal Audio WebZine, Brad Plunkett said, I "came up with a circuit that would allow me to move this midrange boost … As it turned out, it sounded absolutely marvelous while you were moving it. It was okay when it was standing still, but the real effect was when you were moving it and getting a continuous change in harmonic content. We turned that on in the lab and played the guitar through it... I turned the potentiometer and he played a couple licks on the guitar, and we went crazy. A couple of years later... somebody said to me one time, 'You know Brad, I think that thing you invented changed music.'"

The wah-wah effect gives a voice-like quality to an input signal by simulating the formants found in speech [41]. The first formant of the [u] vowel is roughly located around 300 Hz, and the first two formants of the [a] vowel are located at approximately 750 and 1,200 Hz. Therefore, the wah-wah simulates the transitions between vowels by adjusting the center frequency of its filter in roughly this range (the exact range depends on the manufacturer and model of pedal, but a range between 400 and 1,200 Hz is typical). Wah-wah could be considered a simple form of speech synthesizer, though not close enough to be truly mistaken for a vowel sound.

Basic Wah-Wah

Figure 5.6 shows a block diagram of the wah-wah effect. A single second-order filter is typically used, with several possible variations: peaking, band pass or resonant low pass filters. As discussed previously, a *peaking* filter will boost the midrange frequencies while leaving all other frequencies with a gain of 1, and a *band pass* filter will boost the midrange frequencies and gradually roll off the low and high frequencies to 0.

A *resonant low pass* filter will pass the low frequencies with a gain of 1, create a peak with magnitude G_c in the midrange around the crossover frequency ω_c, and gradually roll off the high frequencies to 0. A second-order resonant low pass filter is given below,

$$H_{LP,Res}(z) = \frac{\Omega_c^2\left(z^2+2z+1\right)}{\left[\Omega_c^2+\Omega_c/G_c+1\right]z^2+2\left[\Omega_c^2-1\right]z+\Omega_c^2-\Omega_c/G_c+1}. \tag{5.10}$$

where $\Omega_c = \tan(\omega_c/2)$. The magnitude response of this filter is shown in Figure 5.7.

The *gain* and Q of the filter are generally fixed by design, but the *center frequency* is adjustable under the control of a foot pedal. As mentioned, the centre frequency commonly takes a range of around 400–1,200 Hz. Above and below this range, the effect loses its vocal quality. A *mix* control is sometimes used to vary the intensity of the effect by mixing between the filtered and unfiltered signals. In the case of a peaking filter, this function could be equivalently implemented by changing the *gain* at the center frequency.

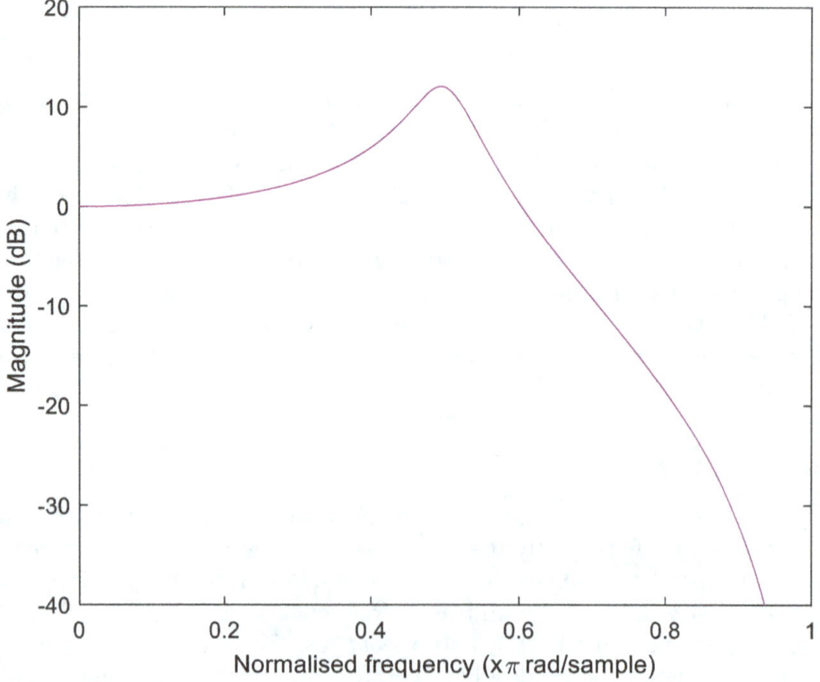

FIGURE 5.7
Magnitude response of a resonant low pass filter, with magnitude 12 dB ($G_c \sim 20$) at $\omega_c = \pi/2$.

Auto-Wah

For the standard wah-wah, as used commonly on guitar pedals, the player manually controls the center frequency. In the *auto-wah* effect, the center frequency is controlled automatically. Two variations of auto-wah are commonly found. In the first, the center frequency sweeps back and forth following a *low-frequency oscillator* (*LFO*) with an adjustable frequency, typically around 1–2 Hz. The range of filter center frequencies is similar to the basic wah-wah, but it is often adjustable by the user.

In the second auto-wah variation, the center frequency is automatically adjusted according to the amplitude of the input signal. This arrangement is sometimes known as an *envelope follower*, and when used on a guitar, it produces a moving resonance on every note. Louder signals push the center frequency upwards (towards the "ah" sound). The sensitivity of this process is typically a user-adjustable parameter, as are the *attack time* and *release time* of the envelope. More details on envelope calculation can be found in Chapter 7.

Tremolo-Wah

If the LFO-controlled auto-wah is combined with a periodic change in amplitude (*tremolo*; Chapter 6), the result is an effect known as *tremolo-wah*. The effect is generally equivalent to placing a tremolo and auto-wah effect in series:

$$y[n] = w[n]g[n] \quad \textit{where} \quad g[n] = 1 + \alpha\cos(\omega n). \tag{5.11}$$

where $y[n]$ is the output of the entire tremolo-wah, $w[n]$ is the output of the auto-wah section and $g[n]$ is a time-varying gain factor, and ω is normalised frequency $2\pi f/f_s$ as usual. $g[n]$ may also be calculated on a logarithmic scale since the human sensitivity of loudness follows a logarithmic relation. The same LFO can be used to control both amplitude and wah center frequency, but often the two move independently with different frequencies, or with the same frequency but out of phase.

Other Variations

The standard wah-wah pedal has only one resonant peak, in contrast to the two or three identifiable formants in most vowel sounds. By adding additional resonances, even more vocal-like sounds from a wah-wah effect are possible. Some pedals use a second peaking filter circuit whose center frequency moves around in a different manner than the main filter, for example, following the second formant in the "oo" and "ah" vowels. This produces an effect much closer to human speech.

WAH-WAH AND WACKA-WACKA

The wah-wah effect is incredibly expressive. It's associated with whole genres of music, and it can be heard on many of the most influential funk, soul, jazz, and rock recordings over the past 50 years.

Jimi Hendrix would sometimes use the wah-wah effect while leaving the pedal in a particular location, creating a unique filter effect that did not change over time. However, in 'Voodoo Child (slight return)', Hendrix muted the strummed strings while rocking the pedal, creating a percussive effect. The sweeping of the wah-wah pedal is more dramatic in the louder verses and the chorus, emphasizing the song's blues styling.

The 'wacka-wacka' sound that Hendrix created soon became a trademark of a whole subgenre of 1970s funk and soul. Melvin 'Wah-Wah Watson' Ragin, a highly respected Motown session musician, is renowned for his use of the wah-wah pedal, especially on The Temptations, 'Papa Was A Rolling Stone'. This distinctive 'wacka-wacka' funk style of soon became a feature of urban black crime dramas, such as in Isaac Hayes' 'Theme from Shaft,' Bobby Womack's score to 'Across 110th Street' and Curtis Mayfield's 'Superfly.'

Another unusual use of the wah-wah pedal can be heard on the Pink Floyd song 'Echoes.' Here, screaming sounds were created by plugging in the pedal back to front; that is, the amplifier was connected to the input, and the guitar was connected to the pedal's output.

Of course, the use of wah pedals is not reserved just for guitar. Bass players have used wah-wah pedals on well-known recordings (Michael Henderson playing with Miles Davis, Cliff Burton of Metallica, ...). John Medeski and Garth Hudson use the pedals with Clavinets. Rick Wright employed a wah-wah pedal on a Wurlitzer electric piano on the Pink Floyd song 'Money,' and Dick Sims used it with a Hammond organ. Miles Davis's ensembles used it to a great extent, both on trumpet and on electric pianos. The wah-wah is frequently used by electric violinists, such as Boyd Tinsley of the Dave Matthews Band. Wah-wah pedals applied to amplified saxophone also feature on albums by Frank Zappa and David Bowie.

Implementation

Filter Design

Like equalization, wah-wah nearly always uses second-order IIR filters. Two of the filter types used in the wah-wah are the same as those found in the various types of equalizer. The *peaking/notch* filter is used in the parametric EQ, and a passable wah-wah effect can be obtained from a parametric EQ by choosing large *gain* (up to 12 dB) and high *Q* (values from 2 to 10 are typical) and varying the *center frequency*. The *band pass* filter, used in some wah-wah implementations, is also found in the graphic EQ. Again, a high *Q* is often used. The design and implementation of these filters are the same as in the equalizers.

Most analog wah-wah pedals for guitar use band pass filters. Analog synthesizers typically use resonant low pass filters to achieve similar effects. Resonant low pass filter coefficients can be calculated similarly to more commonly used second-order low pass filters, but substituting a higher *Q*. Values from 2 to 10 might be found in the wah-wah, compared to 0.71 in a standard Butterworth low pass filter. The resonant low pass filter creates a peak in the frequency response at the *crossover frequency*, which is responsible for the "wah" effect.

As with the equalizers, the filter coefficients should be recalculated if and only if the center frequency (or crossover frequency) has changed. Changing the *mix* control in the basic wah-wah (Figure 5.6) does not require recalculating the filter. In any variation of the auto-wah, the center frequency changes each sample and continuous recalculation of the coefficients is inevitable. However, the computational load can be reduced by recalculating the coefficients less frequently than once per audio sample. For example, at a sample rate of 44.1 kHz, recalculating the coefficients every 16 samples will still update the center frequency over 2,700 times/s, improving efficiency without any significant difference in audio quality.

Low-Frequency Oscillator

In one variant of the auto-wah, a *low-frequency oscillator* (*LFO*) controls the center frequency of the filter. Typical parameters include *LFO frequency*, *LFO waveform*, *minimum frequency*, and *sweep width*. Not all parameters will be user-adjustable; some will be fixed by design. *LFO frequency* (f_{LFO}) is the number of cycles per second the center frequency oscillates; typical values range from 0.2 to 5 Hz. *LFO waveform* controls the shape of the center frequency variation; sinusoidal waveforms are most common in the auto-wah. *Minimum frequency* ($f_{\min}$) sets the lowest center frequency for the filter, typically no less than around 250 Hz and often higher to maintain the vocal effect. *Sweep width* (*W*), expressed in Hz, is the difference between the minimum and maximum center frequencies across an entire oscillation. The center frequency of the filter over time with sinusoidal LFO and sample rate f_s can be written as:

$$f_c[n] = f_{\min} + 0.5W\left(1 + \cos\left(\frac{2\pi f_{\mathrm{LFO}} n}{f_s}\right)\right). \tag{5.12}$$

Further considerations on LFO implementation can be found in Chapter 3.

Envelope Follower

The envelope follower variant of the auto-wah scales the center frequency of the filter proportionally to the level of the input signal. The instantaneous value of each sample is a poor measure of a signal's level, so a *level detector* must be used to calculate a local average value. Level detectors based on the exponential moving average are discussed in detail in Chapter 7, and the same types of level detectors used in the compressor can be used in the envelope follower wah. A common level detector has a variable *attack time* τ_A and *release time* τ_R; its operation is given by:

$$y_L[n] = \begin{cases} \alpha_A y_L[n-1] + (1-\alpha_A) x_L[n] & x_L[n] > y_L[n-1] \\ \alpha_R y_L[n-1] + (1-\alpha_R) x_L[n] & x_L[n] \le y_L[n-1] \end{cases}. \tag{5.13}$$

where $\alpha_A = e^{-1/(\tau_A f_s)}$, $\alpha_R = e^{-1/(\tau_R f_s)}$ and $x_L[n]$ is the input signal. Attack time and release time are often user-adjustable parameters. The other parameters in the envelope follower wah are *minimum frequency* and *sweep width*, which work analogously to the LFO case. We can therefore write the center frequency f_c as a function of the level detector value $y_L[n]$:

$$f_c[n] = f_{\min} + W y_L[n]. \tag{5.14}$$

$y_L[n]$ is taken to always be positive, and assuming the input signal is scaled to have a maximum value of 1, the center frequency will reach a maximum value of $f_{\min} + W$, just as with the LFO auto-wah.

Analog Emulation

Musicians often become attached to particular brands of wah-wah pedal to achieve their signature sound. Even when the gain, center frequency, and Q of a digital filter are tuned identically to the analog case, the sound may still be subtly different. Part of the distinct sound of many analog wah pedals comes from nonlinear distortion introduced by the electronic components, especially the iron-core inductors used in the resonant filters. Nonlinear distortion, discussed in Chapter 8, adds new *harmonic* and *intermodulation* frequency components to the output signal that were not present in the input. Precise replication of these effects requires detailed numerical simulation of the behavior of each circuit element, which is beyond the scope of this text. References for further reading on analog modeling are given in [42].

Phaser

The *phase shifter* (or *phaser*) creates a series of *notches* in the audio spectrum where sound at particular frequencies is attenuated or eliminated. The flanger (Chapter 3) also produces its characteristic sound from notches, and in fact, the flanger can be considered a special case of phasing. However, where the flanger is based on delays, the phaser uses *allpass filters* to create phase shifts in the input signal. When the allpass-filtered signal is mixed with the original, notches result from destructive interference. Where the flanger always generates evenly-spaced notches, the phaser can be designed to arbitrarily control the location of each notch, as well as their number and their width. Like the flanger, though, the phaser's characteristic sound comes from the sweeping motion of the notches over time.

Theory

Basic Phaser

The notches needed for the phaser are most often implemented using allpass filters. Allpass filters, whose design is discussed in Chapter 4, pass all frequencies with no change in magnitude, but they introduce a frequency-dependent phase lag. The output of the allpass filter is then added to the original signal, as in Figure 5.8. The relative level of the filtered signal can be adjusted by a *depth* (or *mix*) control.

Mixing the original and filtered signals creates notches in the frequency response according to the principle of *constructive* and *destructive interference*, just as in the flanger. At certain frequencies, the allpass filters will introduce a phase shift of 180° or an odd multiple thereof (540, 900, etc.). This is equivalent to inverting the input, and when the original and filtered signals are added together with equal weight ($depth = 1$), they will cancel completely, resulting in a notch at those frequencies. Frequencies near the notch, which experience nearly 180° of phase shift, will also be attenuated.

The pure delay used in the flanger can also be considered an allpass filter with *linear phase*: a delay produces no change in magnitude for any

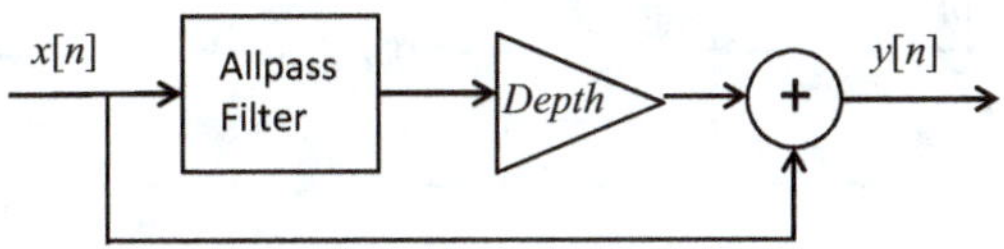

FIGURE 5.8
A phaser, also known as a phase shifter.

frequency, but it produces a phase lag proportional to the input frequency. Therefore, the phase response of a pure delay hits odd multiples of 180° (–180, –540, –900, –1,260; the negative value indicates the output phase *lags* the input) at evenly spaced frequencies, and the notches will also appear at evenly spaced frequencies.

IIR allpass filters, like all IIR filters, are not linear phase, so the phase shift they create is not a linear function of frequency. By changing the order, *Q* and center frequency of the filter, many variations in notch location and width are possible. The number of notches is determined by the number of times the phase crosses an odd multiple of 180 degrees, which in turn is determined by the order of the filter. However, another useful property of allpass filters is that a cascade (series connection) of several allpass filters is itself an allpass filter. Phaser effects thus often consist of several simpler allpass filters in series, where the total phase lag is the sum of each filter's phase response.

Low-Frequency Oscillator

Like the chorus and flanger effects, a low-frequency oscillator (LFO) produces a periodic change in notch location over time. It does this by changing the center frequencies of the allpass filters. But where the chorus and flanger commonly use sinusoidal or triangular LFO waveforms, the phaser often changes the notch frequencies in an *exponential* pattern over time. This more closely corresponds to human hearing, where perceived pitch is an exponential function of frequency. Example LFO waveforms are discussed in the Implementation section.

Phaser with Feedback

Like the flanger, some phasers incorporate feedback (sometimes called *regeneration*) between the allpass filter output and input, as shown in Figure 5.9. Like other effects incorporating feedback, the *feedback gain* of the phaser must be strictly less than 1 to maintain stable operation. The

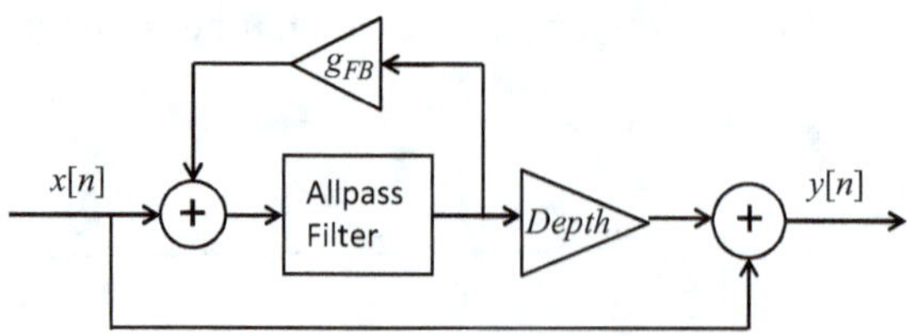

FIGURE 5.9
A phaser with feedback. The feedback gain can be positive or negative, but must have a magnitude strictly less than 1.

effect of feedback is to increase the Q of the allpass filters, making the phase transitions more steep and therefore making the notches sharper.

Stereo Phaser

Just as with stereo flangers and chorus units, a stereo phaser can be created from two monophonic phasers with different filter settings, creating notches at different frequencies. Typically, the notches are controlled by two low-frequency oscillators in *quadrature phase*, where the output of one oscillator trails the other by 90°.

An optional addition is to selectively mix the outputs of each filter, which can create additional notches (Figure 5.10). The phaser effect can even be created acoustically with no mixing at all: each *feed-across* gain is set to zero, and each output is sent to a separate speaker. In this *spatial phaser* arrangement, notches exist at different points in the room where the signal from each speaker cancels out in the air. Moving around the room will change the location of the notches.

Implementation

Allpass Filter Calculation

The phaser uses first-order or second-order IIR allpass filters whose center frequencies vary over time, as shown in Figure 5.11a. A first-order allpass has 0° phase lag at 0 Hz and a total *phase lag* of 180° at high frequencies. By itself, a single first-order section is not enough to create a moving notch in the phaser. A second-order allpass has a total phase of 360° at high frequencies. Since the phase lag increases monotonically with frequency, a single second-order section produces a single notch at the frequency with 180° phase lag (which is the *crossover frequency* of the allpass filter) as shown in Figure 5.12a. To achieve more notches, multiple allpass sections are placed in series.

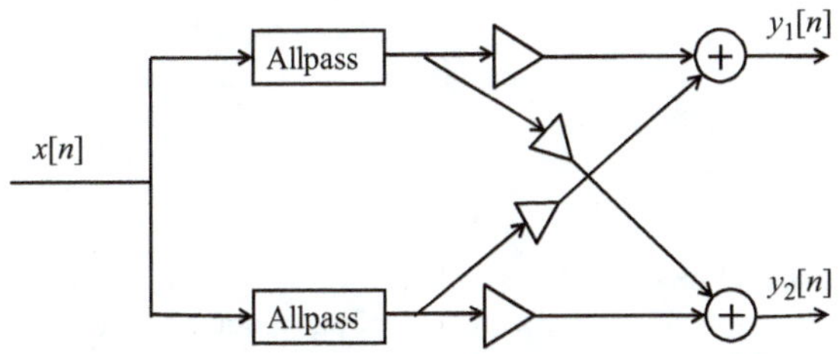

FIGURE 5.10
A generalized diagram of a stereo phaser with feed-across gains.

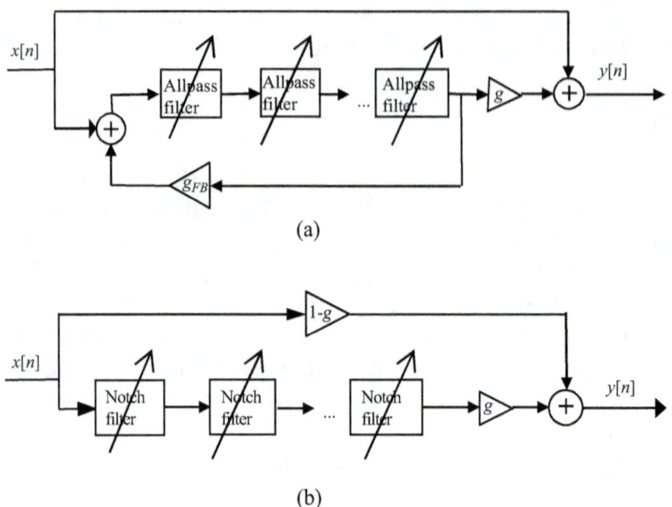

FIGURE 5.11
Two approaches to implementing a phaser: phasing with time-varying allpass filters and optional feedback (a) and phasing with notch filters (b).

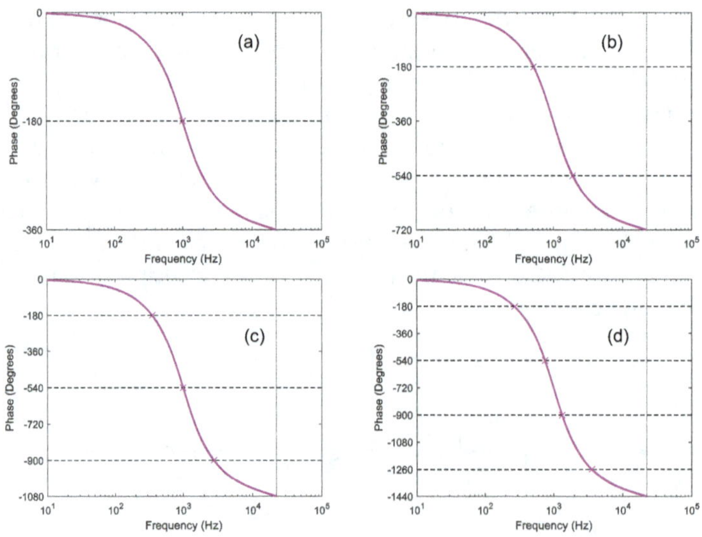

FIGURE 5.12
Allpass filter phase and phaser notch locations for different numbers of second-order allpass filters. Center frequency is 1 kHz and sampling frequency is 44.1 kHz. (a), (b), (c) and (d) represent 1, 2, 3, and 4 second-order sections, respectively.

A common analog phaser design uses four first-order allpass filter sections in series (or, equivalently, two second-order sections). This produces 720° of total phase shift and therefore two notches (at 180° and 540°). In a typical design, all four first-order allpass filters might have the same center frequency (that is, the frequency at which the phase lag is 90°). This does not mean that both notches fall at the same frequency. Rather, the total phase lag is the sum of the contributions of each filter, and the notches occur where this sum reaches 180° and 540°. Therefore, in this four-section design, the notches fall where the phase shift of each individual filter is 45° or 135° (Figure 5.12b).

If six first-order allpass sections are used instead, the total phase lag is 1,080°, and three notches will result (180°, 540°, 900°). Again, each section can be tuned to an identical center frequency, but the notches will occur in different locations than in the four-section design (Figure 5.12c), this time occurring where each filter contributes 30°, 90°, or 150° of phase lag.

The transfer functions for an allpass filter were derived in Chapter 4. We can rewrite these transfer functions, replacing the arbitrary constants with terms relating to how the filter modifies the phase. A first-order allpass filter may be given as,

$$H(z)=\frac{z\left[\tan(\omega_c/2)-1\right]+\tan(\omega_c/2)+1}{z\left[\tan(\omega_c/2)+1\right]+\tan(\omega_c/2)-1}. \tag{5.15}$$

and the crossover frequency ω_c is where the phase response reaches 90º.

For a second-order design, we can write the allpass filter [1] as

$$H(z)=\frac{[1-\tan(B/2)]z^2+2z\cos\omega_c+1+\tan(B/2)}{[1+\tan(B/2)]z^2+2z\cos\omega_c+1-\tan(B/2)}. \tag{5.16}$$

Here, the crossover frequency ω_c is where the phase response reaches 180º °, and the bandwidth B is the difference between the two frequencies where the phase response reaches 90º and 270º. Note that both Eqs. (5.15) and (5.16) conform with the format of any allpass filter given in Chapter 4.

Alternate implementation

An alternate approach to phaser design uses a set of *notch filters* (Figure 5.11b). in place of the allpass sections (Figure 5.11a). The notch filters can be implemented as a *cascade* of second-order IIR sections, similar to a parametric equalizer with high Q and gain 0 at the center frequency of each section. This implementation does not require the filter output to be mixed with the direct sound since the notches are created directly;

however, a mix can still be used to vary the intensity of the effect. Like the allpass implementation, the center frequency of the notch filters is varied with a low-frequency oscillator.

LFO Waveform

Just as with the flanger, the sound of the phaser results from how the notch frequencies change over time. In contrast to sinusoidal low-frequency oscillators commonly found in other effects, *exponential* motion of the notch frequencies is often used in the phaser. Specifically, a waveform that is *triangular* in the log-frequency domain can be used:

$$\begin{aligned} \ln(f_c[n]) &= \ln(f_{\min}) + \ln(W)\text{triangle}(\omega_{\text{LFO}} n) \\ f_c[n] &= f_{\min} W^{\text{triangle}(\omega_{\text{LFO}} n)} \end{aligned} . \tag{5.17}$$

where $f_{\min}$ is the minimum center frequency, $\omega_{\text{LFO}}=2\pi f_{\text{LFO}}/f_s$ is the normalised LFO frequency and $W=f_{\max}/f_{\min}$ is defined as the ratio of the maximum to minimum frequency. Note that this definition of W differs from the wah-wah effect described earlier in the chapter. Here we also define the triangle waveform to take a range of values between 0 and 1 over a complete oscillation:

$$\text{triangle}(t) = \begin{cases} t/\pi & 0 \le t < \pi \\ 2 - t/\pi & \pi \le t < 2\pi \end{cases} . \tag{5.18}$$

Analog and Digital Implementations

Filters implemented digitally (*discrete time*) behave slightly differently than their analog (*continuous time*) counterparts, due to the fact that frequency is unbounded in the continuous-time domain but restricted to the $[0, 2\pi)$ range in discrete time. Techniques exist to convert continuous-time prototype filters into discrete-time equivalents, including the *bilinear transform* with pre-warping. However, even with these methods, the response of the continuous and discrete filters diverge near the Nyquist frequency.

Figure 5.13 shows the differences between the continuous and discrete filter phase responses and the resulting effect on phaser notch location. For an allpass center frequency $f_c=3\,\text{kHz}$ and four second-order sections, the highest notch is 2.5 semitones (15%) lower in discrete time than it would be in an analog implementation. When $f_c=10\,\text{kHz}$, the top notch is 12.8 semitones (52%) lower, over an octave of difference. Fortunately, the range of center frequencies used in the phaser rarely extends much

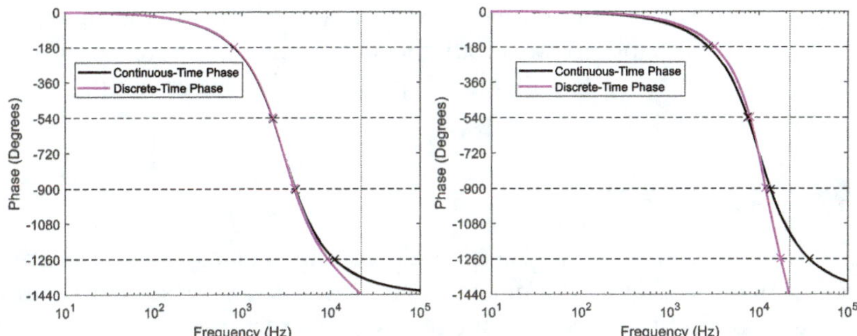

FIGURE 5.13
Continuous-time (analog) and discrete-time (digital) allpass filters produce different notch locations when used in the phaser, particularly at high frequencies. Both plots depict the phase response and notch locations for four second-order allpass sections and 44.1 kHz sampling frequency. On left, center frequency is 3 kHz and on right, center frequency is 10 kHz.

beyond 1 to 2 kHz (though individual notch locations may be higher), so the differences present only a minor concern when emulating analog phasers.

Common Parameters

Depth (mix/level): The depth controls the amount of allpass filtered signal that is added to the output. At a depth of zero, only the original signal appears at the output. At a depth of 1, the mix between original and filtered signals is equal, producing the deepest notches. Some phasers will use "depth" to refer to what this book labels "sweep width", the frequency range in which the notches move, so it is important to know the convention used by any particular phaser unit.

Sweep width (range): This parameter controls the frequency range across which the notches sweep. Possible variations include fixing the minimum frequency location and using the sweep width to change the maximum frequency, or offering separate controls for minimum and maximum frequency.

Feedback/regeneration: This control adjusts the feedback gain between the output and input of the allpass filter section (Figure 5.9). The value must be strictly less than 1 to avoid instability. Using feedback can produce sharper, more pronounced notches.

LFO frequency: Sometimes labeled *speed* or *rate*, this changes the rate at which the notches move up and down in frequency. Its effect and use are similar to the flanger and chorus. Like those effects, the control sets how many times per second the notches sweep across their range. The actual speed at which the notches move (in Hz per second) will also depend on the sweep width and LFO waveform.

Code Example

The following C++ code fragment, adapted from the code that accompanies this book, implements a phaser with feedback, a user-selectable number of allpass sections, and adjustable LFO waveform:

```
int numSamples;      // Indicates how many audio samples to process
float *channelData; // Array of audio samples, length numSamples
float ph;            // Current phase of the LFO (0-1)
float lastFilterOutput; // Output of filter last sample, for
                        // implementing feedback
OnePoleAllpassFilter **allpassFilters; // Objects handling first order
                                       // allpass filter
float inverseSampleRate; // Defined as 1.0/(sample rate)
int sc;                  // Sample count, to decide when to update
                         // coefficients

float depth_;          // Depth of the phaser effect (0-1)
float feedback_;       // Amount of feedback (>= 0, < 1)
float lfoFrequency_;   // Frequency of the LFO
float baseFrequency_;  // Lowest point in the sweep of the allpass center
frequency
float sweepWidth_;     // Width of the LFO (in Hz)
int waveform_;         // What type of waveform to use (sine, triangle,
                       // ...)
int filtersPerChannel_;    // How many allpass filters are used
int filterUpdateInterval_; // How often to update the allpass
                           // coefficients

for (int sample = 0; sample < numSamples; ++sample)
{
  float out = channelData[sample];

  // If feedback enabled, include feedback from last sample in allpass
  // filter
  // chain's input. This is not how analog phasers work because there is
  // sample
  // delay between output and input, which adds further phase shift up
  // to 180
  // degrees at half sampling frequency. To model analog phaser with
  // feedback
  // involves modelling delay-free loop.

  if(feedback_ != 0.0) out += feedback_ * lastFilterOutput;

  for(int j = 0; j < filtersPerChannel_; ++j)
  {
    // First, update current allpass filter coefficients depending on
    // parameter
    // settings and LFO phase.
```

```
    // Recalculating filter coefficients is much more expensive than
    // calculating
    // a sample. Only update coefficients at fraction of sample rate;
    // since
    // the LFO moves slowly, difference won't generally be audible.
    if(sc % filterUpdateInterval_ == 0)
    {
      allpassFilters[j]->makeAllpass(inverseSampleRate,
                       baseFrequency_ + sweepWidth_* lfo
(ph, waveform_));
    }
    out = allpassFilters[j]->processSingleSampleRaw(out);
  }

  lastFilterOutput = out;

  // Add the allpass signal to the output, though maintaining constant
  // level
  // depth = 0 --> input only ; depth = 1 --> evenly balanced input and
  // output
  channelData[sample] = (1.0f-0.5f*depth_)*channelData[sample] +
0.5f*depth_*out;

  // Update the LFO phase, keeping it in the range 0-1
  ph += lfoFrequency_*inverseSampleRate;
  if(ph >= 1.0) ph -= 1.0;
  sc++;
}
```

At the core of this code is an array C++ objects `allpassFilters`, which each implement a single first-order allpass filter. Even if every filter has the same coefficients, we need to maintain separate objects for each filter since the filter must keep track of previous input and output samples in order to calculate its output. The code above passes each sample through each filter in succession, eventually arriving at the sample `out`, which is mixed with the original input `channelData[sample]` to produce the phaser effect. The `depth_` parameter controls the relative balance of these two signals, with `depth_ = 1` producing an even mix between them and therefore the most pronounced phaser effect.

As in the earlier parametric equalizer example, it is not efficient to recalculate the filter coefficients at every single audio sample, and the LFO will change value slowly enough that recalculating once every few samples will be sufficient. The variable `sc` stores how many samples have elapsed since the coefficients were last recalculated, and `filterUpdateInterval_` indicates how often they should be recalculated. A value of 16 or 32 would strike an appropriate balance between efficiency and smoothness of effect.

The function `lfo()` implements one of several LFO waveforms depending on the value of the `waveform_` variable. `waveform_` will take one of several predefined values, for example 0 corresponding to a sine, 1 to a triangle wave, 2 to a square wave, or 3 to a sawtooth wave.

The expression `allpassFilters[j]->processSingleSampleRaw (out)` runs the following code for each allpass filter object:

```
float OnePoleAllpassFilter::processSingleSampleRaw (const float
sampleToProcess)
{
    // Process one sample, storing the last input and output
    y1 = (b0 * sampleToProcess) + (b1 * x1) + (a1 * y1);
    x1 = sampleToProcess;
    return y1;
}
```

Here, `x1` and `y1` keep track of the last input and output $x[n-1]$ and $y[n-1]$ respectively. `b0`, `b1` and `a1` are coefficients of the filter, as calculated by the `makeAllpass()` function in the previous code block.

Further Reading

Audio equalization is a vast topic, and the reader is referred to [28] for an overview of the field, as well as the numerous audio production texts [17,43] for best practices in its application. Recent research has focused on advanced graphic equalizer designs [44], automatic equalization [45], and the application of deep learning to equalization tasks [46]. As for the LFO-driven filter effects, the phaser and wah wah, they are well-established in both analogue and digital forms. So recent work has concentrated on modeling the classic analogue designs, using both spectral [47] and neural [20,48] approaches.

Problems

1. Which one of the three diagrams in Figure 5.14 would *not* produce a working graphic equaliser, and why?
2. Suppose a graphic equalizer is implemented with six filters having an octave spacing between filters. The first (the one with the lowest center frequency) filter has a lower crossover frequency at 375 Hz. Find the lower crossover frequency, upper crossover frequency, bandwidth, and center frequency of the fourth filter.
3.
    i. Draw a magnitude response plot showing the frequency response of each band in a *10-band, 1-octave* graphic EQ with a

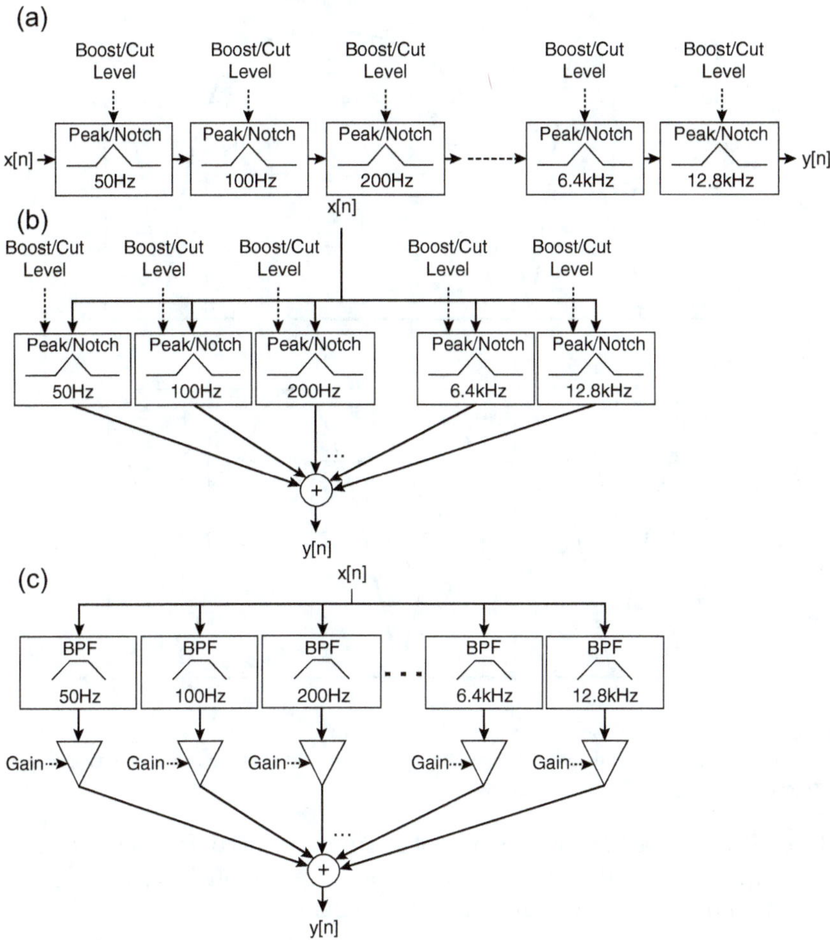

FIGURE 5.14
One of these three does *not* represent a graphic equalizer.

lowest centre frequency of 30 Hz. Assume the controls are all set to flat.

ii. Draw a block diagram of the implementation of this filter.

4.

i. How do graphic equalisers differ from parametric EQs and basic tone controls?

ii. What are the primary controls in a graphic EQ?

iii. What are the primary controls in a parametric equaliser, and what do they do?

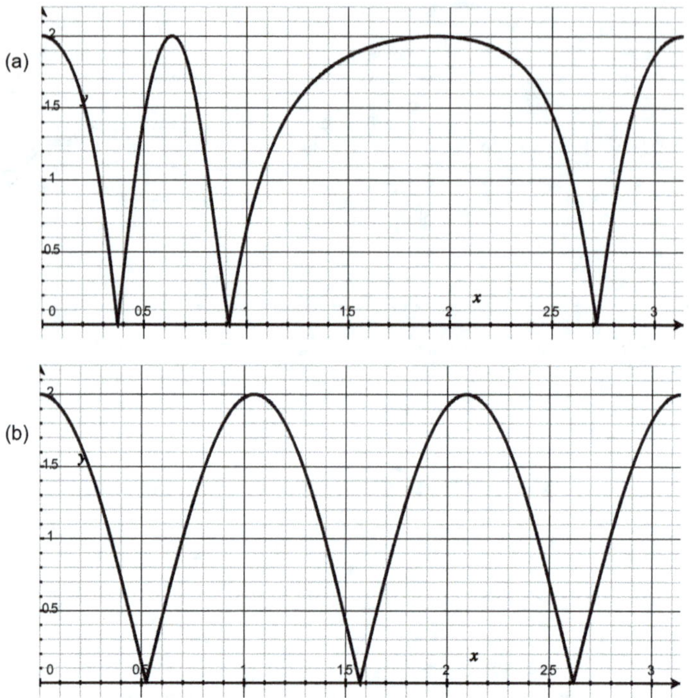

FIGURE 5.15
One of these is the frequency response of a flanger; the other is the frequency response of a phaser.

5. (difficult) In a peaking or notch filter, as used for parametric equalization, find a formula for the lower cutoff frequency as a function of bandwidth and center frequency.
6. Explain why *wah-wah* could be considered a special case of parametric equalisation. (What is the main effect of moving the pedal on a wah-wah box?)
7. Define 3 of the main parameters of a phaser: Depth, Sweep Depth, and Speed, and describe the effect of varying their settings.
8.
   i. How does a flanger differ from a phaser in implementation?
   ii. One of the diagrams in Figure 5.15 is the frequency response of a flanger; the other is the frequency response of a phaser. Identify which is which, and explain why.
9. Explain how a phaser may be implemented using allpass filters to create notches in the frequency spectrum.

6

Amplitude Modulation

The term *modulation* refers to the variation of one signal by another. *Amplitude modulation* specifically refers to one signal changing the amplitude (or gain) of another. There are two common uses of amplitude modulation in digital audio effects: *tremolo* and *ring modulation*. Though mathematically similar, their musical effect is quite different.

Another type of modulation is *frequency modulation*, the periodic variation in frequency (pitch) of a signal. Frequency modulation is used in the *vibrato* effect discussed in Chapter 3. However, this chapter is devoted solely to amplitude modulation effects.

Tremolo

Tremolo is a musical term that literally means "trembling". It typically refers to a style of playing involving fast repeated notes, for example, fast repeated bow strokes on a violin, rolls on a percussion instrument, or continuous rapid plucks on a mandolin. The tremolo audio effect simulates this playing style by periodically modulating the amplitude of the input signal, so a long sustained note comes out sounding like a series of short, rapid notes. The effect is commonly used on electric guitar, and tremolo was built into some guitar amplifiers as early as the 1940s. Tremolo is also one of the more straightforward effects to implement digitally.

Theory

Tremolo, as depicted in Figure 6.1, results from multiplying the input signal $x[n]$ with a periodic, slowly-varying signal $m[n]$:

$$y[n] = m[n]x[n]. \tag{6.1}$$

The simplest interpretation of this equation is to consider the tremolo as a variable gain amplifier whose gain at sample n is given by $m[n]$. When $m[n] > 1$, the output amplitude is greater than the input amplitude, and

DOI: 10.1201/9781003593942-6

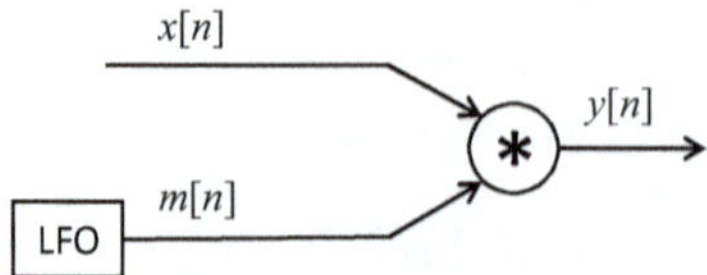

FIGURE 6.1
The tremolo multiplies the input signal by a low-frequency oscillator, which changes its gain.

similarly, $m[n]<1$ reduces the level of the output signal relative to the input.

Low-Frequency Oscillator

The characteristic sound of the tremolo comes from the fact that the gain changes periodically over time. The modulating signal $m[n]$ is generated by a *low-frequency oscillator* (LFO):

$$m[n] = 1 + \alpha \cos \omega_{\text{LFO}}. \tag{6.2}$$

Here, α is the *depth* of the tremolo and $\omega_{\text{LFO}}=2\pi f_{\text{LFO}}/f_s$ is the normalized frequency of the oscillator, where f_s is the sampling frequency. f_{LFO} typically ranges from 0.5 to 20 Hz. Figure 6.2 shows an example of tremolo applied to an audio waveform. A depth of 0 produces no effect; a depth of 1 produces the most pronounced tremolo by having the minima of $m[n]$ reach 0. Both frequency and depth are typically user-adjustable controls.

Some tremolo units also allow the LFO waveform to be selected by the user. In addition to sine waves, triangular and square waveforms are sometimes used. Square waves, with their abrupt transition between high and low, produce a stuttering effect in contrast to the smoother transitions of the sine wave LFO.

Tremolo Properties

Tremolo is a simple effect consisting of a single multiplication per sample. Because multiplication is a *linear* operation, the tremolo is also linear. Tremolo is *time-variant* on account of the low-frequency oscillator: delaying the input signal and passing it to the tremolo will not produce the same result as delaying the output of the tremolo by the same amount, because the gain of the tremolo changes over with the LFO phase. Because a bounded input signal always produces a bounded output, the tremolo is a *stable* effect for all parameter settings. Tremolo is *memoryless* since each output sample only depends on the current input sample and not any previous inputs.

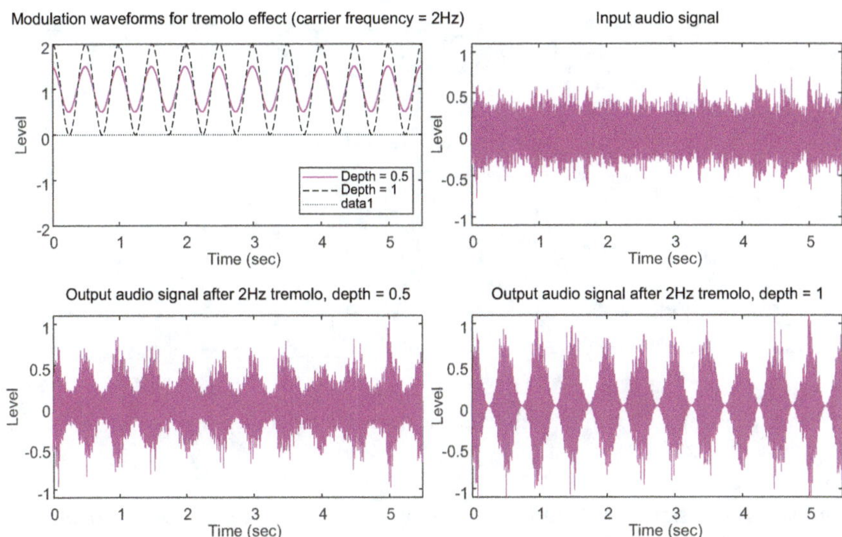

FIGURE 6.2
Tremolo applied to an audio signal, with two different values of the depth parameter.

Implementation

The original tremolo units were based on *voltage-controlled amplifier* (VCA) circuits built from vacuum tubes, in which a time-varying voltage from the LFO would change the gain of the audio signal. Digital implementation of the tremolo is very simple, using only a single multiplication for each audio sample (not including the computation of the LFO). The most complex implementation aspect is the LFO. In particular, keeping track of the *phase* of the LFO is critical. One could imagine the following pseudocode implementation, where `f` and `alpha` are adjustable parameters:

```
for n = 0 to (number of samples):
{
  m[n] = 1.0 + alpha*cos(2*pi*f*n/fs);
  y[n] = m[n]*x[n];
}
```

Assume this code is running in real time (as $x[n]$ arrives). It will produce accurate results unless the user changes the value of `f` midway through. Suppose at time n=N, there is a step change in the frequency `f`. Then the value inside the `cos()` function will jump from sample $N-1$ to N, producing a noticeable artifact in the result (Figure 6.3, left).

Instead, the desired behavior is that the tremolo change frequency but maintain continuity (Figure 6.3, right). We achieve this by recording the phase of the LFO at each sample. Recall that frequency is the derivative of

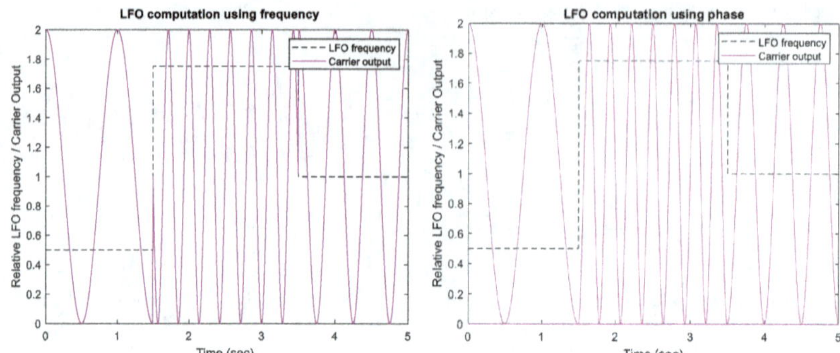

FIGURE 6.3
Tremolo showing LFO output with frequency-based computation (left), and phase-based computation (right).

phase, so by incrementing the phase by the right amount each sample, we cause the LFO to run at the expected frequency while making it robust to frequency changes:

```
phase = 0;
for n = 0 to (number of samples):
{
  m[n] = 1.0 + alpha*cos(phase);
  y[n] = m[n]*x[n];
  phase = phase + 2*pi*f/fs;
}
```

Audio Rate and Control Rate

Because the LFO value changes very little from one audio sample to the next, total computation can be reduced by updating the LFO value less frequently. For example, the LFO value could be recalculated only once every 64 audio samples without noticeable loss of quality. In computer music, signals that update less frequently in this manner are called *control rate* signals (in comparison to *audio rate* signals, which are updated every sample). However, given the simplicity of the tremolo, control rate updating of the LFO is unlikely to be necessary except on the smallest of mobile or embedded processors.

Code Example

The following C++ code fragment, adapted from the code that accompanies this book, implements a basic tremolo effect:

```
int numSamples;          // Indicates how many audio samples to process
float *channelData;      // Array of audio samples, length numSamples
```

```
float ph;                  // Current phase of the LFO (0-1)
float inverseSampleRate; // Defined as 1.0/(sample rate)

float depth_;            // Depth of the tremolo effect (0-1)
float frequency_;        // Frequency of the LFO
int waveform_;           // What type of waveform to use (sine, triangle,
...)

for (int i = 0; i < numSamples; ++i)
{
    const float in = channelData[i];

    // Ring modulation is easy! Just multiply the waveform by a periodic
    // carrier
    channelData[i] = in * (1.0f - depth_*lfo(ph, waveform_));

    // Update the carrier and LFO phases, keeping them in the range 0-1
    ph += frequency_*inverseSampleRate;
    if(ph >= 1.0) ph -= 1.0;
}
```

The code uses the variable `ph` to keep track of the current phase of the LFO. The tremolo effect itself is simply a change in volume modulated by the value of the LFO. This tremolo example includes a choice of LFO styles, as implemented in the `lfo()` function:

```
float lfo(float phase, int waveform)
{
    switch(waveform)
    {
        case kWaveformTriangle:
            if(phase < 0.25f)
                return 0.5f + 2.0f*phase;
            else if(phase < 0.75f)
                return 1.0f - 2.0f*(phase - 0.25f);
            else
                return 2.0f*(phase-0.75f);
        case kWaveformSquare:
            if(phase < 0.5f)
                return 1.0f;
            else
                return 0.0f;
        case kWaveformSquareSlopedEdges:
            if(phase < 0.48f)
                return 1.0f;
            else if(phase < 0.5f)
                return 1.0f - 50.0f*(phase - 0.48f);
            else if(phase < 0.98f)
                return 0.0f;
            else
                return 50.0f*(phase - 0.98f);
        case kWaveformSine:
        default:
            return 0.5f + 0.5f*sinf(2.0 * M_PI * phase);
    }
}
```

Here, `waveform` takes one of several predefined values (`kWaveformTriangle`, `kWaveformSine`, etc.), and the values returned are all scaled to the range 0 to 1.

Ring Modulation

Tremolo is a rhythmic effect, with the LFO inducing periodic amplitude changes in the input signal. The human ear is capable of distinguishing discrete rhythmic events up to a rate of about 10–20 events per second [49]. As the LFO frequency increases above this threshold, the amplitude modulation will cease to be perceived as rhythmic pulses and begin to be heard as a change in timbre.

Ring modulation is an effect that multiplies the input signal by a periodic *carrier* signal, producing unusual, sometimes discordant sounds. Uniquely among common audio effects, the output sounds tend to be *non-harmonic* (not containing multiples of a fundamental frequency) for even the simplest of input sounds. Because of the strangeness of these non-harmonic outputs, the ring modulator is not widely used in music production.

Theory

The basic input-output relation of the ring modulator is identical to the tremolo:

$$y[n] = m[n]x[n], \tag{6.3}$$

where $x[n]$ is the input signal (usually from a musical instrument) and $m[n]$ is the modulating *carrier* signal. All the same properties of the tremolo apply (i.e., $m[n]$ can be thought of as changing the gain of $x[n]$). However, the characteristic sound of the ring modulator results from the particular properties of the input and carrier signals. Consider a simple case, where each of the signals is a single sinusoid,

$$x[n] = \cos(\omega n), m[n] = \cos(\omega_c n). \tag{6.4}$$

The cosine of the sum or difference of two angles may be given by $\cos(A+B)=\cos(A)\cos(B)-\sin(A)\sin(B)$ and $\cos(A-B)=\cos(A)\cos(B) + \sin(A)\sin(B)$. So,

$$m[n]x[n] = \cos(\omega_c n)\cos(\omega n) = \left[\cos((\omega_c - \omega)n) + \cos((\omega_c + \omega)n))\right]/2. \tag{6.5}$$

In other words, multiplying two sinusoids together results in *sum and difference frequencies*. For example, if our input was a 40 0Hz sine wave and the carrier was a 100Hz sine wave, the result would not be the original frequencies at all, but instead would be sinusoids at 300 and 500Hz! This is illustrated in Figure 6.4. Note that using sine instead of cosine in the above equations changes the phase of the results but not the essential sum and difference effect.

Musical instrument signals are composed of many superimposed frequencies. For a single note, these frequencies tend to be harmonically related, or integer multiples of some fundamental frequency. We can thus refine the simple example above to an input $x[n]$ containing k harmonically-related sinusoids, each with a different strength a and a phase offset φ. The carrier $m[n]$ remains the same:

$$x[n] = \sum_{i=1}^{k} a_i \cos(i\omega n + \phi_i), \quad m[n] = \cos(\omega_c n). \tag{6.6}$$

As in Eq. (6.5), trigonometric identities give us the following result:

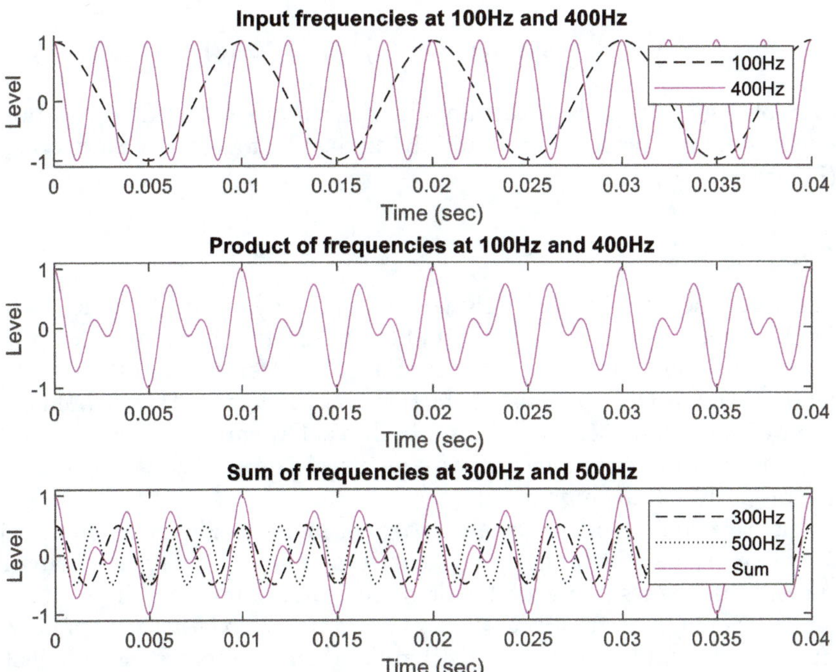

FIGURE 6.4
Multiplying sinusoids at 400 and 100Hz results in sum and difference components at 500 and 300Hz, respectively.

$$m[n]x[n] = \sum_{i=1}^{k} a_i \left[\cos((i\omega + \omega_c)n + \phi_i) + \cos((i\omega - \omega_c)n + \phi_i)\right] / 2. \tag{6.7}$$

In other words, the frequencies present in the modulated output signal will be $\omega - \omega_c$, $\omega + \omega_c$, $2\omega - \omega_c$, $2\omega + \omega_c$, $3\omega - \omega_c$, $3\omega + \omega_c$, etc. Except in the special case where the carrier frequency ω_c is a multiple of the instrument's fundamental frequency ω, these output frequencies will no longer be integer multiples of ω. This shows why the ring modulator produces *non-harmonic* output sounds. The result holds even for more complex real-world musical signals, which typically contain a small amount of inharmonicity.

Modulation in the Frequency Domain

Another way of understanding the ring modulation effect is to consider its behavior in the frequency domain by examining its Z transform. Multiplication in the time domain is equivalent to *convolution* in the frequency domain:

$$y[n] = m[n]x[n] \Rightarrow Y(z) = M(z) * X(z). \tag{6.8}$$

Convolving two signals can be thought of as a process of gradually sliding one signal past the other and multiplying the two together at every offset (Figure 6.5).

Perception

The frequency of the carrier signal $m[n]$ typically ranges from 10 Hz to 1 kHz or more. At the very bottom of the range, the ring modulator produces a rhythmic warbling effect similar to the tremolo, and the sounds tend to become more unusual as the frequency increases. Why should the same effect produce a perception of rhythm at low carrier frequencies and a change in timbre at high carrier frequencies? The answer has to do with the properties of the human ear.

The ear can be modeled as a *filterbank* of many band pass filters, each sensitive to a narrow range of frequencies (similar to a short-time Fourier transform). The resolution of the filterbank is not infinite, and if two tones are close enough in frequency to fall within the *critical bandwidth* of hearing, the ear will be unable to distinguish them as separate signals [50]. In this case, the perceptual result will be a single tone exhibiting *beating* or changing amplitude. This is the effect of the tremolo. At high carrier frequencies, the ring modulator will produce tones that are separated by more than the critical bandwidth, which is typically on the order 15%–20% of each band's center frequency (much more accurate formulae are

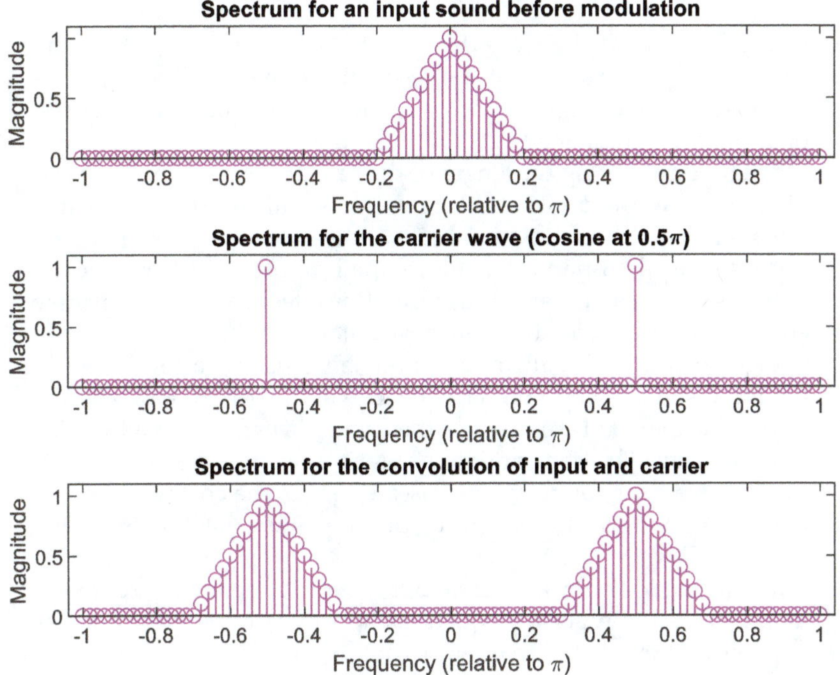

FIGURE 6.5
An example convolution of input and carrier signals.

available in [50]). For moderate carrier frequencies between these extreme cases, where the tones are separated by an amount roughly equal to the critical bandwidth, the perceived result is often described as *roughness*: neither clear beating nor obviously separable tones.

FOURIER IN THE EAR

The cochlea is a coiled cavity within the inner ear that is primarily responsible for much of our auditory system. Within the cochlea is the basilar membrane, a filmy structure that divides the cochlea and vibrates with incoming sound energy. The width of the basilar membrane increases from the base (entrance) to the apex (end), while the width of the cochlear cavity decreases.

The basilar membrane has thousands of basilar fibers embedded within. These fibers affect the local stiffness in the membrane. The membrane starts thick and stiff, but becomes thinner and more flexible toward its apex. When sound waves travel along the basilar membrane, sound energy is dissipated at the place along the membrane that has the same natural resonant frequency. The stiff fibers will resonate with high frequencies, and the more flexible fibers resonate at lower frequencies.

The hair cells along the length of the basilar membrane detect this vibration and convert it into electrical potentials for transmission to the brain. There are also outer hair cells that contract in response to signals from the brain. This allows the brain to adjust or tune the stiffness of the membrane, thus providing a feedback mechanism to enhance the resolution of frequency content.

Together, the basilar membrane and hair cells give a continuum of natural resonance from high frequencies at the base to low frequencies at the apex. This distributes the energy over space as a function of frequency. (The distribution is logarithmic, which also accounts for why we hear sound on a log scale). Thus, the cochlea acts as a filterbank, performing a Fourier decomposition of the incoming sound.

Interestingly, this is a reversal of the normal frequency representation, since the highest frequencies appear at the entrance to the cochlea, and the lowest frequencies at the rear.

As mentioned above, one special case occurs when the carrier and the input fundamental frequency are multiples of one another. In this case, the output frequencies *will* be harmonically related, though with different frequencies emphasized. The result will be a sound with a more clearly identifiable pitch, lacking the strangeness of most ring modulator sounds. However, since the carrier frequency is usually fixed, this effect can only occur for the few notes that match its pitch.

Low-Frequency Oscillator

Like the tremolo, the ring modulator requires a low-frequency oscillator (LFO) to generate the carrier signal $m[n]$. The implementation is similar but not identical to the tremolo:

$$m[n] = 1 - \alpha + \alpha\cos(n\omega_c). \tag{6.9}$$

Here, as in the tremolo, α controls the *depth* of the effect. It takes a range of 0 to 1, with $\alpha=0$ producing no effect, $\alpha=1$ producing a pure ring modulator effect and $\alpha=0.5$ producing an equal balance between the two. Notice that at the maximum depth ($\alpha=1$), the carrier reduces to a simple sinusoid:

$$m[n] = \cos(n\omega_c). \tag{6.10}$$

In comparison to the tremolo, this carrier lacks an offset term. If the offset was present ($m[n]=1+cos(\omega_c n)$), the output $y[n]$ would also contain the original frequency components of $x[n]$. To see why, substitute this alternate definition of $m[n]$ into Eq. (6.3) above. In no case does the carrier frequency itself (ω_c) appear in the output. For this reason, this type of modulation is known as *suppressed carrier* modulation.

Variations

Typical ring modulators take one input for connecting an instrument, and have one control for setting the frequency of the LFO. Sinusoidal LFOs are most commonly used, but more complex waveforms can be used as well. Since complex periodic waveforms can be modeled as the sum of sinusoids, the output quickly becomes quite dense as each term interacts with each other one:

$$x[n] = \sum_{i=1}^{k} a_i \cos(i\omega n + \phi_i), \quad m[n] = \sum_{h=1}^{l} b_h \cos(h\omega_c n + \phi_{c,h})$$

$$m[n]x[n] = \sum_{i=1}^{k} \sum_{h=1}^{l} \frac{a_i b_h}{2} \begin{bmatrix} \cos\left((i\omega + h\omega_c)n + \phi_i + \phi_{c,h}\right) \\ +\cos\left((i\omega - h\omega_c)n + \phi_i - \phi_{c,h}\right) \end{bmatrix}. \tag{6.11}$$

Another variation is to use two input signals in place of a carrier. The two signals are multiplied, producing new frequency components depending on the spectral content of the two inputs. If the same signal is used for both inputs, the output is the square of the input, which produces a type of nonlinear distortion (see Chapter 8) but otherwise lacks the characteristic non-harmonic qualities of the ring modulator.

Implementation

Early ring modulators used "analog multiplier" circuits to modulate the two signals. These were expensive and typically had less-than-perfect accuracy, adding a certain amount of extra non-linearity to the output. Digital implementation of the ring modulator is trivial, consisting of a single multiplication per sample. However, the sound of classic analog units will result in part from the non-linearity of the multiplier circuits, so these effects would need to be simulated if replicating an individual unit's sound was desired.

Another consideration in digital ring modulation is *aliasing*. Depending on the frequency content of the input and the carrier, the sum and difference frequencies at the output may extend beyond the Nyquist frequency $f_s/2$.

If this happens, these frequencies will be aliased, adding yet another layer of inharmonicity to the output. Aliasing can be avoided by strictly band-limiting the input and/or carrier signals such that the maximum frequency term $k\omega+\omega_c$ remains below the Nyquist frequency. If low-pass filtering the input is not desirable, the signals can be *upsampled* in advance and the modulation performed at a higher sampling rate. The output is then low-pass filtered before being *downsampled* back to the original rate.

Code Example

The following C++ code fragment implements a basic ring modulator:

```
int numSamples;          // Indicates how many audio samples to process
float *channelData;      // Array of audio samples, length numSamples
float ph;                // Current phase of the LFO (0-1)
float inverseSampleRate; // Defined as 1.0/(sample rate)

float carrierFrequency_;        // Frequency of the oscillator

for (int i = 0; i < numSamples; ++i)
{
    const float in = channelData[i];

    // Ring modulation is easy! Just multiply waveform by a periodic
    // carrier
    channelData[i] = in * sinf(2.0 * M_PI * ph);

    // Update the carrier phase, keeping it in the range 0-1
    ph += carrierFrequency_*inverseSampleRate;
    if(ph >= 1.0) ph -= 1.0;
}
```

As the code indicates, ring modulation is a simple effect to implement digitally. Each incoming sample is multiplied by the output of an audio-frequency oscillator whose frequency is given by `carrierFrequency_`. The variable `ph` keeps track of the phase of the oscillator at each sample. The main differences between the ring modulation code and the earlier tremolo code example are that the carrier is in the audio frequency range rather than the LFO range, and that waveform takes values between –1 and 1 rather than between 0 and 1.

Applications

The ring modulator's sound is found relatively rarely in music compared to most other well-known audio effects. When it is used, it is often made less strange by mixing in the original instrument sound (low values for the *depth* control). When a small amount of ring-modulated sound is mixed into an otherwise clean instrument, it can add an interesting roughness to the track.

The iconic use of ring modulation was not in music but in television and movies, especially early science fiction. Because the information in speech is contained in the overall shape of the spectrum and not in the fundamental frequency of the voice [51], ring modulation maintains intelligibility while scrambling the identity of the original speaker.

THAT SCI-FI SOUND

Ring modulation was popular in early electronic music, and one of the first examples of its use was in the Melochord, an instrument built by the electronic music pioneer, Harald Bode, in 1947. But ring modulation really came of age as a special effect in science fiction film and television.

In 1956, Louis and Bebe Barron were commissioned to do 20 minutes of sound effects for the Sci-Fi movie *Forbidden Planet*. After creating some initial samples, the producers were sufficiently impressed to ask the husband and wife team to compose the entire score. This was one of the first electronic scores for a major film, and made heavy use of ring modulators that were built by Louis and Bebe. They treated each ring modulator as a different 'actor,' with a unique voice and behavior. As they explained on the sleeve notes for the soundtrack album, "we created individual cybernetics circuits for particular themes and leit motifs, rather than using standard sound generators. Actually, each circuit has a characteristic activity pattern as well as a 'voice'."

Ring modulation was also used heavily by the BBC Radiophonic Workshop, a sound effect unit of the BBC, well known for innovations in sound synthesis and effects. For the voice of the Daleks, first used in 1963, Brian Hodgson applied a 30 Hz modulation, among other effects, to give a harsh buzzing sound to the voices. Ring modulation has been applied to almost every Dalek voice since, and a Moog ring modulator has also been used to generate as a special effect on the voices of another *Doctor Who* villain, the Cybermen.

The ring modulator has become synonymous with sci-fi sound effects. It's easily recognizable in classic productions, such as *The Outer Limits* and *The Hitchhiker's Guide to the Galaxy*. Today, it's an essential sound effect in television, film, and game audio production.

Further Reading

A good review of modulators (and demodulators) from a more technical, research perspective is provided in [52]. As with many other effects, deep learning has been used to model those based on amplitude modulation, especially the ring modulator [48]. Recent work has also been concerned with uncovering the modulation that has been applied to a signal [53]. Two related fields are also of note. Modulation for broadcast (FM and AM radio) is, of course, a huge field. But paralleling audio effects, which process sounds, are virtual instruments that generate the sounds. There, modulation for the synthesis of new audio signals has achieved significant success. Reference [54] is the seminal work on FM synthesis, and [55] provides a review of modulation synthesis techniques.

Problems

1.
   i. When two frequency components of equal amplitude are added together, as could be the case in ring modulation, when are beating, separate tones or roughness perceived in the output?
   ii. Assuming critical bandwidth is 20% of the center frequency, give approximate equations for the conditions for beating, separate tones or roughness when two sinusoids with frequencies f_1 and f_2 are added together?
   iii. Which of the three conditions in part (ii) is most likely to be heard for each of the combinations: 300 and 500 Hz tones, 500 and 495 Hz tones, or 500 and 580 Hz tones?
2. Derive the following result, which was mentioned in the discussion on ring modulation;

$$\text{If } x[n]=\sum_{i=1}^{k} a_i \cos(i\omega n+\phi_i), \quad m[n]=\cos(\omega_c n), \text{ then}$$

$$m[n]x[n]=\sum_{i=1}^{k} a_i\left[\cos((i\omega+\omega_c)n+\phi_i)+\cos((i\omega-\omega_c)n+\phi_i)\right]/2.$$

3. An input signal containing frequencies at 500 and 900 Hz is modulated by a carrier at 200 Hz. What are the frequencies that will appear in the output?
4. A signal containing only two frequency components is modulated by a third carrier frequency. At the end of this process, the output frequencies are 200, 400, 1,100, and 1,300 Hz. What were the original frequencies of the signal and the frequency of the carrier?

7

Dynamics Processing

Dynamic audio effects apply a time-varying gain to the input signal. The applied gain is typically a nonlinear function of the level of the input signal (or a secondary signal). Dynamic effects are most often used in order to modify the amplitude envelope of a signal. They often compress or expand the dynamic range of a signal.

In this chapter, we focus on two of the most common forms of dynamics processing: dynamic range compression and expansion, as well as their more extreme forms, limiting and noise gates.

Dynamic Range Compression

Dynamic range compression (or just compression) is concerned with mapping the perceived dynamic range of an audio signal to a smaller perceived range. Dynamic range compressors achieve this goal by reducing high signal levels while leaving the quieter parts untreated. Dynamic range compression should not be confused with data compression as used in audio codecs, which is a completely different concept.

Our goal here is to describe how a compressor is designed and to discuss how different design choices affect the perceived sonic characteristics of the compressor, considering both technical and perceptual aspects. We first provide an overview of the basic theory, paying special attention to the adjustable parameters used to operate a dynamic range compressor. Then the principles of its operation are described, including detailed discussion of digital implementation based on classic analogue designs. Next, different methods and implementations are given for a complete design. Applications of compressors are then discussed, along with the artifacts that may result from their use.

Theory

Compressor Controls

A compressor has a set of controls directly linked to compressor parameters through which one can set up the effect. The most commonly used compressor parameters may be defined as follows.

DOI: 10.1201/9781003593942-7

A compressor is essentially a variable gain control, where the amount of gain used depends on the level of the input. Attenuation is applied (gain less than one) when the signal level is high, which in turn makes louder passages softer, reducing the dynamic range. The basic scheme, implemented as either a feedforward or feedback device, is shown in Figure 7.1.

Threshold defines the level above which the compressor is active. Whenever the signal level overshoots this threshold, the level will be reduced.

Ratio controls the input/output ratio for signals overshooting the threshold level. It determines the amount of compression applied. A ratio of 3:1 ('three to one') implies that the input level needs to increase by 3 dB in order for the compressor to apply a 1 dB reduction.

A compressor's input/output relationship is often described by a simple graph, as shown in Figure 7.2. The horizontal axis corresponds to the input signal level, and the vertical axis is the output level, where level is

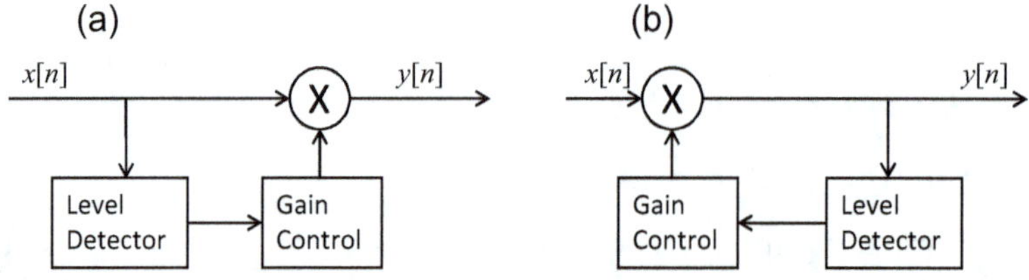

FIGURE 7.1
Flow diagram of a feedforward (a) and feedback (b) compressor.

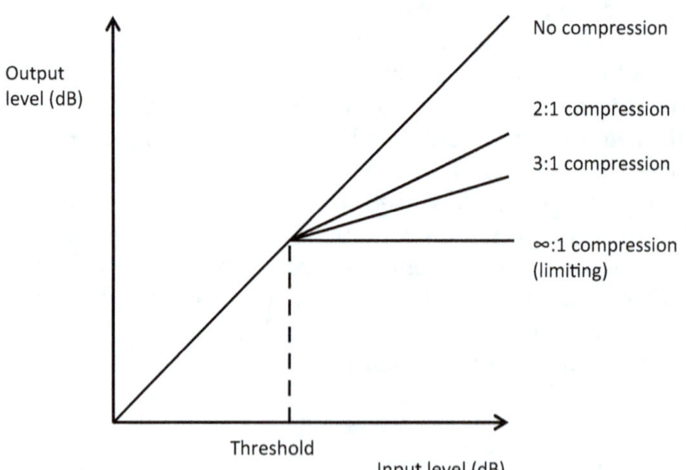

FIGURE 7.2
Compressor input/output characteristic. The compressor weakens the signal only when it is above the threshold. Above that threshold, a change in the input level produces a smaller change in the output level.

measured in decibels. A line at 45° through the origin corresponds to a gain of one, so that any input level is mapped to exactly the same output level. The ratio describes how the compressor changes the slope of that line above the threshold value. The distance between the highest and lowest output levels defines the dynamic range of the output.

Limiting is simply an extreme form of compression where the ratio is very high (often described as 20:1 or higher), and thus, the input/output relationship becomes very flat. This places a hard limit on the signal level.

Attack and release times, also known as time constants, control the speed at which a compressor reacts to a change in signal level. Instantaneous compressor response, as described in [56], is not usually sought because it introduces distortion on the signal.

The attack time defines the time it takes the compressor to decrease the gain to the level determined by the ratio once the signal overshoots the threshold. The release time defines the time it takes to bring the gain back up to the normal level once the signal has fallen below the threshold.

Figure 7.3 shows how the attack and release times affect an example input. There is overshoot when the signal level increases since it takes time for the gain to decrease, and attenuation when the input signal returns to the initial level since it takes time for the gain to increase. As shown, the release time is generally longer than the attack time.

A *Make-Up Gain* control is usually provided at the compressor output. The compressor reduces the level (gain) of the signal, so that applying a make-up gain to the signal allows for matching the input and output loudness level.

The *Knee Width* option controls whether the bend in the response curve has a sharp angle or has a rounded edge. The Knee is the threshold-determined point where the input-output ratio changes from unity to a set ratio. A sharp transition is called a Hard Knee and provides a more noticeable compression. A softer transition where, over a transition region on both sides of the threshold, the ratio gradually grows from 1:1 to a user-defined value is called a Soft Knee. It makes the compression effect less perceptible. Depending on the signal, one can use a hard or soft knee, with the latter being preferred when we want less obvious (transparent) compression.

A compressor has a set of additional controls, most of which are found in most modern compressor designs. These include a *Hold* parameter, *Side-Chain filtering*, *Look-Ahead*, and many more. However, in this chapter, we will focus primarily on the parameters mentioned above.

Signal Paths

The signal entering the compressor is split into two copies. One is sent to a variable-gain amplifier (the gain stage) and the other to an additional path, known as a side-chain, where the gain computer, a circuit controlled by

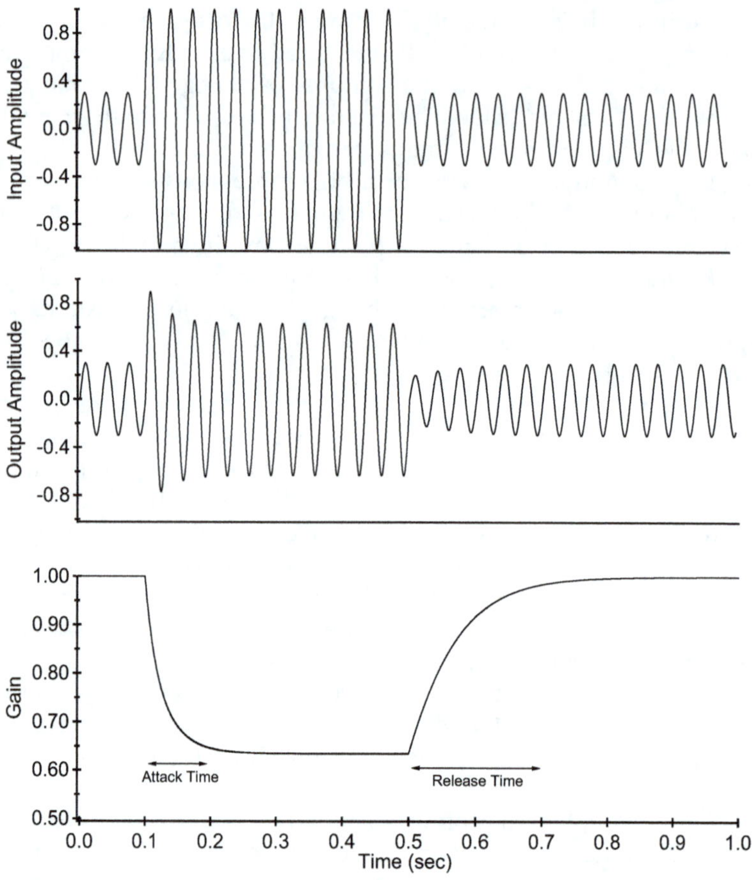

FIGURE 7.3
Effect of a compressor on a signal. Only the middle portion of input is above the compressor's threshold.

the level of the input signal, applies the required gain reduction to the gain stage. The copy of the signal entering the side-chain has its bipolar amplitude converted into a unipolar representation of level. When the level of the signal is determined by its absolute value (instantaneous signal level), this is known as peak-sensing. An alternative is to use RMS-sensing and determine the signal level by its root-mean-square (RMS) value.

The Gain Stage

The gain stage is responsible for attenuating the input signal by a varied amount of decibels (dB) over time. The heart of every compressor is the element that applies this gain reduction: a voltage controlled amplifier

(VCA), which attenuates the input signal according to an external control signal (*c*) coming from the side chain. Building a VCA with analogue components is non-trivial, and thus many approaches have been applied [57]. Optical compressors like the Teletronix LA-2A use a light-dependent resistor (LDR) as the bottom leg of a voltage divider located within the signal path. The control voltage is used to drive a light source, which, with increasing brightness, will lower the resistance of the LDR and therefore, apply the necessary gain reduction. A similar approach is taken by the Universal Audio 1176 compressor, but instead of using an LDR, they used the drain-to-source resistance of a field effect transistor (FET compressor). The control voltage could then be applied to the gate terminal in order to lower the FET's resistance [58]. An even earlier approach used in so-called variable-mu compressors like the Fairchild 670 exploited the fact that altering the grid-to-cathode voltage would change the gain of a tube amplifier. More modern compressor designs make use of specialized integrated VCA circuits. These are much more predictable than the earlier approaches and offer improved specifications (such as less harmonic distortion and a higher usable dynamic range).

In a solely digital design, one can model an ideal VCA as multiplying the input signal by a control signal coming from the *sidechain* (sidechain refers to any signal path in an audio effect that doesn't produce the output signal, see Chapter 13). If $x[n]$ denotes the input signal, $y[n]$ the output signal and $c[n]$ the control signal then $y[n] = c[n]\cdot x[n]$. In addition, make-up gain is often used to add a constant gain back to the signal in order to match output and input levels. In a digital compressor, we can easily implement a make-up gain by multiplying the compressor's output by a constant factor corresponding to the desired make-up gain value, hence representing the signals either in the linear or in decibel domains, where M is the make-up gain,

$$\begin{aligned} y[n] &= x[n]\cdot c[n]\cdot M \\ y_{\mathrm{dB}}[n] &= x_{\mathrm{dB}}[n]+c_{\mathrm{dB}}[n]+M_{\mathrm{dB}} \end{aligned} \quad (7.1)$$

The Gain Computer

The gain computer is the compressor stage that generates the control signal. The control signal determines the gain reduction to be applied to the signal. This stage involves the Threshold T, Ratio R, and Knee Width W parameters. These define the static input-to-output characteristic of compression. Once the signal level exceeds the threshold value, it is attenuated according to the ratio.

The compression ratio is defined as the reciprocal of the slope of the line segment above the threshold,

$$R = \frac{x_G - T}{y_G - T} \quad \text{for} \quad x_G > T, \tag{7.2}$$

where the input and output to the gain computer, x_G and y_G, and the threshold T, are all given in decibels. The static compression characteristic is described by the following relationship:

$$y_G = \begin{cases} x_G & x_G \le T \\ T + (x_G - T)/R & x_G > T \end{cases}, \tag{7.3}$$

In order to smooth the transition between compression and no compression at the threshold point, we can soften the compressor's knee. The width W of the knee (in decibels) is equally distributed on both sides of the threshold. Figure 7.4 presents a compression gain curve with a soft knee.

To implement this, we replace Eq. (7.3) with the following piecewise, continuous function,

$$y_G = \begin{cases} x_G & 2(x_G - T) < -W \\ x_G + (1/R - 1)(x_G - T + W/2)^2/(2W) & 2|(x_G - T)| \le W \\ T + (x_G - T)/R & 2(x_G - T) > W \end{cases}. \tag{7.4}$$

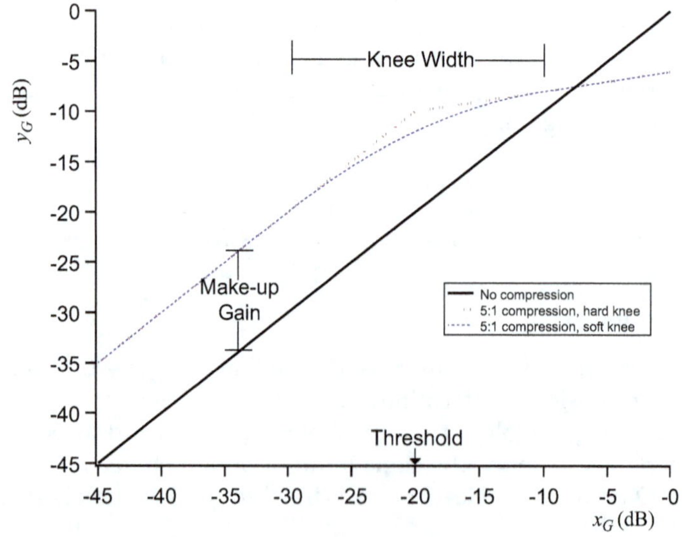

FIGURE 7.4
Static compression characteristic with make-up gain and hard or soft knee.

When the knee width is set to zero, the soft knee becomes a hard knee.

Level Detection

The level detection stage is used to provide a smooth representation of the signal's level, and may be applied at various places in the side-chain. The gradual change of gain is due to the attack and release times. The process of properly setting up these times is crucial to the performance of the compressor since unpleasant artifacts are often associated with the choice of these compressor parameters.

The attack and the release times are usually introduced through a smoothing detector filter. We can simulate the time domain behavior of the filter in the digital domain with a digital one-pole filter,

$$y[n] = \alpha y[n-1] + (1-\alpha)x[n], \tag{7.5}$$

where α is the filter coefficient, $x[n]$ the input, and $y[n]$ the output. The step response of this filter is:

$$y[n] = 1 - \alpha^n \quad \text{for} \quad x[n] = 1, n \geq 0. \tag{7.6}$$

The time constant τ is defined as the time it takes for this system to reach $1-1/e$ of its final value, i.e., $y[\tau f_s] = 1-1/e$. Thus, from (7.6), we have

$$\alpha = e^{-1/(\tau f_s)}. \tag{7.7}$$

Alternate definitions for the time constant are often used [59]. For instance, if one considers the rise time for the step response to go from 10% to 90% of the final value,

$$0.1 = 1-\alpha^{\tau_1 f_s}, 0.9 = 1-\alpha^{\tau_2 f_s} \rightarrow \tau_2 - \tau_1 = \tau \ln 9. \tag{7.8}$$

Another alternative is to relate the attack and release time to the amount of decibel gain change, e.g., if the attack and release times are specified relative to 10 dB, then an 8 ms attack time implies that the compressor will take 8 ms to reduce the gain from 0 to –10 dB.

In analogue detectors, the filter is typically implemented as a simple series resistor capacitor circuit, with τ=RC. Equations (7.5) and (7.7) may then be found by digital simulation using a step invariant transform [1]. The advantage of this approach over other digital simulation methods, like the bilinear transform, is that we preserve the analogue filter topology with the capacitor's voltage as the state variable. Therefore, we will not experience any clicks and pops once we start varying the filter coefficients over time.

RMS Detector

The RMS detector is useful when we are interested in a smoothed average of a signal. Level detection may be based on a measurement of the RMS value of the input signal [60], which is defined as

$$y_L^2[n] = \frac{1}{M}\sum_{m=-M/2}^{M/2-1} x_L^2[n-m]. \tag{7.9}$$

However, this is generally unsuitable for real-time implementations since it enforces a latency of M/2 samples. In the implementation of real-world effects, this measurement is often approximated by filtering the squared input signal with a first-order low-pass IIR filter and taking the square root of the output [61]. This is also commonly found in analogue RMS-based compressors [1]. The difference equation of the RMS detector with smoothing coefficient becomes:

$$y^2[n] = \alpha y^2[n-1] + (1-\alpha)x^2[n]. \tag{7.10}$$

We can find RMS detectors in some compressors at the beginning of the sidechain. In [62], it was shown that Eq. (7.10) produces behavior generally equivalent to Eq. (7.5), save for a scaling of the time constant. Therefore, we will focus on the various options for the peak detector.

Peak Detector

The analogue peak detector circuit, as commonly found in analogue dynamic range controls, is given in Figure 7.5. If we are not required to simulate a particular type of diode, we can idealize it by assuming that it can supply infinite current once the voltage across the diode becomes positive and completely blocks when reverse-biased. This significantly simplifies the calculation [62].

$$\frac{dV_C}{dt} = \frac{\max(V_{in} - V_C, 0)}{R_A C} - \frac{V_C}{R_R C}. \tag{7.11}$$

The capacitor is charged through resistor R_A according to a positive voltage across the diode but continually discharged through R_R. Taking α_A as the attack coefficient and α_R as the release coefficient, calculated according

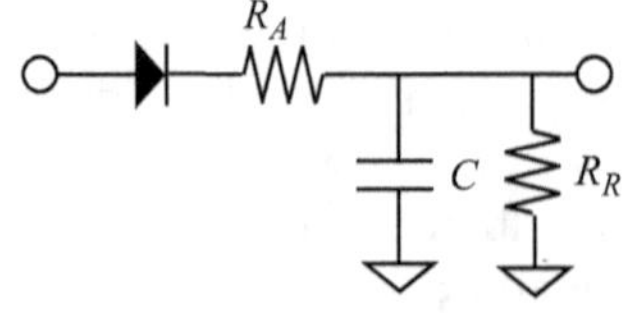

FIGURE 7.5
Peak detector circuit.

to Eq. (7.7) from the attack and release times $\tau_A = R_A C$ and $\tau_R = R_R C$, we can simulate the ideal analogue peak detector with:

$$y_L[n] = \alpha_R y_L[n-1] + (1-\alpha_A)\max(x_L[n] - y_L[n-1], 0). \tag{7.12}$$

Although used in many analogue compressors, and some digital designs [63], this circuit has a few problems. When $x_L[n] \geq y_L[n-1]$, the step response is

$$y[n] = (1-\alpha_A)\sum_{m=0}^{n-1}(\alpha_R + \alpha_A - 1)^m \rightarrow \frac{1-\alpha_A}{2-\alpha_R-\alpha_A} \approx \frac{\tau_R}{\tau_R + \tau_A}, \tag{7.13}$$

where we used the series expansion of the exponential function. Eq. (7.13) implies that we will get a correct peak estimate only when the release time constant is considerably longer than the attack time constant. Another side effect is that the attack time also gets slightly scaled by the release time: there will be a faster attack time than expected when we use a fast release time. Both problems are illustrated in Figure 7.6.

In order to accomplish a program-dependent release behavior (autorelease), some analogue compressors use a combination of two release time constants in their peak detectors. One such design is found in the famous SSL Stereo Bus Compressor. It uses two release networks stacked on top of each other. The much earlier Fairchild compressors and similar tube compressors used similar designs. When used in a compressor, the peak detector with dual time constant automatically increases the release time once the compression continues for a longer

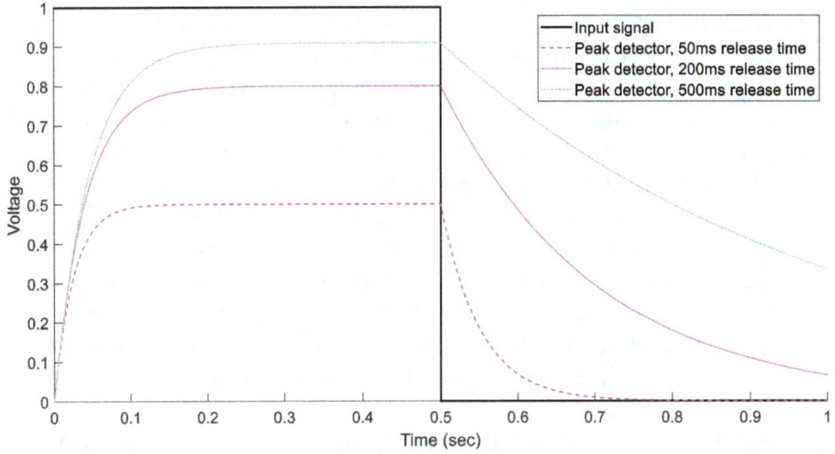

FIGURE 7.6
Output of the analogue peak detector circuit for different release time constants. Attack Time=50 ms.

time period. This gives the desirable property of shorter release times after compressing transients and a longer release time for steady state compression. However, this type of peak detector suffers from the same level problems.

Level Corrected Peak Detectors

A more modern, digital design decouples the attack and release and provides a branch to the difference equation.

$$y_L[n] = \begin{cases} \alpha_A y_L[n-1] + (1-\alpha_A)x_L[n] & x_L[n] > y_L[n-1] \\ \alpha_R y_L[n-1] + (1-\alpha_R)x_L[n] & x_L[n] \le y_L[n-1] \end{cases}. \quad (7.14)$$

This peak detector, also used in [1,63], now simply switches coefficients between the attack and the release phase. As shown in Figure 7.7, it does not suffer from the level differences caused by different time constants exhibited by the standard peak detector circuit.

There are many other choices for the peak detector, both described in the literature and implemented in commercial designs. A detailed description of these, along with an assessment of their performance, is provided in [57].

OVERCOMPRESSION AND THE LOUDNESS WAR

Dynamic range compression is used at almost every stage of the audio production chain. It is applied to minimize artifacts in recording (like variation in loudness as a vocalist moves towards or away from a microphone), to reduce masking and to bring different tracks into a comparable loudness range. Compression is also applied in mastering to make the recording sound 'loud.' since a loud recording will be more noticeable than a quiet one, and the listener will hear more of the full frequency range. This has resulted in a trend to more and more compression being applied, a 'loudness war.'

Broadcasting also has its loudness wars. Dynamic range compression is applied in broadcasting to prevent drastic level changes from song to song and to ensure compliance with standards regarding maximum broadcast levels. But competition for listeners between radio stations has resulted in a trend to very large amounts of compression being applied.

So a lot of recordings have been compressed to the point where dynamics are compromised, transients are squashed, clipping occurs, and there can be significant distortion throughout. The end result is that many people think that, compared to what they could have been, a lot of modern recordings sound terrible. And broadcast compression only adds to the problem.

Who is to blame? There is a belief among many that 'loud sells records.' This may not be true, but believing it encourages people to participate in the loudness war. And each individual may think that what they are doing is appropriate. Collectively, the musician who wants a loud recording, the record producer who wants a wall of sound, the engineers dealing with artifacts, the mastering engineers who prepare content for broadcast, and the broadcasters themselves are all acting as soldiers in the loudness war.

THE TIDE IS TURNING

The loudness war may have reached its peak shortly after the start of the new millennium. Audiologists became concerned that the prolonged loudness of new albums might cause hearing damage. Musicians began highlighting the sound quality issue, and in 2006, Bob Dylan said, "... these modern records, they're atrocious, they have sound all over them. There's no definition of nothing, no vocal, no nothing, just like static. *Even these songs probably sounded ten times better in the studio.*" Also in 2006, a vice-president at a Sony Music subsidiary wrote an open letter decrying the loudness war, claiming that mastering engineers are being forced to make releases louder in order to get the attention of industry heads.

In 2008, Metallica released an album with tremendous compression, and hence clipping and lots of distortion. But a version without overuse of compression was included in downloadable content for a game, *Guitar Hero III,* and listeners all over noticed and complained about the difference. Again in 2008, Guns N' Roses producers (including the band's frontman Axl Rose) chose a version with minimal compression when offered three alternative mastered versions. Recently, an annual Dynamic Range Day has been organized to raise awareness of the issue, and the nonprofit organization Turn Me Up! was created to promote recordings with more dynamic range.

The European Broadcasting Union addressed the broadcast loudness wars with EBU Recommendation R 128 and related documents that specify how loudness and loudness range can be measured in broadcast content, as well as recommending appropriate ranges for both.

Together, all these developments may go a long way to establishing a truce in the loudness war.

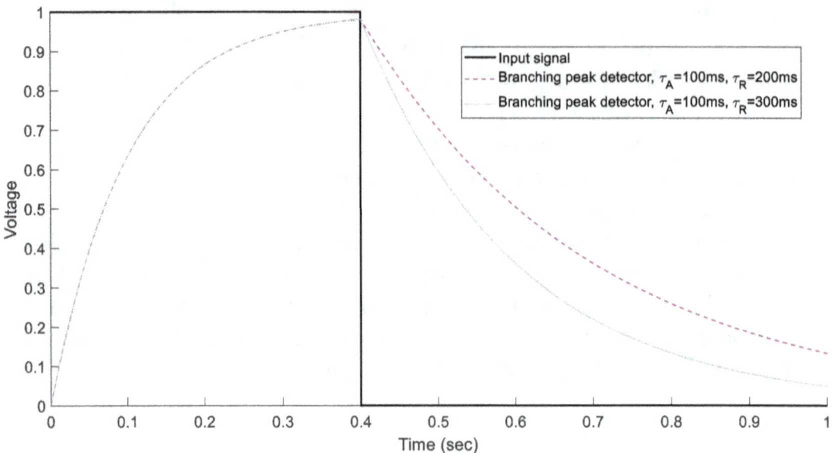

FIGURE 7.7
Output of the branching peak detector for different release time constants.

Implementation

Considering the classic audio effects (equalization, delay, panning, etc.), the dynamic range compressor is perhaps the most complex one. There is no single correct form or implementation, and there are a bewildering variety of design choices. This explains why every compressor in common usage behaves and sounds slightly different and why certain compressor models have become audio engineers' favorites for certain types of signal. Analysis of compressors is difficult because they represent nonlinear time-dependent systems with memory. The gain reduction is applied smoothly and not instantaneously as would be the case with a simple static nonlinearity. Furthermore, the large number of design choices makes it nearly impossible to draw a generic compressor block diagram that would be valid for the majority of real world compressors. Some differ in topology, others introduce additional stages, and some simply differ from the precise digital design since these deviations add character to the compressor. However, we can describe the main parameters of a compressor unit and specify a set of standard stages and building blocks that are present in almost any compressor design.

Feedback and Feedforward Design

There are two possible topologies for the sidechain: a feedback or feedforward topology. In the feedback topology, the input to the side-chain is taken after the gain has been applied. This was used in early compressors and had the benefit that the side-chain could rectify possible inaccuracies of the gain stage.

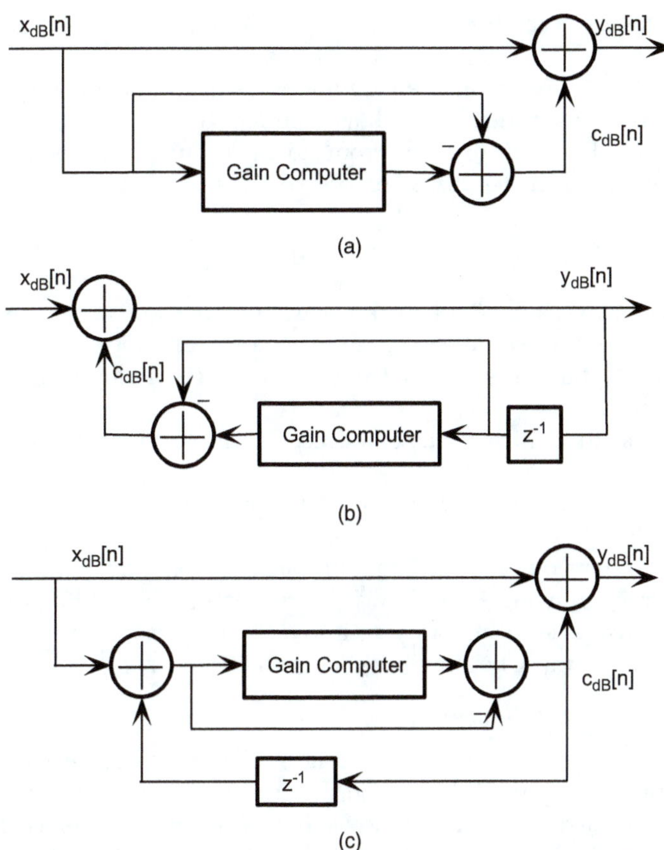

FIGURE 7.8
Static compressor block diagrams with linear/decibel conversion not depicted. Feedforward design on top, feedback design in the middle and alternate feedback design on bottom.

The feedforward topology has the side-chain input before the gain is applied. This means that the side-chain circuit responsible for calculating the gain reduction, the gain computer, will be fed with the input signal. Therefore, it will have to be accurate over the whole of the signal's dynamic range, as opposed to a feedback type compressor where it will have to be accurate over a reduced dynamic range since the side-chain is fed with the compressors' output. This bears no implications for digital design, but it should be taken into consideration if designing an analogue feedforward compressor. Most modern compressors are based on the feedforward design.

By combining Eqs. (7.1) and (7.3), we can calculate the control signal either from the input or from the output of the compressor, leading to the two topologies of feedforward and feedback compression, Figure 7.8, (a) and (b).

The feedback design has a few limitations, such as the inability to allow a look-ahead function or to work as a perfect limiter due to the infinite negative amplification needed. From Eqs. (7.1) and (7.3), when, $x_{L>}T$, the control signal for the feedback compressor is calculated as:

$$c_{\mathrm{dB}}[n]=(1-R)\left(y_G[n-1]-T\right), \tag{7.15}$$

where we have assumed a hard knee and no attack or release. A limiter (with a ratio of ∞:1) would need infinite negative amplification to calculate the control signal. This is why feedback compressors are not capable of perfect limiting.

In contrast, the control signal of the feedforward compressor, for $x_L>T$, is:

$$c_{\mathrm{dB}}[n]=(1/R-1)\left(x_G[n]-T\right). \tag{7.16}$$

Here, since the slope for a limiter approaches –1, implementing limiting is not a problem using a feedforward compressor. The feedforward compressor is also able to smoothly go into over-compression (with $r<0$); the slope variable simply becomes smaller than minus one.

An Alternate Digital Feedback Compressor

As an alternative to the feedback design depicted in Figure 7.8b, we can use the previous gain multiplier on the current input sample in order to estimate the current output sample, as shown in Figure 7.8c. Since the compressor's gain is likely to change only by small amounts from sample to sample, the estimation is a fairly good one. Since the sidechain is now fed by the input signal, we could instead feed in an arbitrary signal in the side-chain, in order to achieve side-chaining operation (Chapter 13). A similar technique has been used in analogue compressors such as the SSL Bus Compressor (by using a slave VCA that mirrors the main VCA's action).

Detector Placement

There are various choices for the placement of the detector circuit inside the compressor's circuitry, as depicted in Figure 7.9 for feedforward compressors. In [1,61,62], the suggested position for the detector is within the linear domain and before the log converter and gain computer. That is,

$$\begin{aligned}x_L[n]&=\left|x[n]\right|\\x_G[n]&=20\log_{10}y_L[n]\cdot\\c_{\mathrm{dB}}[n]&=y_G[n]-x_G[n]\end{aligned} \tag{7.17}$$

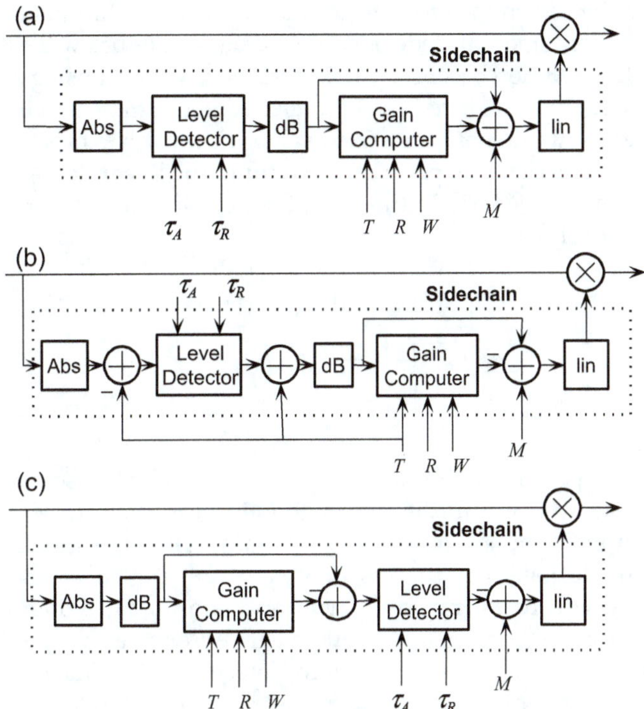

FIGURE 7.9
Block diagrams of the compressor configuration. (a) The return-to-zero detector; (b) the return-to-threshold detector; (c) the log domain detector.

However, the detector circuit then works on the full dynamic range of the input signal, while the gain computer only starts to operate once the signal exceeds the threshold (the control signal will be zero for a signal below the threshold). The result is that we again experience a discontinuity in the release envelope when the input signal falls below the threshold. It also generates a lag in the attack trajectory since the detector needs some time to charge up to the threshold level even if the input signal attacks instantaneously.

A way to overcome this and achieve a smooth release trajectory and avoid the envelope discontinuity is by leaving the detector in the linear domain but biasing it at the threshold level. That is, subtract the threshold from the signal before it enters the detector and add the threshold back in again after the signal has left it.

$$
\begin{aligned}
x_L[n] &= |x[n]| - 10^{T/20} \\
x_G[n] &= 20\log_{10}(y_L[n] + 10^{T/20}). \\
c_{dB}[n] &= y_G[n] - x_G[n]
\end{aligned}
\tag{7.18}
$$

This turns the return-to-zero detector into a return-to-threshold type, and the envelope will smoothly fade out once the signal falls below the threshold. However, this solution becomes problematic once we start using a soft knee characteristic. Since the threshold becomes a smooth transition instead of a hard boundary, we do not have a fixed value with which to bias the detector.

A similar approach is to place the detector in the linear domain but after the gain computer [64,65]. In this case, the detector circuit works directly on the control signal.

$$\begin{aligned} x_G[n] &= 20\log_{10}|x[n]| \\ x_L[n] &= 10^{y_G[n]/20} \\ c[n] &= y_L[n] \end{aligned} \quad . \tag{7.19}$$

For each of Eqs. (7.17)–(7.19), since the release envelope discharges exponentially towards zero in the linear domain, this is equivalent to a linear discharge in the log (or decibel) domain. Although the release rate in terms of decibels per time is now constant, it also means that the release time will be longer for heavier and shorter for lighter compression. Unfortunately, this is not what the ear perceives as a smooth release trajectory, and thus the return-to-zero detector in the linear domain is mostly used by compressors where artefacts are generated on purpose [66]. Therefore, the preferred position for the detector is within the log domain and after the gain computer [57].

$$\begin{aligned} x_G[n] &= 20\log_{10}|x[n]| \\ x_L[n] &= x_G[n] - y_G[n] \\ c_{dB}[n] &= -y_L[n] \end{aligned} \quad . \tag{7.20}$$

Now, the detector directly smoothes the control signal instead of the input signal. Since the control signal automatically returns to zero when the compressor does not attenuate, we do not depend on a fixed threshold, and a smooth release envelope is guaranteed. The trajectory now behaves exponentially in the decibel domain, which means that the release time is independent of the actual amount of compression. This behavior seems smoother to the ear since the human sense of hearing is roughly logarithmic [50]. It is therefore used in most compressors that want to achieve smooth and subtle (artefact-free) compression characteristics for use with complicated signals (such as program material).

Code Example

The following C++ code fragment, adapted from the materials that accompany the book, implements a dynamic range compressor:

```
AudioSampleBuffer inputBuffer;  // Working buffer to analyse the input
                                // signal
int bufferSize;                 // Size of the input buffer
float samplerate;               // Sampling rate
float yL_prev;                  // Previous sample used for gain
                                // smoothing
float *x_g, *x_l, *y_g, *y_l;   // Buffers used in control voltage
                                // calculation
float *c;                       // Output control voltage used for
                                // compression

float threshold_;               // Compressor threshold in dB
float ratio_;                   // Compression ratio
float tauAttack_, tauRelease_;  // Attack and release time constants
float makeUpGain_;              // Make-up gain of the compressor

inputBuffer.clear(0,0,bufferSize);

// Mix down left-right to analyse the input
inputBuffer.addFrom(0,0,buffer,0,0,bufferSize,0.5);
inputBuffer.addFrom(0,0,buffer,1,0,bufferSize,0.5);

// Compression : calculates the control voltage
float alphaAttack = exp(-1/(0.001 * samplerate * tauAttack_));
float alphaRelease = exp(-1/(0.001 * samplerate * tauRelease_));

for (int i = 0 ; i < bufferSize ; ++i)
{

    // Level detection- estimate level using peak detector
    if (fabs(inputBuffer.getSampleData(0)[i]) < 0.000001) x_g[i] = -120;
    else x_g[i] = 20*log10(fabs(inputBuffer.getSampleData(0)[i]));
    // Gain computer- static apply input/output curve
    if (x_g[i] >= threshold_) y_g[i] = threshold_ + (x_g[i] -
    threshold_) / ratio_;
    else y_g[i] = x_g[i];
    x_l[i] = x_g[i] - y_g[i];
    // Ballistics- smoothing of the gain
    if (x_l[i] > yL_prev)
      y_l[i] = alphaAttack * yL_prev + (1 - alphaAttack ) * x_l[i] ;
    else
      y_l[i] = alphaRelease* yL_prev + (1 - alphaRelease) * x_l[i] ;
    // find control
    c[i] = pow(10,(makeUpGain_ - y_l[i])/20);
    yL_prev=y_l[i];
}

// apply control voltage to the audio signal
for (int i = 0 ; i < bufferSize ; ++i)
{
    buffer.getSampleData(0)[i] *= c[i];
    buffer.getSampleData(1)[i] *= c[i];
}
```

This code first implements a level detector, converting the level of the input signal to decibels. It then applies the compression curve to find the required gain for the given signal level. Finally, it applies the calculated gain to the sample data in the buffer. This code assumes a stereo input, and the level detector uses the mix of the two channels.

Application

As the name implies, dynamic range compression reduces the dynamic range of a signal. It is used extensively in audio recording, noise reduction, production work and live performance applications. But it does need to be used with care, since its overuse or use with nonideal parameter settings can introduce distortion and unforeseen side effects, reducing sound quality.

Compressors have a wide variety of applications. When recording onto magnetic tape, distortion can result from high signal levels. So compressors and limiters, with no or minimal make-up gain, are used to prevent sudden transient sounds causing the distortion, while still ensuring that the signal level is kept above the background noise. Compression is also used when editing and mixing tracks to correct for issues that arose in the recording process. For example, a singer may move towards and away from a microphone, so a small amount of compression could be applied to reduce the resultant volume changes. And once tracks have been recorded, a compressor provides a means to adjust the dynamic range of the track. In some cases, compression may be used as an alternative to equalization, as we shall see in Chapter 13.

A popular use of dynamic range compression is to increase the sustain of a musical instrument. By compressing a signal, the output level is made more constant by suppressing the louder portions while amplifying the overall signal. For example, after a note is played, the envelope will decay towards silence. A compressor with appropriate settings will slow this decay, and hence preserve the instrument's sound. Alternatively, a transient modifier could be used [67], attenuating or boosting just the transient portions (as opposed to the sustain), which include the attack stage of a musical note.

Artifacts

There are a large number of artifacts associated with compressors. These are primarily to do with parameter settings that result in unwanted modification of the signal. To name just a few:

- **Dropouts**: these are 'holes,' or periods of unwanted near silence, in the output signal that result from a strong attenuation immediately after a short, intense sound;

- **Pumping**: perceived variation of signal level, as a result of reducing gain as the signal crosses the threshold and turning up as the signal dips below;
- **Breathing**: similar to pumping, this is a perceived variation of background noise level due to rapid gain changes in conjunction with high background noise;
- **Modulation distortion**: distortion that is caused by the gain control changing too rapidly;
- **Spectral ducking**: broadband gain reductions in the processed signal that occur as a result of a narrow-band interfering signal; and
- **SNR reduction**: reduction in the signal-to-noise ratio caused by boosting low-level (noise signals) signals and/or attenuating high-level sources.

Drop-outs, pumping, and breathing are functions of the attack and release time changes, and to some extent, the threshold. Pumping and breathing, in particular, are quite well-known and sometimes used intentionally. Sometimes, especially with dance music, the producer may want an audible change every time the compressor 'kicks in.' A short attack or release time can be used in order to achieve a quick change in the gain and achieve modulation of the overall signal level. This gives a 'pumping' sound.

When the sound level drops below the threshold, the gain increases with a rate dependent on the release time. With an appropriate choice of time constants, the signal level can remain reduced even though the input level is low. Thus, the noise can be made audible, giving the impression of breathing. A more sophisticated compressor may watch the input closely and adjust the gain when the input hits zero momentarily to reduce the 'breathing' effect. Since breathing involves raising the noise floor, it could be considered a subset of SNR reduction. Spectral ducking is dealt with through sideband filtering or multiband compression, and is discussed in Chapter 13.

Generally, a compressor should leave the power spectrum unchanged. If our input is a pure sinusoid, with a sudden change in level, then the spectrum should be a single peak, with a tiny bit of harmonic distortion in just the window where the transition occurred.

Summary

Based on the analysis and design choices presented, we can now present a compressor configuration that serves the goals of ease of modification and avoidance of artifacts.

Feedforward compressors are preferred since they are more stable and predictable than the feedback type ones, and high dynamic range problems do not occur with digital designs. The detector is placed in the log domain after the gain computer, since this generates a smooth envelope, has no attack lag, and allows for easy implementation of variable knee width. For the compressor to have smooth performance on a wide variety of signals, with minimal artifacts and minimal modification of timbral characteristics, the smooth, decoupled peak detector should be used. Alternately, the smooth, branching peak detector could be used in order to have more detailed knowledge of the effect of the time constants, although this may yield discontinuities in the slope of the gain curve when switching between attack and release phases.

Noise Gates and Expanders

The expander is a dynamic processor that attenuates the low-level portions of a signal and leaves the rest unaffected. A noise gate is an expansion taken to the extreme, where it will heavily attenuate the input or eliminate it entirely, leaving only silence. The expander and noise gate, which act on low-level signals, are analogous to the compressor and limiter, which act on high-level signals.

Theory and Implementation

The expander can be viewed as a variable gain amplifier, where the gain is controlled by the level of the input signal and always set less than or equal to one. When the signal level is high, the expander has unity gain. But when the signal drops below a predetermined level, the gain will decrease, making the signal even lower. The basic structure of an expander is a level-dependent, feedforward gain control, and identical to the feedforward compressor block diagram (Figure 7.1a).

The input/output relationship is represented with a simple graph, like the one in Figure 7.10. The input signal level is given in dB by the horizontal axis, and the output level is given by the vertical axis. When the input/output curve has slope 1 (i.e., angled at 45°) and intersects the origin, the gain of the expander is one, and the output level is identical to the input level. A change in the height of this curve corresponds to a change in the expander's gain. For the expander, part of the line will have a slope greater than 1. The point where the slope of the line changes is the *threshold*, which is generally a user-adjustable parameter. When the input signal level is

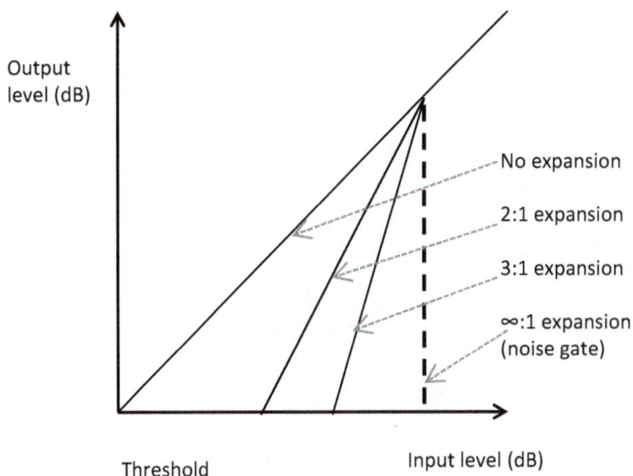

FIGURE 7.10
The expander reduces the signal level only when it drops below the threshold.

above the threshold nothing happens, and the output equals the input. But a gain reduction results whenever the level drops or stays below the threshold. This gain reduction lowers the input level, thus expanding the dynamic range. For a typical expander, the signal levels are taken from an average level based on a root-mean-square (RMS) calculation, rather than from instantaneous measurements, since even a signal with high average level may produce low instantaneous measurements.

As with compression, the amount of expansion that is applied is usually expressed as a ratio between the change in input level (expressed in dB) and the corresponding change in output level. So with a 4:1 expansion ratio (with the input level below the threshold), a dip of 2 dB in the input will produce a drop of 8 dB in the output.

Since the level detection is based on a short time average, it takes some time for a change in the input level to be detected, which then triggers the change in the gain. Like the compressor, the ballistics are given by attack and release times, which may optionally be used in the level sensing or in smoothing the applied gain. The time required for the expander to restore the gain to one once the input level rises above the threshold is the *attack time*. Likewise, the time taken for the expander to reduce its gain after the input drops below the threshold is the *release time*. The attack and release times give the expander a smoother change in the gain, rather than abrupt changes that may produce pops and other noise. Note that this is different from in the compressor, where attack relates to the time to reduce the gain and release to the time to restore to the original levels. Figure 7.11 shows how the attack and release times in an expander affect an example input signal.

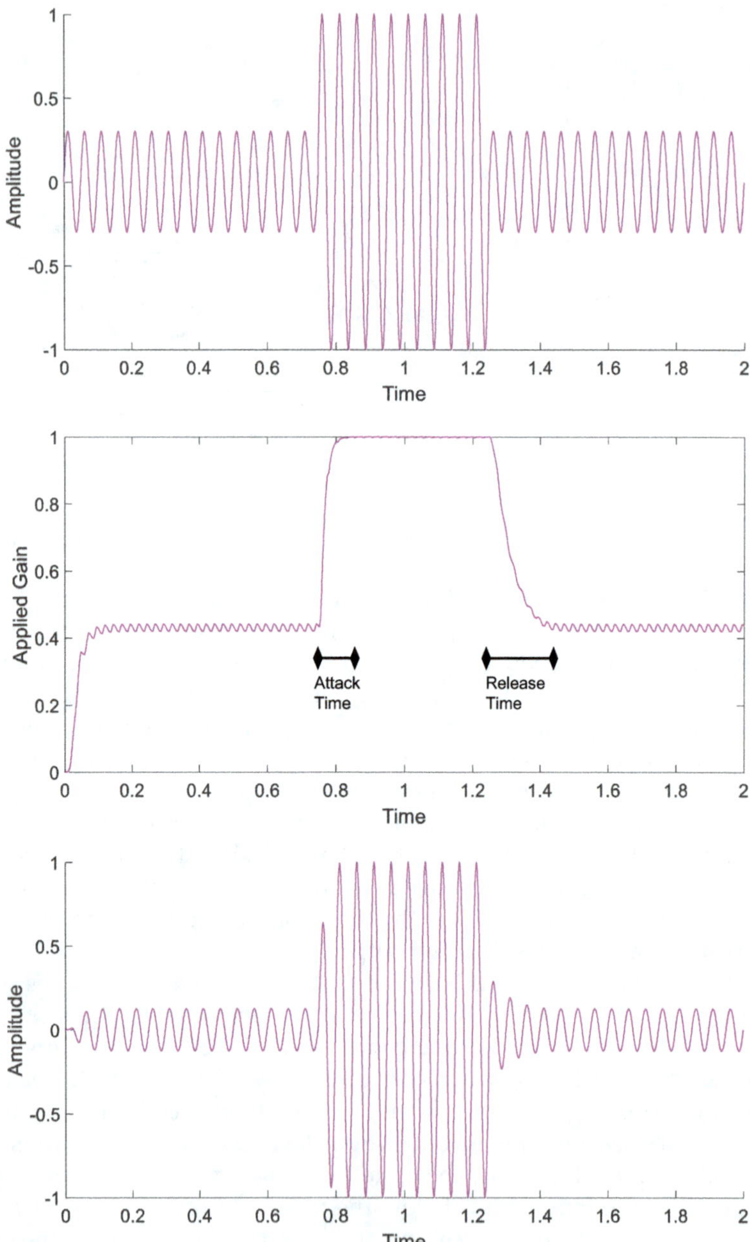

FIGURE 7.11
Effect of an expander. Only the middle portion of input is above the expander's threshold value. It takes time for the expander to increase gain when the input rises above the threshold. When input drops below the threshold, the expander slowly reduces gain.

If an expander is used with extreme settings where the input/output characteristic becomes almost vertical below the threshold, e.g., expansion ratio larger than 8:1, it is known as a *noise gate*. In this case, the signal may be very strongly attenuated or even completely cut. The noise gate will act like an on/off switch for an audio signal. When the signal has a large amplitude, the switch is on and the input reappears at the output, but when it drops below the threshold, the switch is off and there is no output. As the name suggests, noise gates are used to reduce the level of noise in a signal.

A noise gate has five main parameters: threshold, attack, release, hold, and gain [68]. Threshold and gain are measured in decibels, and attack, release, and hold are measured in milliseconds. The threshold is the level above which the signal will open the gate and below which it will not. The gain is the attenuation applied to the signal when the gate is closed. The attack and release parameters control how quickly the gate opens and closes. The attack is a time constant representing the speed at which the gate opens. The release is a time constant representing the speed at which the gate closes. The hold parameter defines the minimum time for which the gate must remain open. It prevents the gate from switching between states too quickly, which can cause modulation artifacts.

Applications

There are many audio applications of noise gates. For example, noise gates are used to remove breathing from vocal tracks, hum from distorted guitars, and bleed on drum tracks, particularly snare and kick drum tracks.

Consider a drum kit containing snare, kick drum, hi-hats, cymbals, and any number of tom toms. An example microphone setup will include a kick drum microphone, a snare microphone (possibly two), a microphone for each tom tom, and a set of stereo-overheads to capture a natural mix of the entire kit. In some instances, a hi-hat microphone will also be used. When mixing the recording, the overheads will be used as a starting point. The signals from the other microphones are mixed into this to provide emphasis on the main rhythmic components, that is, the kick, snare, and tom toms. Processing is applied to these signals to obtain the desired sound. Compression is invariably used on kick drum recordings. A compressor raises the level of low amplitude regions in the signal, relative to high amplitude regions which has the effect of amplifying the bleed. Noise gates are used to reduce (or remove) bleed from the signal before processing is applied.

Figure 7.12a shows an example snare drum recording containing bleed from secondary sources, and Figure 7.12b shows the same signal after a noise gate has been applied. The amplitude of the spill from other sources is very low and will have minimal effect on the gate settings.

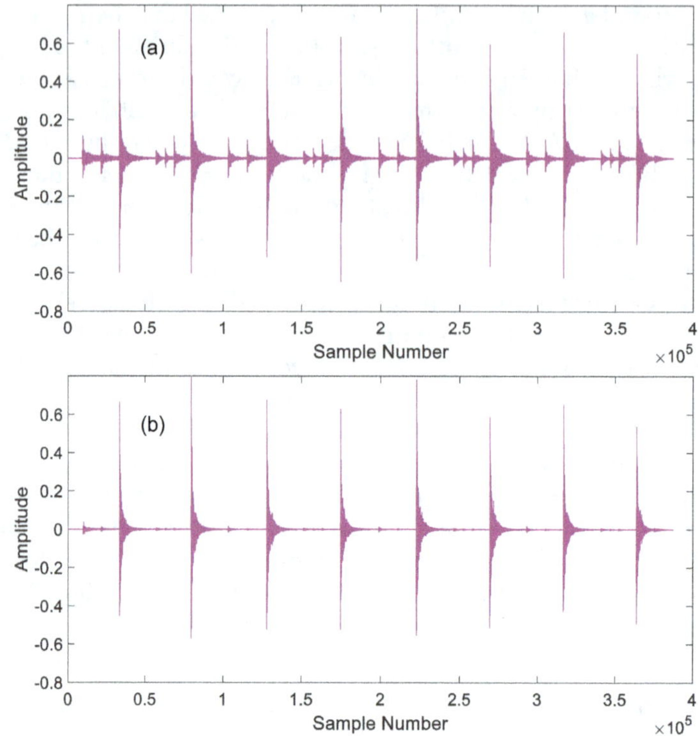

FIGURE 7.12
(a) Audio recorded from a snare drum microphone, with spill from the cymbals and bass drum in a drum kit; (b) the same recording, but with a noise gate applied to remove the low-level bleed from other sources.

Components of the bleed signal that coincide with the snare cannot be removed by the gate, because the gate is opened by the snare.

If the release time is short, the gate will be tightly closed before the spill from other sources, but the natural decay of the snare will be cut off. If the release time is long, the gate will remain partially open, and the spill will be audible to some extent, but the snare will be allowed to decay more naturally. If the threshold is below the peak amplitude of any part of the bleed signal, then the bleed will open the gate and will be audible. It is necessary to strike a balance between reducing the level of bleed and minimizing distortion of the wanted source (in this case, the snare).

Expanders have applications in consumer electronics. They are often used to produce more extremes on recordings that have a limited dynamic range, such as vinyl or cassettes. An expander will make the dynamics much more dramatic during playback.

One of the most significant applications for expanders is noise reduction. They help reduce feedback and unwanted audio content, such as background sounds or bleed and interference from other sources. Noise gates are often used to eliminate noise or hiss, which may otherwise be amplified and heard when an instrument is not being played. In which case, the threshold should be set high enough such that the ambient noise falls below it, but not set so high that the instrument's sound and sustained notes are prematurely cut off.

Expanders can also be used in conjunction with compressors to reduce the effects of noise when transmitting a signal, audio or otherwise. A transmission channel typically has limited dynamic range capacity. If the signal is compressed before transmission, one can increase the average level of the signal with respect to the noise in the system, thus reducing the effect of the noisy channel. An expander is then used on the receiving end to return the transmitted signal back to its original dynamic range (though it does not undo or invert the compression [69]). This process of compressing and then expanding a signal is called *companding*, and is an important component of the Dolby A noise reduction technique. Compressing the signal when recording and then expanding it on playback reduces the overall noise level.

Further Reading

Virtually all textbooks on mixing or mastering will devote significant discussion to dynamics processing, with Alex Case's text [17] being an excellent introduction, and Roey Izhaki's text [43] being a standout example that combines practical and theoretical explanation. The analysis in this chapter was limited to the design of standard dynamics processors. Their uses and recommended parameter settings are also discussed in detail in texts such as [70,71]. Variations and more advanced designs, such as sidechain filtering or multiband compressors, are discussed in Chapter 13.

Recent research into dynamics processing has been concerned with extracting the parameters of a compressor [72,73], or even reversing the effect entirely [74]. And since dynamics compressors have a large number of parameters, yet their use often has a clear objective, research has also concentrated on automating their parameter values [75–77]. As with other audio effects, modeling of analogue devices for dynamics processing is now often performed with neural networks [78–80]. Research and development in this field, especially in relation to multiband dynamic range compression, is also driven by the hearing aid community as much as the audio production community [81].

Problems

1. Define the following parameters for a dynamic range compressor: threshold, ratio, attack, release, and knee.
2. Explain whether or not the compressor has the following properties, and why:
   i. Linearity; $f(x+y)=f(x)+f(y)$ and $f(ax)=af(x)$
   ii. Shift or time invariance; if $f(x[n])=y[n]$, then $f(x[n-k])=y[n-k]$
   iii Memory independence f is not an explicit function of time.
3. The diagrams in Figure 7.13 show two different compressor input-output characteristics.
   i. Label the following quantities on each diagram: *threshold, ratio, knee*, i.e., show what aspect of the diagram corresponds to each quantity.
   ii. Which curve more closely corresponds to a limiter? Why?
4. Define two side effects of compression, *pumping* and *breathing*. Explain the difference between them and how they can be addressed.
5. Suppose a compressor is given by the input signal depicted in Figure 7.14. Draw a plot of the compressor gain over time for the following two configurations (exact numerical accuracy is not expected, but keep approximately the right scale in time and amplitude):

   i. Attack time = 20 ms, Release time = 500 ms
   ii. Attack time = 100 ms, Release time = 100 ms

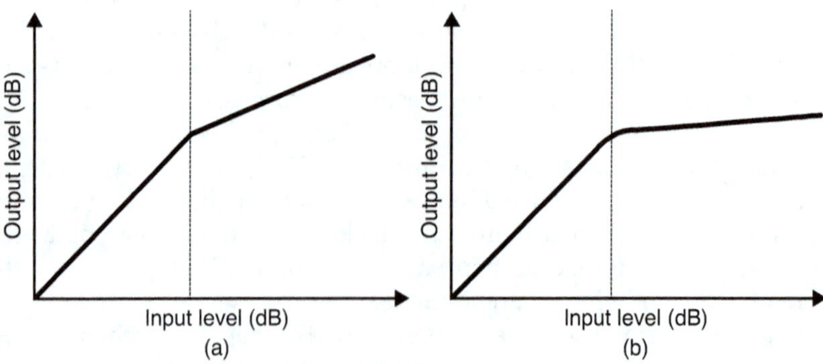

FIGURE 7.13
Possible compressor input/output curves.

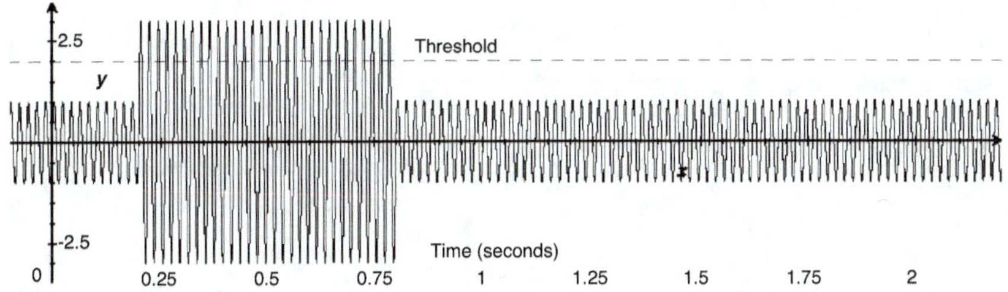

FIGURE 7.14
Example waveform.

iii. For case i), draw a plot of the *output* audio signal after the compressor is applied. In your diagram, label where *pumping* occurs and explain what can be done to reduce its effect.

6. Consider a dynamic range compressor with hard knee, threshold at –12 dB and ratio of 3:1. What is the output signal level if the input signal level is –15, –12, or 3 dB?

8

Distortion

Since the earliest days of the electric guitar, guitarists have been adding *distortion* to the sound of their instruments for expressive purposes. Distortion effects can create a wide palette of sounds ranging from smooth, singing tones with long sustain to harsh, grungy effects. Distortion can be introduced deliberately by the amplifier (especially in vacuum tube guitar amplifiers) or by a self-contained effect, and the particular choice of distortion effect is often part of a player's signature sound. An overview of many distortion-based effects, along with new implementations, is provided in [82].

This chapter covers *overdrive, distortion,* and *fuzz* effects, which are all based on the same principle of *nonlinearity*. Indeed, the three terms are sometimes used interchangeably. When a distinction is made, it is largely one of degree: *overdrive* is a nearly linear effect for low signal levels, which becomes progressively more nonlinear at high levels; *distortion* operates mainly in a nonlinear region for all input signals; *fuzz* is a completely nonlinear effect, which creates more drastic changes to the input waveform, resulting in a harder or harsher sound.

Theory

Characteristic Curve

Most common audio effects are *linear* systems. In a linear system, if two inputs are added together and processed, the result is the same as processing each input individually and adding the results. And if an input is multiplied by a scalar value before processing, the result is the same as processing the input and then multiplying by that same scalar value. That is, for a linear audio effect, the following holds;

$$\begin{gathered} f\left(x_1[n]+x_2[n]\right)=f\left(x_1[n]\right)+f\left(x_2[n]\right) \\ f\left(ax[n]\right)=af\left(x[n]\right) \end{gathered} \quad (8.1)$$

DOI: 10.1201/9781003593942-8

Overdrive, distortion, and fuzz (hereafter collectively referred to as "distortion") are always *nonlinear* effects for at least some input signals, meaning that Eq. (8.1) does not hold.

The simplest class of distortion effects are those described by a *characteristic curve,* a mathematical function relating the output sample $y[n]$ to the input sample $x[n]$. The following equation is one of many possible characteristic curves to produce a distortion effect [63]:

$$f(x) = \begin{cases} 2x & 0 \le x < 1/3 \\ 1-(2-3x)^2/3 & 1/3 \le x < 2/3 \\ 1 & 2/3 \le x \le 1 \end{cases}. \tag{8.2}$$

In this example, for input samples with magnitude less than 1/3, the effect operates in a linear region, but as the magnitude of x increases, it becomes progressively more nonlinear until *clipping* occurs above $x = 2/3$ and the output no longer grows in magnitude. A plot is shown in Figure 8.1. This particular equation is best classified as an *overdrive* effect since it contains a linear and a nonlinear region with a gradual transition between them. It is the nonlinear region that will give this effect its distinct sound.

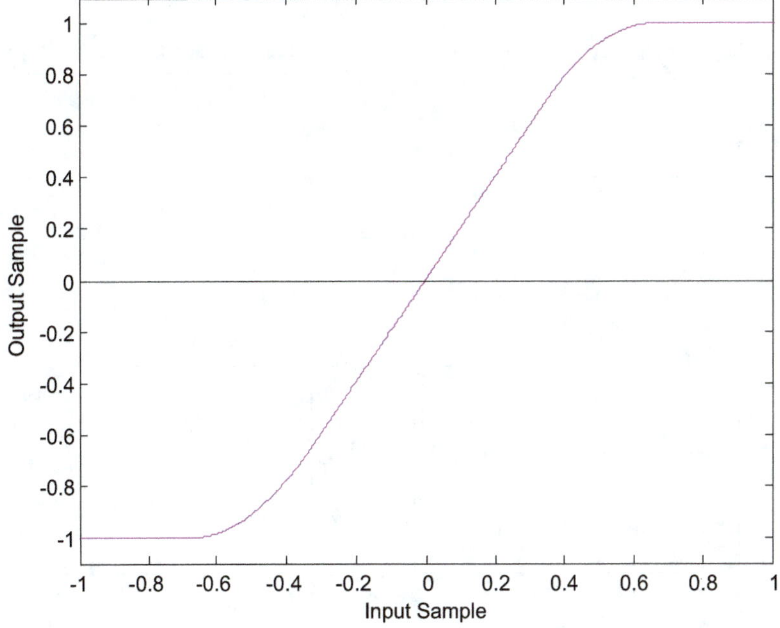

FIGURE 8.1
The characteristic input/output curve for a quadratic distortion.

The characteristic curve defines a *memoryless* effect: the current output sample $y[n]$ depends only on the current input sample $x[n]$ and not on any previous inputs or outputs. This is a reasonable approximation to how analog distortion circuits operate [83], though not an exact one, as we will see in a later section. Distortion is also a *time-invariant* effect in that the output samples depend only on the input samples and not the time at which they are processed.

Hard and Soft Clipping

Both digital and analog systems have limits to the magnitude of signal they can process. For analog systems, these limits are typically determined by the power supply voltages and architecture of each amplifier stage. In digital systems, the limits are usually determined by the number of bits in the analog-to-digital converter (ADC) and digital-to-analog converter (DAC). When a signal exceeds these limits, *clipping* occurs, meaning that a further increase in input does not produce any further increase in output. Clipping is an essential feature of distortion effects, and the way that an effect approaches its clipping point is a crucial part of its sound.

Distortion effects are often classified by whether they produce *hard clipping* or *soft clipping*. Figure 8.2 compares the two forms. Hard clipping is characterized by an abrupt transition between unclipped and clipped

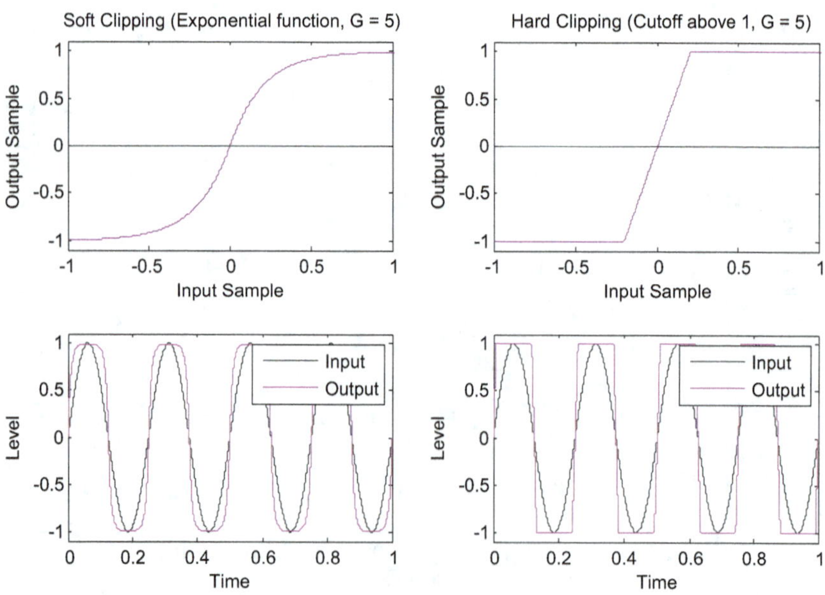

FIGURE 8.2
Comparison of hard and soft clipping.

regions of the waveform, which produces sharp corners in the waveform. Soft clipping is characterized by a smooth approach to the clipping level, creating rounded corners at the peaks of the waveform. In general, soft clipping produces a smoother, warmer sound, whereas hard clipping produces a bright, harsh, or buzzy sound. Hard versus soft clipping is not a binary decision, and any given characteristic curve will fall on a continuum between the two.

The simplest, purest form of hard clipping simply caps the input signal above a certain magnitude threshold:

$$f(x) = \begin{cases} -1 & Gx \leq -1 \\ Gx & -1 < Gx < 1 \\ 1 & Gx \geq 1 \end{cases}. \tag{8.3}$$

where G is an *input gain* applied to x before comparing to the threshold, explained in the next section. Digital systems produce this result when overloaded. Musicians generally find it to be an unpleasant, overly harsh sound. In the analog domain, many amplifiers based on transistors produce hard clipping, and hard clipping can also be created in an effect pedal through the use of silicon diodes.

The characteristic curve in Eq. (8.2) produces a form of soft clipping since the transition from unclipped to clipped is gradual. The equation below [63] also produces soft clipping:

$$f(x) = \operatorname{sgn}(x)\left(1 - e^{-|Gx|}\right). \tag{8.4}$$

In this equation, the output asymptotically approaches the clipping point as the input gets larger but never reaches it.[1] The amount of distortion added to the sound increases smoothly as the input level increases. Soft clipping occurs in analog vacuum tube amplifiers and certain effects pedals based on germanium diodes. It is not a natural occurrence in digital systems unless deliberately created by a suitable characteristic curve.

Input Gain

The term G in Eqs. (8.3) and (8.4) is a *gain* term applied to the input signal x before it passes through the nonlinear function. Because distortion is a nonlinear effect, the gain (or amplitude) of the input signal changes how the effect sounds. For nearly all practical characteristic curves, higher gain produces more distortion in the output. Notice that applying more gain to the input signal does not substantially affect the amplitude of the output, since the clipping level remains in the same place.

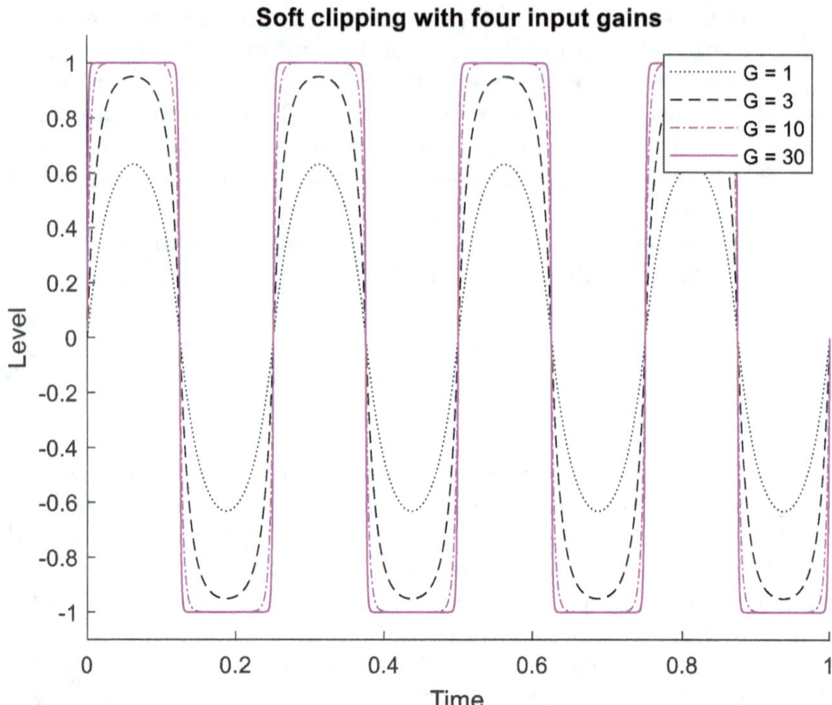

FIGURE 8.3
Soft clipping of a sine wave with four different input gains.

Figure 8.3 shows a sine wave subjected to soft clipping, Eq. (8.4), with four different input gains. In the extreme case, the output approaches a square wave with amplitude equal to the clipping level. An extremely large input gain with a hard clipping effect would also produce an output approaching a square wave, showing that the differences between hard and soft clipping are less pronounced for very large gains.

Symmetry and Rectification

The equations presented in the preceding sections were all *symmetrical* in that they applied the same nonlinear function to the positive and negative halves of the waveform. Real analog guitar amplifiers, especially those based on vacuum tubes, do not always behave this way. Instead, the clipping point might differ for positive and negative half-waves, or the curve for each half-wave could be entirely different. As we will see in the next section, symmetrical and asymmetrical characteristic curves produce different effects in the frequency domain, which are responsible for distinctive differences in sound.

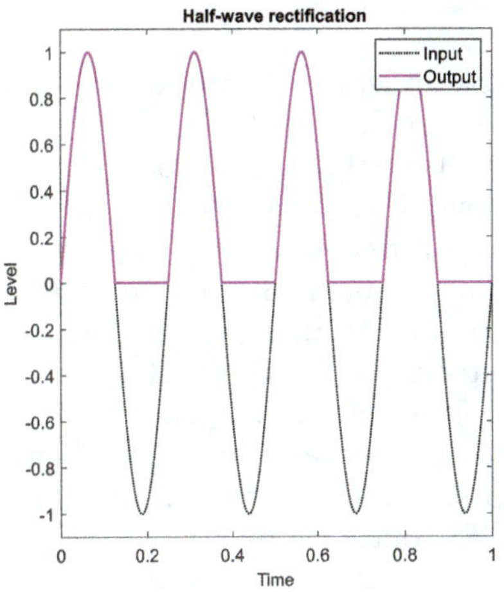

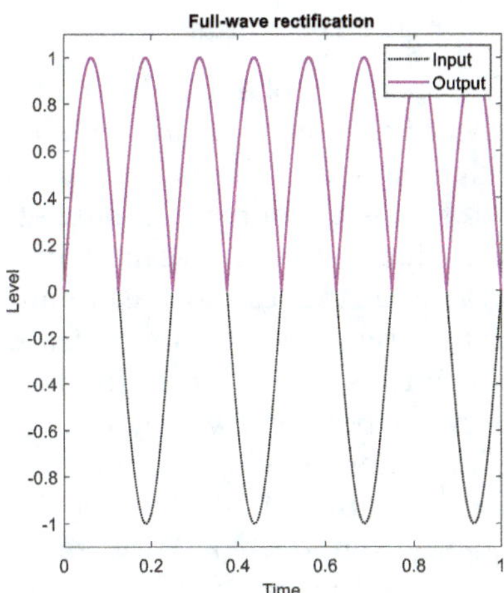

FIGURE 8.4
Half-wave and full-wave rectification.

Rectification is a special case of an asymmetrical function used in distortion effects. Rectification passes the positive half-wave unchanged but either omits or inverts the negative half-wave. It comes in two forms, shown in Figure 8.4. *Half-wave rectification* sets the negative half-wave to 0,

$$f_{\text{half}}(x) = \max(x, 0). \tag{8.5}$$

Full-wave rectification, equivalent to the absolute value function, inverts the negative half-wave:

$$f_{\text{full}}(x) = |x|. \tag{8.6}$$

Rectification is often combined with another nonlinear transfer function in a distortion effect. It adds a strong *octave harmonic* (twice the fundamental frequency) to the output signal. The reason can be seen in Figure 8.4. In the full-wave rectifier, the period of the waveform is half the original input since the negative half-waves have been inverted. The half-wave rectifier is mathematically equivalent to the average of the input and its full-wave rectified version:

$$f_{\text{half}}(x) = \left(x + f_{\text{full}}(x)\right)/2. \tag{8.7}$$

For this reason, the half-wave rectifier contains both the original fundamental frequency and its octave harmonic.

Harmonic Distortion

The operation of a distortion effect is best understood as applying a nonlinear function to the input signal in the time domain. However, its characteristic sound comes from the artifacts the nonlinear function creates in the frequency domain. Linear effects have the property that while they may change the relative magnitudes and phases of frequency components in a signal, they cannot create new frequency components that did not exist in the original signal. By contrast, the nonlinear functions used in distortion effects produce new frequency components in the output according to two processes: *harmonic distortion* and *intermodulation distortion.*

Consider applying a distortion effect to a sine wave input with frequency f and sample rate f_s: $x[n] = \sin(2\pi f/f_s)$. A sine wave contains a single frequency component at f (Figure 8.5, left). The output of the effect may have a different magnitude and phase at f, but it may also contain energy at every *multiple* of f: $2f$, $3f$, $4f$, etc. (Figure 8.5, right). These frequencies, which were not present in the input, are known as the *harmonics* of the fundamental frequency f, and the process that creates them is known as *harmonic distortion.* Every nonlinear function will introduce some amount of harmonic distortion. As a rough guideline, the more nonlinear the function, the greater the relative amplitude of the harmonics. Where multiple input frequencies are present, as in most real-world instrument signals, harmonics of each input frequency will appear in the output. In general, the magnitude of each harmonic decreases toward zero as frequency increases, but there is no frequency above which the magnitude of every harmonic is exactly zero. In other words, harmonic distortion will create *infinitely many* harmonic frequencies of the original input. This result can create problems with *aliasing* in digital implementations of distortion effects.

Detailed nonlinear analysis on the origins of harmonic distortion and its relation to specific characteristic curves is beyond the scope of this text, but in general, the more nonlinear the characteristic curve, the greater the magnitude of the harmonic distortion products that are introduced (Figure 8.6, left). Hard clipping also produces a different pattern of distortion products than soft clipping (Figure 8.6, right).

Another important relationship should be highlighted: *odd symmetrical* distortion functions produce only *odd* harmonics and *even symmetrical* distortion functions produce only *even* harmonics, where *asymmetrical* functions can produce both *even* and *odd* harmonics. Odd functions are those that obey the relationship $f(-x) = -f(x)$, known as *odd symmetry* in mathematical terminology. Similarly, even functions obey the relationship $f(-x) = f(x)$. For example, Eqs. (8.3) and (8.6) are both symmetrical by this definition. For a fundamental frequency f, the odd harmonics are the odd multiples of f: $3f$, $5f$, $7f$, etc. Similarly, the even harmonics are the even multiples of f: $2f$, $4f$, $6f$, etc. The *octave harmonic* created by rectification is

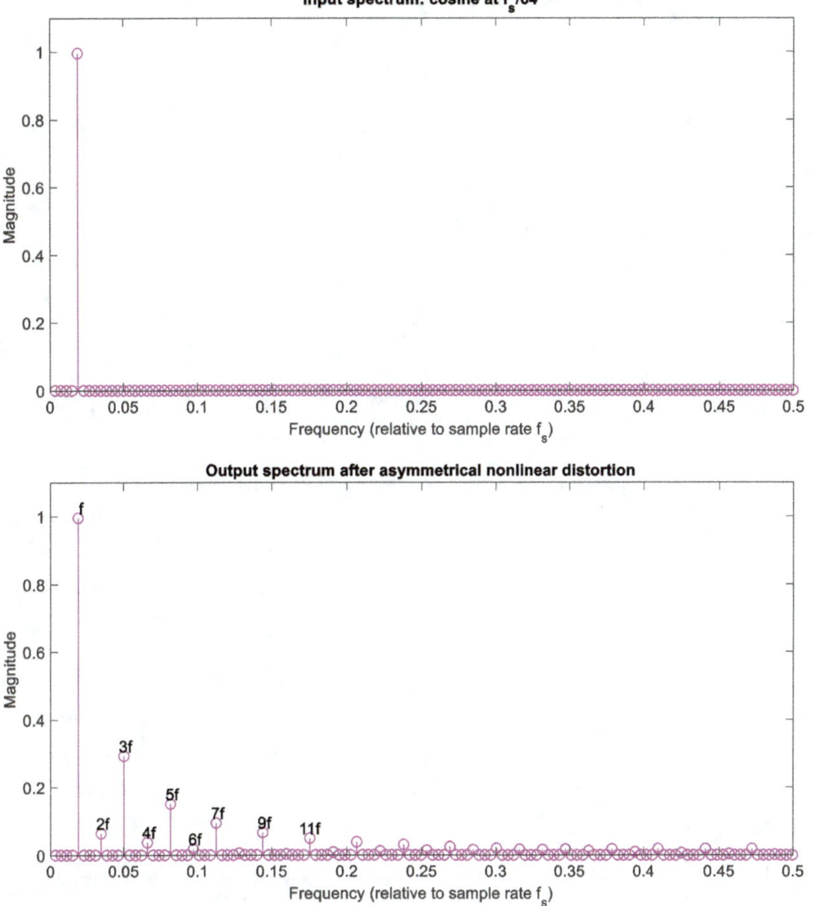

FIGURE 8.5
The spectrum of a single sine wave before and after asymmetric distortion has been applied.

an even harmonic: 2*f*. Musicians often prefer the combination of both even and odd harmonics, and consequently, asymmetrical functions are often preferred to symmetrical ones.

To see why an odd symmetrical function produces only odd harmonics, consider a sine wave input signal written in the complex domain, $x(t) = e^{j\omega t}$. We can perform a polynomial expansion of $f(x)$,

$$\begin{aligned} f\left(x(t)\right) &= c_0 + c_1 x(t) + c_2 x^2(t) + \ldots \\ &= c_0 + c_1 e^{j\omega t} + c_2 \left(e^{j\omega t}\right)^2 + \ldots \\ &= c_0 + c_1 e^{j\omega t} + c_2 e^{2j\omega t} + \ldots \end{aligned} \tag{8.8}$$

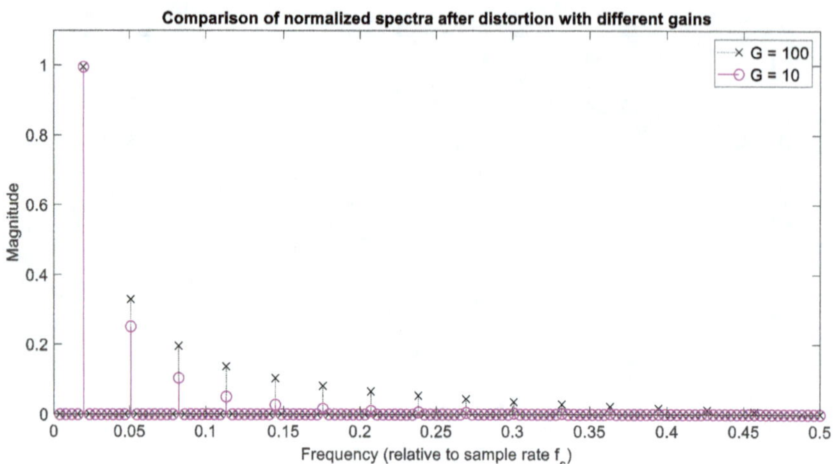

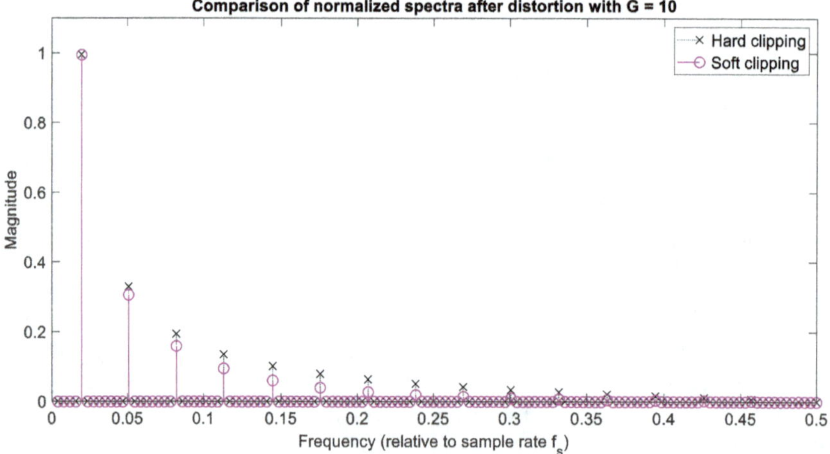

FIGURE 8.6
The output spectrum after distortion has been applied for sinusoidal input, comparing two values of distortion level for soft clipping (top), and comparing soft and hard clipping (bottom).

And in this complex exponential form, we can easily see that the odd powers, like $\left(e^{j\omega t}\right)^3$, are also odd harmonics, like $e^{3j\omega t}$. And similarly for the even powers and even harmonics. So we can rewrite this as the sum of even terms (even harmonics) and odd terms (odd harmonics):

$$\begin{aligned} f\left(x(t)\right) &= c_0 + c_2 e^{2j\omega t} + c_4 e^{4j\omega t} + \ldots && \text{even terms} \\ &= c_1 e^{j\omega t} + c_3 e^{3j\omega t} + c_5 e^{5j\omega t} + \ldots && \text{odd terms} \end{aligned} . \tag{8.9}$$

It can easily be shown that the sum of even functions is an even function, the sum of odd functions is an odd function, and the sum of even and odd functions is neither. So if $f(x)$ is odd, then all the even terms in the polynomial expansion must be zero. Therefore, an odd symmetrical function can produce only odd harmonics of the original input frequency. A similar argument can be used to show that even symmetrical functions only have even harmonics. Creating both even and odd harmonics requires that the function is neither even nor odd, so the positive and negative half-waves must be treated asymmetrically.

Intermodulation Distortion

Harmonic distortion is a desirable property of overdrive, distortion, and fuzz effects. Another result, *intermodulation distortion*, is also a direct consequence of any nonlinear transfer function, but this result is generally undesirable in musical situations. Suppose the input signal contains two frequency components at f_1 and f_2:

$$x(t) = \sin(2\pi f_1 t) + \sin(2\pi f_2 t). \tag{8.10}$$

A general analysis of all nonlinear functions is beyond the scope of this text, but to see the mechanism behind intermodulation distortion, consider the simple nonlinear function $f(x) = x^2$:

$$f\left(x(t)\right) = \sin^2(2\pi f_1 t) + 2\sin(2\pi f_1 t)\sin(2\pi f_2 t) + \sin^2(2\pi f_2 t). \tag{8.11}$$

By trigonometric identity, the square terms produce an octave-doubling effect (twice the frequency):

$$\sin^2(2\pi f t) = \left[1 - \cos(2\pi\left(2f\right)t\right)]/2. \tag{8.12}$$

This property offers another way to understand the operation of the *full-wave rectifier*, which produces a similar (though not identical) output. However, it is the term in the middle, the product of two sines at different frequencies, which is responsible for the intermodulation distortion. Also by trigonometric identity:

$$\sin(2\pi f_1 t)\sin(2\pi f_2 t) = \left[\cos(2\pi\left(f_1 - f_2\right)t\right) - \cos(2\pi\left(f_1 + f_2\right)t)]/2. \tag{8.13}$$

The output will therefore contain *sum and difference frequencies* between the two frequency components at the input. Unless f_1 an f_2 are multiples of one another, these sum and difference frequencies will not be *harmonically-related* to either one (that is, not a multiple of either f_1 or f_2). This in turn means that these new frequencies will sound discordant and often unpleasant. This intermodulation process happens with *every* pair of frequencies in the input signal, so the more complex the input, the greater the number and spread of intermodulation products. An example of intermodulation distortion is shown in Figure 8.7.

Sum and difference frequencies are also found in the *ring modulator* (Chapter 6), with similarly non-harmonic results. But in that case, the frequency products are sums and differences with the *carrier* frequency and not between the frequencies of the input signal itself. Furthermore, in ring modulation, the original frequency is not present, while in intermodulation distortion, it is still the most prominent.

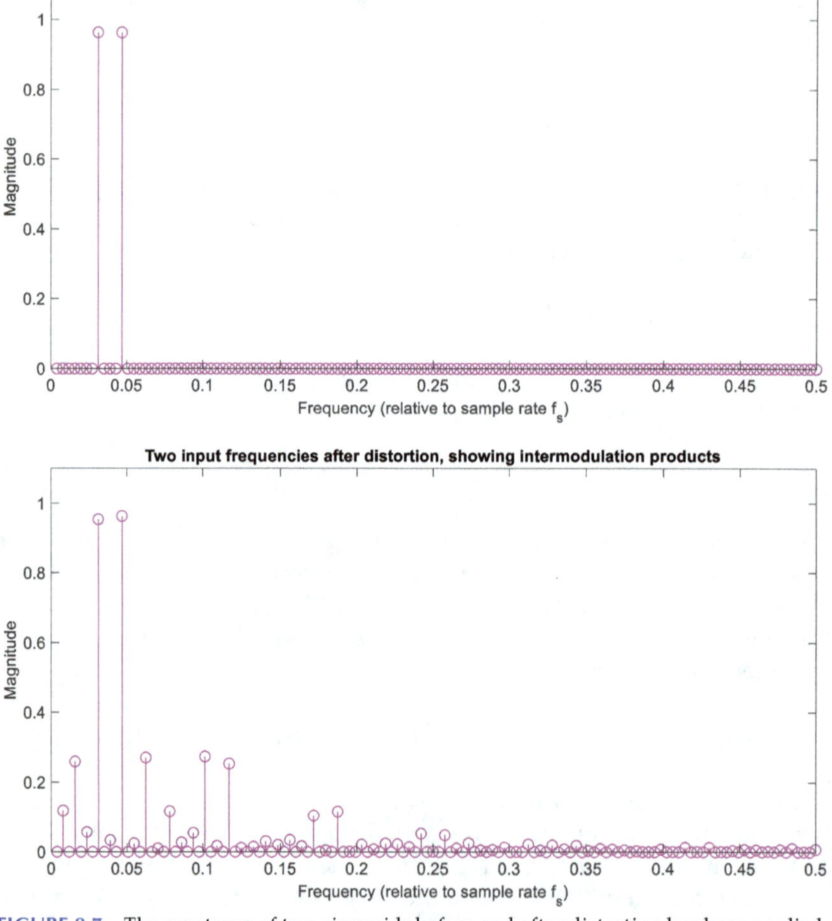

FIGURE 8.7 The spectrum of two sinusoids before and after distortion has been applied.

Highly nonlinear characteristic curves, such as those found in fuzz effects, will have a higher amplitude of intermodulation products, just as they produce higher amplitudes of harmonic distortion. This is the reason that single notes and "power chords" (combinations of octaves and perfect fifths) often work best with fuzz boxes. These inputs typically contain only harmonically related frequencies, so all the intermodulation products remain harmonic. Unfortunately, we cannot choose only harmonic distortion products without intermodulation; the right type of distortion effect must be chosen for each musical application, which balances these two qualities.

Waveshaping

The characteristic curve of a memoryless distortion effect is closely related to waveshaping, which is frequently found as a distortion effect in synthesizers to turn simple waveforms into more complex and spectrally rich signals. The basic premise of the waveshaper is to implement a nonlinear transfer function $y = f(x)$ which maps each input sample to an output sample. $f(x)$ can be implemented analytically, as in the characteristic curve examples earlier in this chapter, but in digital audio systems, a common approach is to implement it as a look-up table.

A look-up table $C[k]$, $0 \le k < N$, is an array of N elements specifying the output value corresponding to each value of an input signal. $C[k]$ represents a sampled version of $f(x)$ for $-1 \le x \le 1$. In other words, $C[0]$ represents the output value for an input of –1, $C[N-1]$ represents the output value for an input of +1, and $C[k]$ corresponds to the value at $\dfrac{2k}{N-1} - 1$.

For each input sample $x[n]$, to implement the waveshaper, we find the nearest index in the look-up table according to the formula $k = (x[n]+1)(N-1)/2$. This value of k may be fractional, in which case interpolation can be used (see Chapter 3) to obtain a more accurate approximation of the original function $f(x)$. When implementing a look-up table in code, the array index must always be restricted to the range $[0, N-1]$ to avoid reading an invalid index from the array, which might cause the code to crash.

An advantage of the look-up table approach to waveshaping, compared to the characteristic curve, is that more complex and elaborate transfer functions can be implemented without the need for an explicit formula. The shape of the look-up table could also be presented to the user as an adjustable parameter.

Analog Emulation

Many guitarists hold the vacuum tube to be the ideal device for constructing amplifiers and distortion effects. Its soft, asymmetrical clipping creates sounds that have defined generations of blues, jazz, and rock musicians. Unfortunately, tube amplifiers are expensive, heavy, noisy and require replacement parts every few years. Digital emulation of vacuum tube circuits has thus become an active and profitable area of development, and these emulation techniques can extend to other analog circuits as well, including diode- and transistor-based distortion effects.

Accurate emulation of even the simplest analog distortion effects rapidly becomes mathematically complex on account of the nonlinearity of each circuit element and the way the elements affect one another. Consider the *diode clipper* circuit in Figure 8.8. Each diode turns on when the voltage between its anode and cathode exceeds a fixed threshold voltage V_d. Once the diode turns on, it begins to conduct current and prevents the output voltage from rising much further. Placing the diodes back-to-back in opposite directions therefore clips the output to approximately the range $[-V_d, V_d]$. However, real-world diodes do not exhibit perfectly sudden turn-on behavior, so there will be some degree of rounding of the waveform peaks before hard clipping sets in.

As this chapter has shown, hard and soft clipping can be easily simulated with *characteristic curves*, but there is a subtle aspect of this circuit that is not simulated. The characteristic curves implement a *memoryless* nonlinear system whose output depends only on the current input sample. While the diodes in Figure 8.8 by themselves might reasonably approximate a memoryless system, the capacitor C_1 and resistor R_1 behave quite differently. Specifically, these two components together form a low pass filter with cut-off frequency $f_L = 1/(2\pi R_1 C_1) = 8$ kHz for the values given. But as the diodes turn on, they act for small signals like resistors whose resistance changes with the overall signal level [84]. As the diodes approach a completely on state, their resistance moves toward 0, and the cut-off frequency of the filter rises accordingly. Since this process depends on the signal level, the cut-off frequency of the filter will change dramatically over the course of a single waveform period! But since the filter is not a memoryless system, it cannot be accounted for in the characteristic curve.

The mathematical techniques to solve this problem are beyond the scope of the book. Most of them involve iterative approximation methods. *Wave digital filters* [85] are one popular technique for handling interconnected analog circuit elements. Interested readers are referred to [86–88].

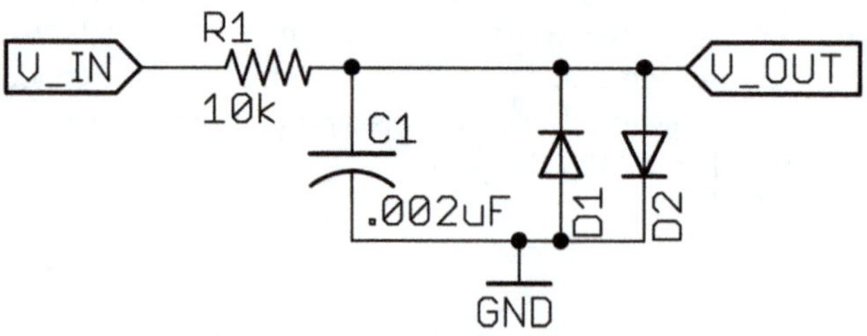

FIGURE 8.8
A diode clipper circuit.

Implementation

Basic Implementation

In their basic form, overdrive, distortion, and fuzz are among the simplest effects to implement. Since the nonlinear characteristic curve is a *memoryless* effect (at least in the simpler cases), these effects can be calculated on a sample-by-sample basis by applying the nonlinear function to each input sample.

Aliasing and Oversampling

We have seen that the nonlinear functions used in overdrive, distortion, and fuzz create an infinite number of *harmonics*, frequency components which are integer multiples of an original frequency in the input signal. In the analog domain, this is not a problem: eventually, the frequency limits of the electronic devices or filters deliberately added to the effect attenuate the higher-frequency components to a level at which they are not heard. Even when they do appear in the output sound, the highest harmonics will not be perceived if they are above the range of human hearing.

In the digital domain, the unbounded series of harmonics creates a problem with *aliasing*. Harmonics that are above the *Nyquist frequency* will be aliased, appearing in the output as lower-frequency components (Figure 8.9a). The aliased components are no longer harmonically related

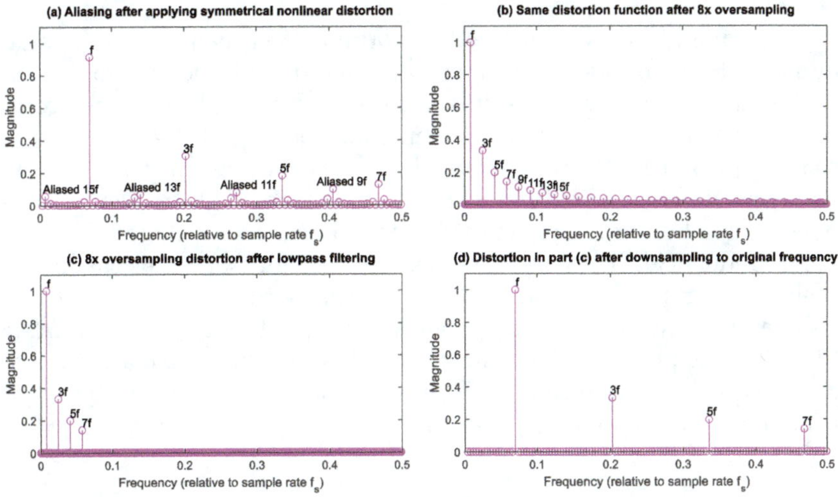

FIGURE 8.9
Output spectrum with aliasing due to distortion, and the output spectrum after oversampling, low pass filtering, and downsampling.

to the original sound, nor can they be filtered out once they appear. Unless aliasing is avoided, the quality of digital distortion effects will suffer compared to their analog counterparts.

The best way to reduce aliasing in a distortion effect is to employ *oversampling*. Prior to applying the nonlinear function, the input signal is *upsampled* to several times the original sampling frequency. Oversampling by a factor of N can be accomplished by inserting $N - 1$ zeroes between each input sample. This signal is then filtered to remove frequencies above the original Nyquist frequency (Figure 8.9b and c).

Once the signal has been upsampled, the nonlinear characteristic curve can be applied. This will still generate an infinite series of harmonic products, but now considerably more of them will fit within the new, higher Nyquist frequency. Furthermore, even though aliasing still occurs, the first aliased components are still in the higher frequency regions above the original Nyquist frequency. Since the harmonic distortion components decrease in amplitude with increasing frequency, those components that are aliased back into the audible range will be greatly reduced in amplitude (Figure 8.9d).

After the nonlinear curve is applied, the signal is again filtered to remove frequencies above the original Nyquist frequency. The signal is then *downsampled* back to the original rate. Downsampling by a factor of N can be accomplished by choosing every Nth sample and discarding the rest.

Filtering

Even when aliasing is minimized, distortion effects can produce a large number of new frequency components extending all the way to the top of the human hearing range. In some cases, these high-frequency components create an undesirable harshness in the output. For this reason, some distortion effects incorporate a *low pass filter* or *shelving filter* after the nonlinear function to reduce the magnitude of the high-frequency components. First-order filters are often used to create a gentle rolloff in the upper frequencies. The corner frequency of this filter or, in the case of the shelving filter, the gain, may be a user-adjustable control.

Other distortion effects incorporate a low pass filter *before* the nonlinear function. The purpose of this arrangement is to reduce the magnitude of high-frequency components in the input signal, which reduces their contribution to intermodulation distortion.

Common Parameters

The *characteristic curve* of a distortion effect is typically fixed by design. Most analog distortion effects have a particular circuit containing diodes, transistors, or tubes which determines the characteristic curve. Simple digital distortion effects generally follow analog effects in having a single curve; however, sophisticated multi-effects units may let the user choose from a range of options to simulate different classic analog sounds.

The *input gain* (or just *gain*) is a user-adjustable parameter on most effects. This control changes the gain of the input signal before it passes through the nonlinear transfer function. Implementation is simply a matter of multiplying the input sample by a constant before putting it into the nonlinear function. A higher gain produces a more distorted sound.

Since a large input gain is usually required to produce a heavily distorted sound, the output of the nonlinear function might be much higher in level than the input. Thus, most distortion effects produce an output that is consistently near the *clipping level*. Thus, it can be useful to incorporate a post-effect *volume* (or *output gain*) control, which scales the level of the output to more closely match the input. This is accomplished by multiplying the output of the nonlinear function by a constant, which typically ranges from 0 (muted) to 1 (full volume). In a purely linear effect, such as a filter or delay, the *gain* and *volume* controls would produce the same result, since it does not matter whether a scaling operation takes place before or after a linear effect. In the distortion effects, the two controls have different results, so it is useful to include them both in a practical effect.

Some effects also feature a *tone* control, which affects the timbre or brightness of the output. This control can be implemented in several ways, but it typically involves a low pass filter placed before or after the nonlinear transfer function. The control can affect the *cut-off frequency* of the filter or, if a low shelving filter is used, the *shelf gain*. Placing a low pass filter before the nonlinear function can help reduce intermodulation distortion by eliminating high-frequency components from the input signal. Placing a low pass filter after the nonlinear function will attenuate the resulting high-frequency distortion products.

Tube-Sound Distortion

As discussed in earlier sections, guitarists often seek digital alternatives that recreate the sound of classic vacuum tube amplifiers. More accurate emulation of every component of a tube amplifier, including not just tubes but also transformers and speakers, is an active area of academic and industrial research [87,89]. Emulation techniques are often mathematically complex, but the following choices in a basic distortion effect will help approach a tube-like sound:

1. Use a *soft clipping* characteristic curve that rounds the corners of the waveform as it approaches the clipping level.

2. Choose the curve to be at least mildly *asymmetrical*, which will produce even and odd harmonics. For example, the top and bottom half-waves in Eq. (8.4) could use a different input gain.
3. Use *oversampling* to control non-harmonic products from aliasing. If the sound is still too harsh, consider adding a gentle *low pass filter* before or after the nonlinear function.

Code Example

The following C++ code fragment implements several types of basic distortion effects:

```
int numSamples;           // Indicates how many audio samples to process
float *channelData;       // Array of audio samples, length numSamples
float inputGain;          // Input gain (linear scale), before distortion
float inputGainDecibels_; // Gain in decibels, from the user control
int distortionType_;      // Index containing the type of distortion

// Calculate input gain once to save calculations
inputGain = powf(10.0f, inputGainDecibels_ / 20.0f);

for (int i = 0; i < numSamples; ++i) {
    const float in = channelData[i] * inputGain;
    float out;

    // Apply distortion based on type
    if(distortionType_ == kTypeHardClipping) {
        // Simple hard clipping
        float threshold = 1.0f;
        if(in > threshold) out = threshold;
        else if(in < -threshold) out = -threshold;
        else out = in;
    }
    else if(distortionType_ == kTypeSoftClipping) {
        // Soft clipping based on quadratic function
        float threshold1 = 1.0f/3.0f;
        float threshold2 = 2.0f/3.0f;
        if(in > threshold2) out = 1.0f;
        else if(in > threshold1)
            out = (3.0f - (2.0f - 3.0f*in) * (2.0f - 3.0f*in))/3.0f;
        else if(in < -threshold2) out = -1.0f;
        else if(in < -threshold1)
            out = -(3.0f - (2.0f + 3.0f*in) * (2.0f + 3.0f*in))/3.0f;
        else out = 2.0f* in;
    }
    else if(distortionType_ == kTypeSoftClippingExponential) {
        // Soft clipping based on exponential function
        if(in > 0) out = 1.0f - expf(-in);
        else out = -1.0f + expf(in);
    }
    else if(distortionType_ == kTypeFullWaveRectifier) {
        // Full-wave rectifier (absolute value)
        out = fabsf(in);
    }
```

```
    else if(distortionType_ == kTypeHalfWaveRectifier) {
        // Half-wave rectifier
        if(in > 0) out = in;
        else out = 0;
    }

    // Put output back in buffer
    channelData[i] = out;
}
```

The code first applies an input gain to the samples in the buffer. It then applies one of several characteristic curves based on the value of `distortionType_`. The curves follow the formulas given earlier in this chapter.

Applications

Expressivity and Spectral Content

Distortion is most commonly used with the electric guitar, though it is sometimes applied to other instruments including the bass and even the voice. For some guitarists, the particular choice of distortion effect is as much a matter of personal identity as the choice of guitar. On the flip side, the designer seeking to recreate a particular player's distortion sound should remember that the tone ultimately depends not only on the distortion effect, but also on the amp, the guitar, and the manner of playing.

By adding harmonic distortion to the signal, the distortion effect creates a spectrally richer, "fatter" sound that can help an instrument achieve prominence in a mix. For example, a plucked guitar string may contain energy primarily in the bass or midrange frequencies, but adding distortion will create harmonics stretching all the way up the audio spectrum. Because of this extra spectral energy, distortion can be usefully paired with filter effects like *wah-wah* and *equalization,* which boost some frequencies while attenuating others. For example, if a wah-wah pedal is placed after a distortion effect, the wah effect can be much more pronounced than if using it on a clean guitar signal. Jimi Hendrix is said to have used the reverse arrangement, placing a "Fuzz Face" distortion pedal after his wah-wah pedal [90], such that the frequencies boosted by the wah effect were particularly distorted.

Sustain

Like the compressor (Chapter 7), distortion effects can increase an instrument's apparent sustain. When a guitar string is plucked, the signal begins strongly but rapidly decays. However, when heavy distortion is

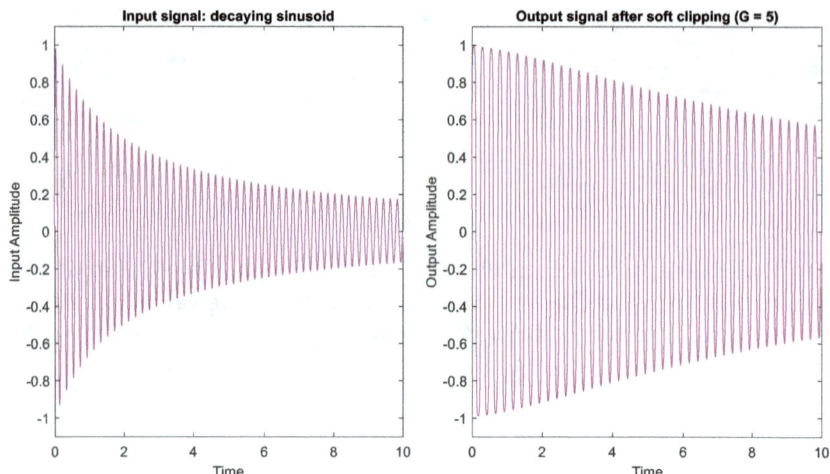

FIGURE 8.10
The effect of soft clipping on a decaying sinusoid.

used, the signal will be amplified to the point where *clipping* occurs (either hard or soft, depending on the type of distortion effect). Since the clipping point does not change over time, the output level will remain at or near this point until the original signal has decayed so much that, even with input gain applied, the amplified signal no longer reaches the clipping point. Figure 8.10 shows this process for a soft clipping distortion effect. Figure 8.10a shows the input, a gradually decaying tone. The output in Figure 8.10b stays near the clipping point and decays much more slowly, which creates the perception of longer sustain.

Comparison with Compression

Distortion and dynamic range compression both have the effect of attenuating or limiting the loudest signals. In fact, the nonlinear function in a distortion effect can be considered a nonlinear gain control where the gain depends on the level of the input signal. For example, we can calculate the gain for the hard clipping function from Eq. (8.3),

$$gain(x) = f(x)/x = \begin{cases} -1/x & Gx \le -1 \\ G & -1 < Gx < 1 \\ 1/x & Gx \ge 1 \end{cases}. \tag{8.14}$$

We can see that the hard clipping function is linear with gain G when the level of the scaled input signal is less than 1, and that the gain progressively decreases as the input level increases. This is identical to the behavior of the *limiter* (compressor with a very high or infinite ratio). Why, then, do distortion effects produce audible harmonic distortion when compressors and limiters generally do not?

The difference between distortion, compression, and limiting has to do with the design of the *level detector*. In the compressor and limiter, the gain is smoothed, and for RMS level detectors, the signal level is determined by the local average level of the input signal and not by the instantaneous sample value. If the *attack time* and *decay time* parameters in a limiter were both set to 0 (instantaneous response) and a peak detector was used, the result would be similar to the distortion effect.

Further Reading

There is a lot of nostalgia for classic, analogue distortion, and this is still often preferred over digital implementations. Thus, there has been much work on modeling those analogue designs with digital implementations. However, up until about 2017, this was primarily achieved through established circuit modeling and signal processing techniques. Recent work has instead used multilayer neural networks for modeling distortion [80,83,91]. This is quite a significant research area, with an open source toolbox dedicated to supporting the field [92]. Another focus of research is the recovery of the clean signal from the distorted signal [93], a task that could be solvable when the input-output function is invertible. There has been far less work on the creation and analysis of new distortion effects, which suggests that there may be an opportunity for new technologies in that area.

Problems

1. Define the terms *overdrive, distortion,* and *fuzz* (in the context of audio effects). How are they similar, and how are they different?
2.
   i. Using equations, demonstrate the concept of *intermodulation distortion* for modulation with two sine waves. What are the sidebands that result? Hint: you may use the following formula, $\cos(A + B) = \cos(A)\cos(B) - \sin(A)\sin(B)$.
   ii. Why is intermodulation distortion unpleasing to hear, and why is it particularly problematic for distortion, overdrive, and fuzz (compared to other audio effects)?

3. Signal-to-noise ratio (SNR) can be defined as $10\log_{10}(P_I/P_N)$, where P_I is the input signal power and P_N is the noise (difference between input and output) power. Calculate the SNR of hard clipping for a square wave input. You can assume that the square wave has some amplitude *A*, and the signal is clipped at some threshold *T*.
4. Draw plots of *hard clipping* and *soft clipping* of a sine wave input. Which kind of clipping do vacuum tubes generally produce? Which kind do digital systems (by default) produce? Why?
5. Draw and explain the difference between *half wave rectification* and *full wave rectification*. Explain the musical use of rectification (either version) and what it does to the sound.
6. Which of the following distortion functions have even symmetry, $f(-x) = f(x)$, which have odd symmetry, $f(-x) = -f(x)$, and which are asymmetric?

$$f(x) = \begin{cases} -1 & Gx \leq -1 \\ Gx & -1 < Gx < 1 \\ 1 & Gx \geq 1 \end{cases}$$

$$f(x) = \text{sgn}(x)\left(1 - e^{-|Gx|}\right)$$

$$f(x) = \max(x, 0)$$

$$f(x) = |x|$$

7. Show that even functions will produce only even harmonics. A similar approach can be taken to the one used earlier to show that odd functions only produce odd harmonics.
8. Explain why *aliasing* can be a problem in digital distortion implementations, and what can be done to minimize its effects.
9. A distortion effect produces odd harmonics of a 2 kHz input. Assuming a sampling frequency $f_s = 44.1$ kHz, what is the first harmonic to alias back to below $f_s/2$, and at what frequency? You may wish to refer back to the discussion of aliasing in Chapter 1.

Note

1 The expression *sgn(x)* takes the value 1 for $x \geq 0$, –1 otherwise.

9

The Phase Vocoder

The term 'phase vocoder' is used to describe a group of sound analysis-synthesis techniques where the processing of the signal is performed in the *frequency domain*. Most audio effects, including delays, filters, compression, and distortion, are implemented directly on the incoming signal in the time domain. By contrast, phase vocoder effects use frequency and phase information calculated from *Fourier transforms* (described in Chapter 1) to implement a variety of audio effects, including *time stretching*, *pitch shifting*, *robotization*, and *whisperization*.

The basic phase vocoder operation involves segmenting the incoming signal into discrete blocks, converting each block to the frequency domain, performing amplitude and phase modification of specific frequency components, and finally converting each block back to the time domain to obtain the final output [94,95]. There is now a widely recognized 'standard' implementation [96,97], which is well documented, and for which the term 'phase vocoder' is most commonly used. The specific effect that is produced depends on the type of processing done on the frequency domain signal.

A detailed introduction to Fourier transform theory can be found in several excellent texts [2,3], and this chapter cannot substitute for a complete digital signal processing course. Instead, it will provide an overview of how Fourier transforms are used to create the overall phase vocoder structure, with a focus on practical implementation strategies and specific audio effects that make use of the phase vocoder.

AUTO-TUNE

From 1976 through 1989, Dr. Andy Hildebrand worked for the oil industry, interpreting seismic data. By sending sound waves into the ground, he could detect the reflections and map potential drill sites. Dr. Hildebrand studied music composition at Rice University and then developed audio processing tools based on his knowledge in seismic data analysis. He was a leading developer of a variety of plug-ins, including MDT (Multiband Dynamics Tool), JVP (Jupiter Voice Processor), and SST (Spectral Shaping Tool). At a dinner party,

DOI: 10.1201/9781003593942-9

a guest challenged him to invent a tool that would help her sing in tune. Based on the phase vocoder, Hildebrand's Antares Audio Technologies released Auto-Tune in late 1996.

Auto-Tune was intended to correct or disguise off-key vocals. It moves the pitch of a note to the nearest true semitone (the nearest musical interval in traditional, equal temperament Western tonal music), thus allowing the vocal parts to be tuned. The original Auto-Tune had a speed parameter that could be set between 0 and 400 ms, and determined how quickly the note moved to the target pitch. Engineers soon realized that by setting this 'attack time' very short, Auto-Tune could be used as an effect to distort vocals, and make it sound as if the voice leaps from note to note in discrete steps. It gives it an artificial, synthesizer-like sound that can be appealing or irritating depending on taste. This unusual effect was the trademark sound of Cher's 1998 hit song, 'Believe.'

Like many audio effects, engineers and performers found a creative use, quite different from the intended use. As Hildebrand said, "I never figured anyone in their right mind would want to do that." Yet Auto-Tune and competing pitch correction technologies are now widely applied (in amateur and professional recordings, and across many genres) for both intended and unusual, artistic uses.

Phase Vocoder Theory

Overview

The phase vocoder is based on the *short-time Fourier transform* (STFT). What sets the STFT apart from other Fourier transform techniques is that it deliberately operates only on a small time segment of the input signal, giving a snapshot of the frequency content of the signal at a particular moment in time. By dividing the signal into a series of discrete frames of samples and performing the STFT on each frame, we get a picture of how the frequency content of the signal evolves over time. By modifying the frequency content in each frame, many new effects are possible that are not easily implemented in the time domain.

Figure 9.1 shows a diagram of the phase vocoder process. The following steps are common to every phase vocoder effect. Each step is discussed in detail in a subsequent section.

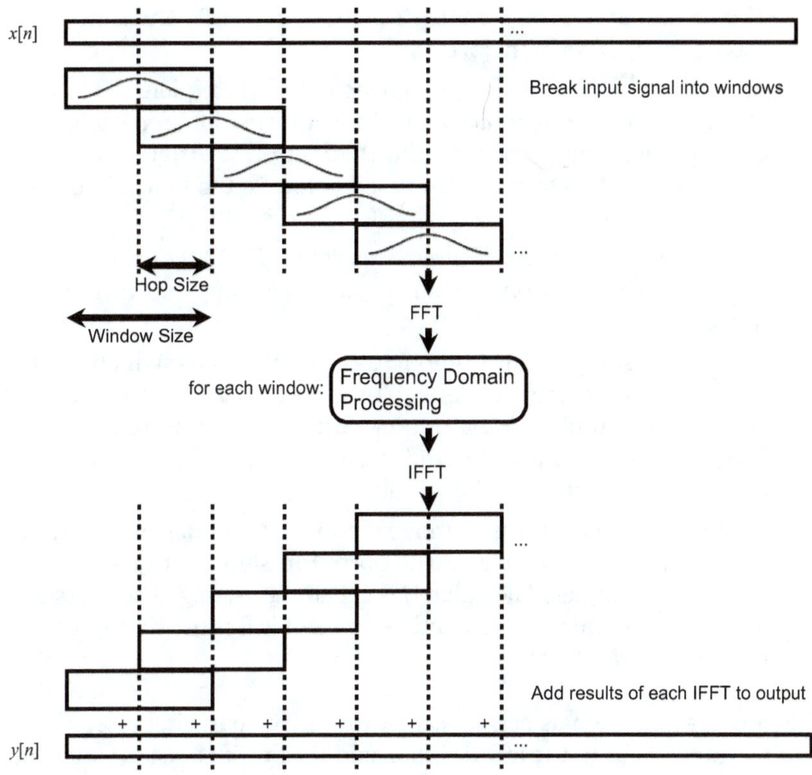

FIGURE 9.1
Overview of the overlap-add process for phase vocoder effects.

1. Given an input signal of arbitrary length, choose a *frame* of N consecutive samples. The value N is known as the *frame size* or *window size.*
2. Multiply the signal a *window function* of length N. The window function is defined to have a nonzero value for N consecutive samples and a value of zero everywhere else. By multiplying the input signal and the window function, *only* the N samples in the frame will remain; the rest will be set to zero.
3. After the signal has been *windowed*, apply a *fast Fourier transform* (FFT). Since the windowed signal contains N points, an FFT of size $M \geq N$ must be used. Typically, the window size and FFT size are identical; however, in some cases, shorter windows can be used with larger FFTs. In practice, the FFT size M is almost always a power of 2 for reasons of computational efficiency.

The combination of a window function and FFT constitutes the short-time Fourier transform.

4. The output of the FFT will be a collection of M frequency-domain *bins* containing *magnitude* and *phase* information for each frequency. Each phase vocoder effect will apply a different type of processing in this step, as discussed below in the Phase Vocoder Effects section.
5. Apply the *inverse fast Fourier transform* (IFFT) to the output of step 4. This will once again produce M samples in the time domain.
6. Add the M samples from step 5 to the *output buffer,* which holds the output signal from the effect. In some configurations, prior to adding the samples to the output buffer, a window function of length N is applied again in the same manner as Step 2, to reduce artefacts at the edges of the windows.
7. Move on to the next frame: move forward H samples in the input signal and return to step 1. The output in step 6 will also move forward H samples. The value H is called the *hop size*. The hop size is sometimes equal to the window size but is frequently a fraction of it (e.g., $H=N/2$ or $H=N/4$).

This process is known as *overlap-add;* it works by analyzing a set of overlapping frames within the input signal, and the output of each frame after processing, properly aligned in time, is added to form the output signal. The following sections discuss each aspect of the overlap-add process.

Windowing

It is important to note that the phase vocoder does not analyze the frequency content of the entire signal at once. Instead, the STFT analyzes only the frequency content present in a particular frame located at a specific time within the signal. A window function of length N will contain N nonzero samples starting from $n=0$, with a value of 0 at all other samples. The simplest type is the *rectangular* window:

$$w_{\text{rect}}[n] = \begin{cases} 1 & 0 \le n < N \\ 0 & |n| \ge N \end{cases}. \tag{9.1}$$

Many types of windows are possible. Other common variants are the *Bartlett* (triangular) window, the *Hann* window, and the *Hamming* window (Figure 9.2):

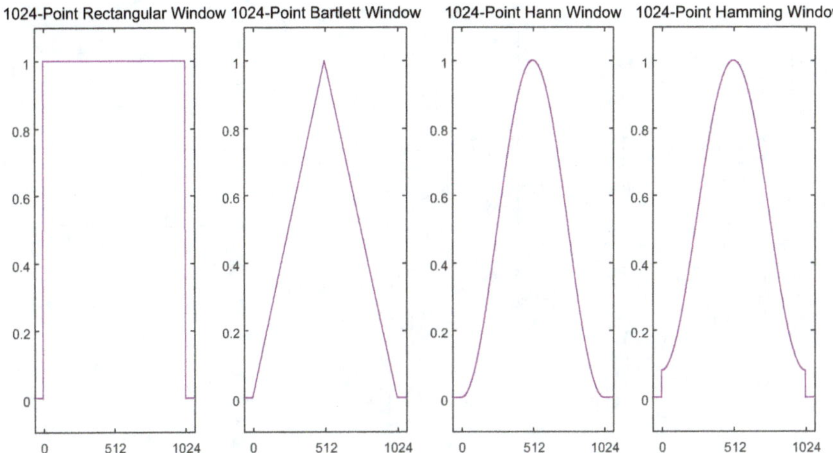

FIGURE 9.2
Four popular window functions.

$$w_{\text{bart}}[n] = \begin{cases} 1-\left|\dfrac{2n}{N-1}-1\right| & 0 \le n < N \\ 0 & |n| \ge N \end{cases}$$

$$w_{\text{hann}}[n] = \begin{cases} \left[1-\cos\left(\dfrac{2\pi n}{N-1}\right)\right]/2 & 0 \le n < N \\ 0 & |n| \ge N \end{cases} \qquad (9.2)$$

$$w_{\text{hamm}}[n] = \begin{cases} 0.54-0.46\cos\left(\dfrac{2\pi n}{N-1}\right) & 0 \le n < N \\ 0 & |n| \ge N \end{cases}$$

Intuitively, the motivation for choosing different types of windows relates to smoothing the edges of the windowed signal. As Figure 9.3 shows, the rectangular window cuts off abruptly at the start and end, and these discontinuities will create undesirable effects known as *sidelobes* in the frequency domain, where energy at one frequency will bleed into other frequency bins. The triangular, Hann and Hamming windows produce more gradual transitions at the edges, which reduce the sidelobe amplitude. However, they do so at the cost of reducing the precision with which any given frequency component can be identified. In signal processing, the trade-off is described as *main lobe width* versus *sidelobe height*; a more

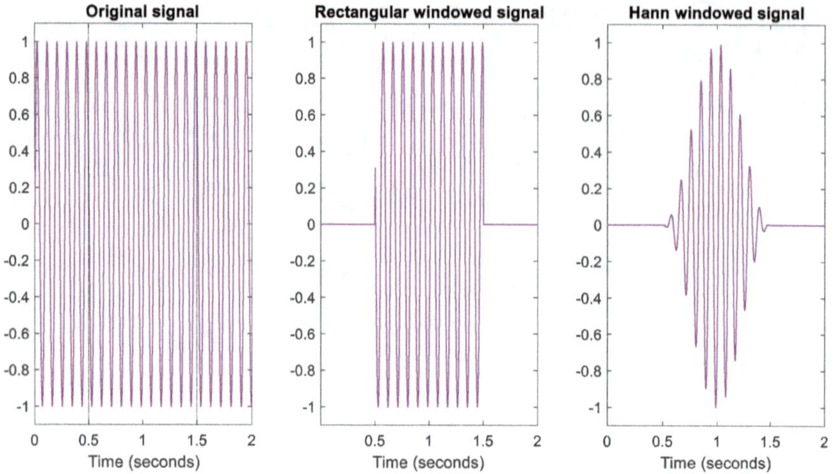

FIGURE 9.3
The effect of a rectangular window and a Hann window applied to a signal.

complete discussion can be found in [4]. Though there is no single best solution for all cases, in phase vocoder audio effects, Hann and Hamming windows are frequently used.

To apply a window function, the input signal is multiplied by the window. The window function can be offset in time to isolate different regions of the input signal:

$$x_{\text{window}}[m,n] = x[n]w[n-m], \tag{9.3}$$

where m is the offset of the window. The samples from m to $m+N-1$ will be isolated by the window function, with all other samples set to zero.

Analysis: Fast Fourier Transform

The *fast Fourier transform* (FFT) is an algorithm for calculating the *discrete Fourier transform* (DFT) of a signal. In turn, the DFT takes M time-domain points as input and produces M regularly-spaced frequency domain bins as output. Converting from the time domain to the frequency domain in this way is known as the *analysis* step of the phase vocoder (in contrast to the *synthesis* step, which goes from frequency to time domain). It is important that the size M of the FFT be at least as large as the length of the input signal. Since real-world audio signals can be of arbitrary length, this illustrates the importance of the window function in isolating only a small number of points for the FFT.

We can represent the combination of window and FFT (together, the short-time Fourier transform) as follows:

$$X[m,k] = \sum_{n=-\infty}^{\infty} x[n]h[m-n]e^{-j2\pi nk/M}, \tag{9.4}$$

where n, k represent the short time frame start and frequency bins, respectively. Here k ranges from 0 to $M-1$, representing evenly-spaced frequency bins between 0 and 2π.

Interpreting Frequency-Domain Data

Audio signals are always real numbers, but the values of the frequency bins are complex, containing a real and an imaginary component, $x + jy$. Both components are crucial to making sense of the frequency data, but their meaning is easier to understand in *polar* (or *magnitude-phase*) representation, $Ae^{j\phi}$. We can convert the polar representation and the *Cartesian* real-imaginary form,

$$\begin{aligned} &x + jy = Ae^{j\phi} \rightarrow \\ &A = |x + jy| = \sqrt{x^2 + y^2}; \phi = \arg(x + jy) = atan2(y, x) \end{aligned}. \tag{9.5}$$

where *atan2* is a version of the arctangent function, available in most programming languages, which produces values in the range $(-\pi, \pi]$ depending on the quadrant of x and y. So, the time frequency bins in Eq. can be written as:

$$X[m,k] = |X[m,k]|e^{j\phi[m,k]}. \tag{9.6}$$

One important use of phase information is to improve the resolution of frequency detection. The FFT produces only a fixed number of frequency bins, but signals may contain frequencies that fall between the bin frequencies. The following section shows how comparing phase from two consecutive frames can be used to recover exact frequencies between the bins.

Target Phase, Phase Deviation, and Instantaneous Frequency

For frequency bin k, a 'target phase', ϕ_t, can be calculated using the bin frequency and unwrapped bin phase of the previous hop. This target phase is the sum of the previous unwrapped phase and the expected phase increment. The expected increment is the frequency of the sinusoid multiplied by the hop size. So we have,

$$\phi_t[n,k] = \phi[n-1,k] + \omega_k h, \tag{9.7}$$

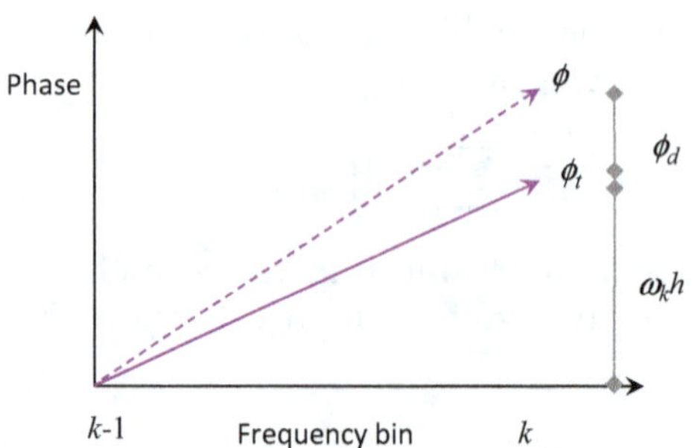

FIGURE 9.4
Relationship between actual phase and target phase. The instantaneous frequency is represented by the gradient of the dashed line, and the bin frequency is the gradient of the solid line.

where ω_k is the frequency of bin k, n is the hop number, and h is the hop size. This target phase represents the perfect case of a steady state sinusoid fitting exactly into a frequency bin, i.e., no spectral leakage. Since this is almost never the case, we have some phase deviation ϕ_d, as shown in Figure 9.4 and given by,

$$\varphi\phi_d[n,k] = \phi[n,k] - \phi_t[n,k]. \tag{9.8}$$

The corresponding value in the range (–π:π], known as the principal argument, of this deviation phase is then used to calculate the unwrapped phase increment per hop, $\Delta\phi_h$:

$$\Delta\phi_h[n,k] = \omega_k h + \arg\phi_d[n,k]. \tag{9.9}$$

The instantaneous frequency, f_i, can then be calculated:

$$f_i[n,k] = \frac{\Delta\phi_h[n,k]}{2\pi h} f_s, \tag{9.10}$$

where f_s is the sample frequency. This instantaneous frequency can then be viewed as a more accurate frequency measurement than the frequency resolution offered by the filterbank or STFT.

Estimating Frequency from Phase Deviation

Frequency is the derivative of phase, $\omega[n] = \Delta\phi/\Delta n$ where n here is measured in samples. This basic fact is the key to estimating exact frequencies from FFT bins. Estimating the frequency of a signal in an FFT bin always requires two consecutive FFTs (i.e., two frames or hops of the phase vocoder algorithm). Since we know how far apart the hops are in time, we can use the amount that the phase in a bin has changed from one hop to the next to estimate its frequency. First, we can write out a first-order difference equation:

$$\omega[n] = \frac{\phi[n]-\phi[n-1]}{n-(n-1)} = \phi[n]-\phi[n-1]. \tag{9.11}$$

Generalizing over a hop size of H samples, we get:

$$\omega[n] = \frac{\phi[n]-\phi[n-H]}{n-(n-H)} = \frac{\phi[n]-\phi[n-H]}{H}. \tag{9.12}$$

A caveat of the above equation is that the phases $\phi[n]$ must be unwrapped. Rather than being bounded within the range $(-\pi, \pi]$, the values should be strictly increasing from one hop to the next, so that $\omega[n]$ is always positive. Unwrapping phase can be accomplished by adding multiples of 2π to the phase $\phi[n]$ on the second hop. But how many multiples of 2π should we add for a given FFT bin?

It is easier to consider the problem the other way around. Inverting the above formula, we can calculate the expected phase shift between hops based on signal frequency $\phi[n]-\phi[n-H]=H\omega[n]$.

FFT bin k will have a center frequency of $2\pi k/N$. Therefore, a sinusoid exactly centered on the bin frequency would give the following expected phase shift:

$$\phi[n]-\phi[n-H] = \frac{2\pi kH}{N}. \tag{9.13}$$

Given measurements of $\phi[n]$ and $\phi[n-H]$ for our actual signal, we can subtract the expected phase shift above to give a phase remainder:

$$\phi_r[n] = \left(\phi[n]-\phi[n-H]\right) - \frac{2\pi kH}{N}. \tag{9.14}$$

The phase remainder tells us whether the frequency in bin k is higher or lower than the bin's center frequency. If $\phi_r[n]$ is positive, this indicates the phase advanced more than expected between hops, which indicates an actual frequency $\omega[n]$ that is higher than the center of the bin. If $\phi_r[n]$ is negative, it indicates a lower frequency than the center of the bin.

To estimate the actual frequency from the phase remainder, we wrap it to stay within the range $(-\pi, \pi]$ (also known as calculating the principal argument), and we add the bin center frequency $\omega_k=2\pi k/N$:

$$\omega[n] = \frac{\text{wrap}(\phi[n])}{H} + \omega_k. \tag{9.15}$$

This instantaneous frequency $\omega[n]=2\pi f[n]$ is more precise than the center frequency of the bin, though the ideal case assumes a signal consisting of a signal sinusoid: in practice, spectral leakage from adjacent bins will affect the accuracy of the frequency estimate.

Later in this chapter, we will see how pitch shifting effects can work by inverting this process, modifying the frequency of a signal by changing the phase of each bin from one hop to the next.

Synthesis: Inverse Fast Fourier Transform

The specific processing done in the frequency domain depends on the particular phase vocoder effect. Once this processing is complete, however, all phase vocoder effects will convert the signal back to the time domain using the *inverse fast Fourier transform* (IFFT):

$$x[m]=\frac{1}{M}\sum_{k=0}^{M-1}\left|X[m,k]\right|e^{j(2\pi km/M+\phi[m,k])}. \tag{9.16}$$

This process is known as *synthesis* or *reconstruction*. It produces M points in the time domain, which can be added to the output buffer. Depending on the type of frequency-domain processing performed, it is possible that the output frame may not have smooth edges even if a Hamming window (or similar) was used for analysis. Occasionally then, applying a second window before adding the result to the output buffer will help reduce audible artifacts.

Overlap-Add

The purpose of the phase vocoder process is to analyze the frequency content of each short frame, modify it, and reconstruct it in the time domain. The manner in which the frames advance from one to the next is important to the proper operation of phase vocoder. Once a frame has been analyzed, processed, and synthesized, the effect should advance by the *hop size*. Suppose the hop size is given by H. Then if the window in frame i previously began at sample m, the window in frame $i+1$ will begin at sample $m+H$. The hop size is always less than or equal to the window size; if it were greater than the window size, there would be a gap between successive windows. Smaller hop sizes (more overlap) will often produce better-quality output in many phase vocoder effects; however, they require more computation.

Hop sizes of one-half, one-fourth, or one-eighth the window size are common. A useful guideline for choosing hop size is the *constant overlap-add* or *COLA criterion*. Suppose that in the steps listed above for the phase vocoder theory overview, no processing at all is done in the frequency domain (step 4), and that instead, the FFT is immediately followed by the

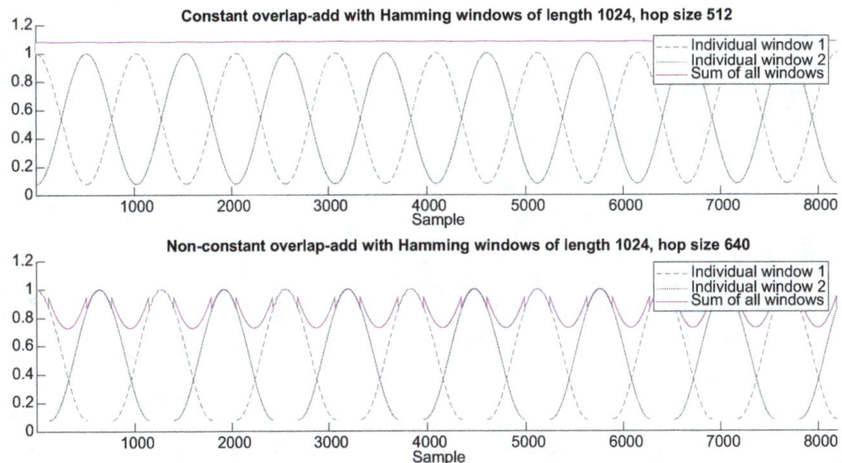

FIGURE 9.5
The constant overlap-add criterion requires that the window functions, when overlapped, add to a constant value. It holds in the top plot, but not the bottom plot.

inverse FFT. For windows and hop sizes meeting the COLA criterion, the output signal at the end of the overlap-add process will be identical to the input signal, with no distortion or modulation introduced by the windowing process. Essentially, this requires that the window functions, when overlapped, add to a constant value, as depicted in Figure 9.5. Rectangular windows meet the COLA criterion for a hop size equal to the window size; Bartlett, Hann, and Hamming windows require a hop size of at most *half* the window size. Integer divisions of this maximum hop size are also possible: for example, a hop size of 1/8 the window size will meet the COLA criterion for all four windows.

To summarize, it is important to choose a window and hop size such that the windowing process itself does not introduce unnecessary artifacts into the output. The exact choice of window type, length, and hop size depends in large part on the specific effect, several of which are discussed below.

Code Example

The following C++ code fragment is adapted from the materials that accompany this book. This example implements the basic structure of an overlap-add phase vocoder system. JUCE classes are used for audio buffers and FFT calculations, the details of which are not included here.

```
// Persistent state across each audio callback
// Sample buffers (mono in this example) for input to FFT and output of
// IFFT
```

```
AudioSampleBuffer inputBuffer, outputBuffer;
// Location of pointers in circular buffers for input and output
unsigned int inputWritePointer, outputReadPointer, outputWritePointer;
// How many samples have gone by since the last hop
unsigned int hopCounter;
// FFT (window) and hop sizes, and scale factor depending on their overlap
unsigned int fftSize, hopSize;
float windowScaleFactor;
// Variables for calculating FFTs in JUCE
std::unique_ptr<dsp::FFT> fft;
HeapBlock<float> fftWindow;
HeapBlock<dsp::Complex<float>> timeDomtainBuffer;
HeapBlock<dsp::Complex<float>> frequencyDomainBuffer;

// Process an overlap-add phase vocoder step for a single channel of
// audio data
// of length numSamples, calling FFT and IFFT when a hop size has
// elapsed
void overlapAdd(float *audioData, unsigned int numSamples)
{
  for(unsigned int n = 0; n < numSamples; n++) {
    // Write input sample into input buffer
    const float inputSample = audioData[n];
    inputBuffer.setSample(0, inputWritePointer, inputSample);
    if(++inputWritePointer >= inputBuffer.getNumSamples())
    inputWritePointer = 0;

    // Take audio output from output buffer (holding previous IFFTs)
    audioData[n] = outputBuffer.getSample(0, outputReadPointer);
    // Clear output sample after we read it, in preparation for future
    // windows
    outputBuffer.setSample(0, outputReadPointer, 0);
    if(++outputReadPointer >= outputBuffer.getNumSamples())
  outputReadPointer = 0;

    // Execute an FFT / IFFT pair once per hop
    if(++hopCounter >= hopSize) {
      hopCounter = 0;

      // Pre-analysis step: unwrap circular buffer, apply window
      int inputBufferIndex = inputWritePointer;
      for(unsigned int index = 0; index < fftSize; ++index) {
        // Time domain data is always real
        timeDomainBuffer[index].real(fftWindow[index]
                                * inputBuffer.getSample(0,
      inputBufferIndex));
        timeDomainBuffer[index].imag(0.0f);
        // Wrap around circular buffer
        if(++inputBufferIndex >= inputBuffer.getNumSamples())
        inputBufferIndex = 0;
      }

      // Analysis: calculate FFT
      fft->perform(timeDomainBuffer, frequencyDomainBuffer, false);

      modification(); // What happens in frequency domain is
                      // effect-specific

      // Synthesis: calculate IFFT
      fft->perform(frequencyDomainBuffer, timeDomainBuffer, true);
```

```
      // Post-synthesis step: add this IFFT to others in the buffer
      // Notice: output buffer has separate read and write pointers
      int outputBufferIndex = outputWritePointer;
      for(unsigned int index = 0; index < fftSize; ++index) {
        float outputSample = outputBuffer.getSample(0,
  outputBufferIndex);
        outputSample += timeDomainBuffer[index].real() *
  windowScaleFactor;
        outputBuffer.setSample(0, outputBufferIndex, outputSample);
        if(++outputBufferIndex >= outputBuffer.getNumSamples())
  outputBufferIndex = 0;
      }

      // Increment write pointer and keep in range
      outputWritePointer += hopSize;
      if(outputWritePointer >= outputBuffer.getNumSamples())
        outputWritePointer -= outputBuffer.getNumSamples();
    }
  }
}
```

This code fragment does not include the initialization, which includes several important steps such as calculating the window, calculating windowScaleFactor, setting initial values for the read and write pointers, and preparing the FFT library. A complete implementation might also use multichannel buffers to process stereo or multichannel audio signals, where a further for() loop would iterate over the number of channels.

Phase Vocoder Effects

The phase vocoder and its underlying theory have a wide range of applications. In the fields of machine listening and music informatics, these include signal content analysis, pitch detection, and steady-state/transient separation. In the domain of audio effects, important applications include *time-scaling* and *pitch-shifting*. Normally, when an audio file is played at a higher sample rate than it was recorded at, faster playback rate and higher pitch go together. The phase vocoder allows playback speed and pitch to be varied independently of one another. Other common phase vocoder effects include *robotization* and *whisperization* voice effects.

Robotization

The *robotization* effect is most commonly used on voice signals. It applies a constant pitch to the signal while preserving the vocal formants that determine vowel and consonant sounds, resulting in a robot-like monotone voice that is nonetheless very intelligible.

Within the context of the overlap-add phase vocoder procedure described in Phase vocoder theory, Analysis: Fast Fourier Transform, the robotization effect itself is very simple. Following the FFT at each frame, the phase of every frequency bin is set to zero, while the magnitude is left unchanged. Setting the phase to zero is equivalent to making each frequency bin a real number (imaginary component set to 0). Suppose the value of frequency bin k is a_k+jb_k. Then the output value following robotization will be $\sqrt{a_k^2+b_k^2}$.

This procedure should be repeated identically for each bin of each frame. By preserving the magnitude, the overall shape of the spectrum remains the same, preserving the vocal formants. But by regularizing the phase information, each frequency component will effectively restart from zero phase on each hop rather than connecting smoothly from one hop to the next. This causes a constant audible pitch which depends on the *hop size*. In general, the pitch of the robot voice can be determined by:

$$f_{\text{robot}} = f_s/H, \tag{9.17}$$

where f_s is the sample rate and H is the hop size in samples. The sound of the robotization effect also depends on the *window size*. In general, moderately sized windows (roughly 256 to 1,024 samples) produce the most striking effect. Very small windows reduce the clarity of the output, where longer windows attenuate the robot-like quality by passing through more of the signal's original pitch. Intuitively, we can understand this by noting that phase information is used to resolve small frequency differences between bins. If there are a large number of bins, each pitch will be closely determined even without phase information, but if fewer bins are used, then regularizing the phase will better obscure the original frequencies.

Note that for both robotization and whisperization, the *constant overlap-add criterion* for choosing window and hop size is not an important consideration, since both use deliberate artifacts of the phase vocoder analysis and resynthesis process to achieve their signature sounds. Therefore, the hop size H will usually not be an integer division of the window size as in other phase vocoder effects but rather will be chosen based on the desired pitch of the robot voice.

The quality of the robotization effect is often improved by multiplying the output of each IFFT by a window function as described in Step 6 of the Overview section of this chapter.

Robotization Code Example

Given a system for performing overlap-add calculations with FFTs and inverse FFTs, described earlier in the chapter, the robotization effect itself is very simple: set the phase of each bin to 0 while keeping its magnitude.

```
// Iterate over all the FFT bins
for(int index = 0; index < fftSize; ++index) {
    // Get magnitude of the bin
    float magnitude = std::abs(frequencyDomainBuffer[index]);

    // Set phase to 0 (all real, no imaginary component)
    frequencyDomainBuffer[index].real(magnitude);
    frequencyDomainBuffer[index].imag(0);
}
```

This example relies on the variable definitions from the overlap-add code example. Each FFT bin is of type `std::complex<float>`, a class which provides methods for measuring magnitude and phase and accessing real and imaginary components, amongst other things. This code example does not include a second synthesis window after performing the IFFT.

Whisperization

The *whisperization* effect, like robotization, is also most commonly used on voice signals. It maintains the vocal formants while completely eliminating any sense of pitch. The resulting effect sounds similar to a person whispering: the contents of speech remain clear while any sense of voicing is lost.

Whisperization is implemented similarly to robotization, but instead of setting the phase of each frequency bin to zero at every frame, the phase is set to a random value in the range $\phi = [0, 2\pi)$. Assuming the original contents of bin k can be written $a_k + i^* b_k$ and ϕ_k holds the random phase value for bin k, the output is:

$$\sqrt{a_k^{\ 2} + b_k^{\ 2}}\left(\cos\phi_k + j\sin\phi_k\right). \tag{9.18}$$

Notice that a different random value is used for each frequency bin, each frame. As with robotization, the magnitude of each bin is preserved, which maintains the overall shape of the spectrum. But scrambling the phase erases any sense of periodicity from one frame to the next.

The most important parameter determining the whisperization effect is *window size*. In this case, shorter windows (e.g., 64 to 256 samples) are typically most effective at eliminating any pitch content. If the window is too short, the effect becomes unintelligible, but for longer windows, the original pitch of the signal remains apparent. Hop size is a less important parameter that can be adjusted experimentally, though generally it should be no longer than the window length or other artifacts will result. As with robotization, applying a window to each IFFT output may improve the quality of the effect.

Whisperization Code Example

To adapt the previous robotization code example to instead perform whisperization, the basic overlap-add structure stays the same and we only need to change the lines following the FFT calculation:

```
// Iterate over the first half of the FFT bins
for(int index = 0; index < fftSize / 2 + 1; ++index) {
    // Get magnitude of the bin
    float magnitude = std::abs(frequencyDomainBuffer[index]);
    // Calculate a random phase
    float phase = 2.0f * M_PI * (float)rand() / (float)RAND_MAX;

    // Calculate real and imaginary from magnitude and phase
    frequencyDomainBuffer[index].real(magnitude*cosf(phase));
    frequencyDomainBuffer[index].imag(magnitude*sinf(phase));

    // Spectrum is conjugate symmetric; calculate the upper half here
    if (index > 0 && index < fftSize / 2) {
        frequencyDomainBuffer[fftSize - index].
    real(magnitude*cosf(phase));
        frequencyDomainBuffer[fftSize - index].
    imag(magnitude*sinf(-phase));
    }
}
```

This code calculates the amplitude and phase of each bin based on the real and imaginary components. However, instead of using the phase of the original signal, the phase is replaced with a random number between 0 and 2π. Amplitude and phase are converted back to real and imaginary values using `sinf` and `cosf`.

Another change between robotization and whisperization code occurs in the first line of the `for()` loop. Notice that in this example, the loop only goes through the first half of the bins. This is because the FFT of a real-valued signal is always *conjugate-symmetric*: $F(k)=F(N-k)^*$ where k is the bin and N is the total transform size. If we want the output to remain as a real signal, we need to maintain conjugate symmetry even as we randomize the phase. The code inside the last `if()` statement fills in the second half of the frequency bins using the complex conjugates of the first half.

Time Scaling

The phase vocoder allows a signal to be rescaled in time without any change in pitch. To perform time scaling, the basic idea is to take advantage of the basic relationship between time, frequency, and phase, $\phi = \omega t$. Looking at this equation, we can see that frequency can be preserved, and time can be varied, at a cost of the original analysis phase being lost. Despite this loss of phase preservation, results are surprisingly good for many phase vocoder applications, especially when a short analysis hop size, such as 1/8 of the window length, is used.

When time scaling using the phase vocoder, the *analysis* stage (steps 1–3 in Phase vocoder theory, Overview) is performed identically to any other phase vocoder application. The *synthesis* stage (steps 5–7) is then varied to facilitate time compression and expansion. The analysis and synthesis hop sizes are no longer equal.

If R is the stretching ratio (e.g., for 20% expansion, R equals 1.2) applied to the signal, the synthesis hop size, h_s, is related to the analysis hop size, h_a, by $h_s = Rh_a$.

Recall that the phase vocoder analysis offers a method for calculation of the instantaneous frequency using phase information. This instantaneous frequency is proportional to the phase increment over a single analysis hop. This phase increment is the parameter that is used in this time scaling application. Consider a single frequency bin k, representing a sinusoidal track over time. Dividing Eq. (9.9) by the analysis hop size h_s, and making an analogous calculation for amplitude, we see that the phase and amplitude increment per sample within hop n are given by:

$$\Delta\phi_i(n,k) = \frac{\omega_k h_a + \arg[\phi_d(n,k)]}{h_a}$$
$$\Delta A_i(n,k) = \frac{A_i(n,k) - A_i(n-1,k)}{h_a}. \tag{9.19}$$

Refer to Eqs (9.7)–(9.9) for explanation of the terms.

Time Scaling Resynthesis

These values may now be used in the resynthesis stage. For the same sinusoidal track, the phase could be calculated at each sample m using:

$$\theta(m,k) = \theta(m-1,k) + \Delta\phi_i(n,k). \tag{9.20}$$

However, we do not need to perform this calculation at every sample. We only need to know the total phase increment after the number of samples in the *synthesis* hop size h_s. This leads to a difference in phase by the end of the hop (recall the phase deviation mentioned earlier), as long as the synthesis hop size differs from that of the analysis hop size, which is a definition of time scaling. In the case of amplitude, the analysis increment cannot be used in the same way, as the same cumulative errors applied to amplitude would lead to severe artifacts compared to phase.

The amplitude increment is calculated instead using the synthesis hop size, to maintain identical analysis and synthesis amplitudes at both the beginning and end of the hop.

$$\Delta A_{is}(n,k) = \frac{A_i(n,k) - A_i(n-1,k)}{h_s}. \tag{9.21}$$

The amplitude can then be calculated for the duration of the synthesis hop size using:

$$A_k(m,k) = A_k(m-1,k) + \Delta A_{ks}(n,k). \tag{9.22}$$

These sinusoidal components are then summed to synthesize the time-scaled audio:

$$y(x) = \sum_{k=0}^{N/2} A(m,k)\cos(\phi(m,k)). \tag{9.23}$$

Clear problems arise from time scaling using a spectral method such as this, discussed in a later section. Also, by definition, time-scaling cannot be a real-time audio effect except for a limited period of time. If the output is compressed in time compared to the input (i.e., plays faster), then at some point the effect would become *non-causal* since the output would depend on future input samples. On the other hand, if the output is stretched in time (plays slower), the difference in time between input and output will steadily accumulate and an ever-increasing amount of memory will be needed to buffer the audio waiting to play. However, the converse case of pitch-shifting without time-scaling, discussed in the following section, is a common real-time audio effect.

Pitch Shifting

Pitch Shifting by Time Scaling and Resampling

There are several ways to shift the pitch of a signal without changing its speed using the phase vocoder. One of the most straightforward approaches uses the time-scaling algorithm presented in the previous section. Suppose we apply a time stretch factor of R; then every block of N input samples will produce $R{\cdot}N$ output samples after time-scaling. If we then changed the sample rate to play R times faster, then the output would have the same speed as the input but with a pitch R times higher.

In practice, we don't actually change the sample rate at the output, but we achieve a similar effect using interpolation to fit $R{\cdot}N$ samples in the space of N. If $x[n]$, $0<n<R{\cdot}N-1$, represents the output of a single frame of the phase vocoder, then we can calculate the pitch-shifted output using linear interpolation:

$$y[n] = \left(1+\lfloor Rn\rfloor - Rn\right)x\left[\lfloor Rn\rfloor\right] + \left(Rn-\lfloor Rn\rfloor\right)x\left[\lfloor Rn\rfloor+1\right]. \tag{9.24}$$

This interpolation is performed for each frame of the phase vocoder output. With time-scaling, the synthesis hop size was $R{\cdot}h$ samples, but in the pitch-shifting, each of the interpolated buffers advances by the analysis hop size h, ensuring that input and output signals remain at the same speed despite the pitch shift.

To summarize, raising the pitch of the output requires first stretching the signal in time ($R > 1$), then compressing the longer output buffer with interpolation to fit in the original length. Lowering the pitch requires compressing the signal in time, producing fewer output samples, which are then stretched with interpolation to fit the original length.

Pitch Shifting Without Resampling

It is also possible to implement a pitch shifter by directly manipulating the phase of each FFT bin, without the time scaling and resampling in the previous approach. Both approaches are based on the sample fundamental relationships between frequency and phase described earlier in this chapter.

Equation (9.15) estimated the frequency of a signal in an FFT bin from the phase remainder between two successive analysis hops. The same relationship can be expressed in fractional FFT "bins" rather than frequency, using the relationship between bin number k and frequency $\omega_k = 2\pi k/N$:

$$b[n] = \frac{\mathrm{wrap}(\phi_r[n])N}{2\pi H} + k. \tag{9.25}$$

The formulation in terms of fractional bins can be more convenient to implement in code, so we will use it for the remainder of this section.

Pitch shifting involves multiplying every frequency component of the signal by a value R. Here, $R > 1$ will raise the pitch of the signal, while $R < 1$ will lower it. Suppose that for each bin of the FFT, we calculated the frequency using the above formula to produce an analysis frequency $b_a[n]$. From there, we can calculate a set of synthesis frequencies $b_s[n] = Rb_a[n]$ representing the pitch-shifted components of the signal.

Because of the pitch shifting by R, the synthesis frequencies $b_s[n]$ will not necessarily occupy the same FFT bins as their corresponding analysis frequencies $b_a[n]$. Therefore, the next step is to identify which (integer) FFT bin each synthesis frequency belongs to. We can do this by rounding $b_s[n]$ to the nearest integer to get the new bin k':

$$k' = \lfloor Rk + 0.5 \rfloor. \tag{9.26}$$

Next, we need to calculate a new phase remainder for bin k'; this corresponds to the amount by which the phase should advance between hops to generate the precise frequency $R\omega[n]$ that we want to synthesize.

$$\phi_{rs}[n] = \frac{2\pi H}{N}\left(b_s[n] - k'\right). \tag{9.27}$$

Intuitively, the meaning of $\phi_{rs}[n]$ matches what we saw earlier in the chapter ("Interpreting Frequency-Domain Data"). Positive values indicate a frequency higher than the bin center frequency $2\pi k'/N$, and negative values indicate a lower frequency than the bin center frequency.

We then convert the phase remainder into the final phase value for this hop by adding the phase calculated at the last hop, the phase remainder, and the expected amount that the phase would advance between hops, given the bin center frequency:

$$\phi_s[n] = \text{wrap}\left(\phi_s[n-H] + \phi_{rs}[n] + \frac{2\pi k'H}{N}\right). \tag{9.28}$$

Notice that implementing this equation requires that we remember the phase of each bin on the previous hop. This is generally a requirement for any implementation of a phase vocoder pitch-shifting effect.

We now have the phase $\phi_{rs}[n]$ for bin k'. The magnitude of this bin can be copied from the corresponding analysis bin k. Then, the magnitude and phase can be converted to real and imaginary components in the usual way. The last remaining consideration is what to do with duplicated or unused bins: when $R > 1$, there may be the synthesis bins $b_s[n]$ that are more widely spaced than the analysis bins $b_a[n]$, and some integer bins of the FFT may be skipped in synthesis. The magnitude of these bins should be set to 0 before performing the IFFT. When $R < 1$, then two analysis bins might map to the same (integer) synthesis bin. Various approaches could be used to resolve the duplication, but for a basic effect, it suffices just to choose one or the other of the analysis frequencies and ignore the other.

Code Example

The following C++ code fragment uses much of the same phase vocoder structure as the robotization and whisperization effects, but instead performs real-time pitch shifting. Some initial code, which is identical to the robotization example, has been omitted.

```
// Persistent state across each audio callback
std::vector<float> lastInputPhases(fftSize / 2 + 1);
//Phases from previous hop analysis
std::vector<float> lastOutputPhases(fftSize / 2 + 1);
// ...and synthesis

// These containers hold the converted representation from
// magnitude-phase
// into magnitude-frequency, used for pitch shifting
std::vector<float> analysisMagnitudes(fftSize / 2 + 1);
std::vector<float> analysisFrequencies(fftSize / 2 + 1);
std::vector<float> synthesisMagnitudes(fftSize / 2 + 1);
std::vector<float> synthesisFrequencies(fftSize / 2 + 1);
```

```
float pitchShift; // Ratio of pitch shift: >1 raises the pitch,
                  // <1 lowers it

// ------

// Analysis step: iterate over the first half of the FFT bins
for(int index = 0; index < fftSize / 2 + 1; ++index) {
    // Turn real and imaginary components into amplitude and phase
    float amplitude = std::abs(frequencyDomainBuffer[index]);
    float phase = std::arg(frequencyDomainBuffer[index]);

    // Calculate the phase difference in this bin between the last
    // hop and this one, which will indirectly give us the exact
    // frequency
    float phaseDiff = phase - lastInputPhases[index];

    // Subtract the amount of phase increment we'd expect to see based
    // on the centre frequency of this bin (2*pi*n/gFftSize) for this
    // hop size, then wrap to the range -pi to pi
    float binCentreFrequency = 2.0 * M_PI * (float)index / (float)
  fftSize;
    phaseDiff = wrapPhase(phaseDiff - binCentreFrequency * hopSize);

    // Find deviation in (fractional) number of bins from the centre
    // frequency
    float binDeviation = phaseDiff * (float)fftSize / (float)hopSize /
  (2.0 * M_PI);

    // Add the original bin number to get the fractional bin where this
    // partial belongs
    analysisFrequencies[index] = (float)index + binDeviation

    // Save the magnitude for later and for the GUI
    analysisMagnitudes[index] = amplitude;

    // Save the phase for next hop
    lastInputPhases[index] = phase;
}

// Zero out the synthesis bins, ready for new data
for(int index = 0; index < fftSize / 2 + 1; ++index) {
    synthesisMagnitudes[index] = synthesisFrequencies[index] = 0;
}

// Handle the pitch shift, storing frequencies into new bins
for(int index = 0; index < fftSize / 2 + 1; ++index) {
    // Find the nearest bin to the shifted frequency
    int newBin = floorf(index * pitchShift + 0.5);

    // Ignore any bins that have shifted above Nyquist
    if(newBin <= fftSize / 2) {
        synthesisMagnitudes[newBin] += analysisMagnitudes[index];
        synthesisFrequencies[newBin] = analysisFrequencies[index] *
  pitchShift;
    }
}
```

```
// Synthesise frequencies into new magnitude and phase values for FFT
// bins
for(int index = 0; index < fftSize / 2 + 1; ++index) {
    float amplitude = synthesisMagnitudes[index];

    // Get the fractional offset from the bin centre frequency
    float binDeviation = synthesisFrequencies[index] - index;

    // Multiply to get back to a phase value
    float phaseDiff = binDeviation * 2.0 * M_PI * (float)hopSize /
   (float)fftSize;

    // Add the expected phase increment based on the bin centre
    // frequency
    float binCentreFrequency = 2.0 * M_PI * (float)index / (float)
   fftSize;
    phaseDiff += binCentreFrequency * hopSize;

    // Advance the phase from the previous hop
    float outPhase = wrapPhase(lastOutputPhases[index] + phaseDiff);

    // Calculate real and imaginary from magnitude and phase
    frequencyDomainBuffer[index].real(amplitude*cosf(outPhase));
    frequencyDomainBuffer[index].imag(amplitude*sinf(outPhase));

    // Also store the complex conjugate in the upper half of the
    // spectrum
    if(index > 0 && index < fftSize / 2) {
        frequencyDomainBuffer[index].real(amplitude*cosf(outPhase));
        frequencyDomainBuffer[index].imag(amplitude*sinf(-outPhase));
    }

    // Save the phase for the next hop.
    lastOutputPhases[index] = outPhase;
}
```

After performing the FFT, each bin is given in magnitude-phase representation. At that point, the phase is updated based on Eq. (9.19) , so the windows can be overlapped at a different synthesis hop size. Normally, making analysis and synthesis hop sizes different would result in a time stretch. However, the output of the inverse FFT is resampled and then overlapped at the original (analysis) hop size. This results in a pitch shift without changing the timing of the signal.

Phase Vocoder Artifacts

Amplitude estimation is not easily achieved with linear interpolation, especially when using a long synthesis window. Consider the idea of a piano piece being played more slowly: the attacks would have much the same temporal envelope, but steady state and decay regions would be longer before the start of the next note although the decay *rate* would be the same. Linear interpolation implies that all regions of the envelope are stretched equally.

Also, problems arise in phase relationships, as the phase error is cumulative. If phase coherence is lost at transients (note attacks), transient smearing occurs, which softens the attacks and loses the natural sound of these regions. Loss of phase coherence across harmonic partials of the same note can lead to loss of natural sounding steady-state regions.

Finally, a 'phasiness' is introduced, which adds a reverb-like quality to the sound. One reason for this 'phasiness' is the loss of coherence of phase within the same sinusoidal component. If the phase is not maintained across all components that correspond to a windowed sinusoid, although frequency is maintained, severe amplitude distortions can occur. This produces a 'robot-like' effect when applied to signals such as voice.

Further problems arise when chirp-like signals are applied to phase vocoders. When the signal crosses over to the next frequency bin, it takes the phase of the new frequency track, leading to discontinuities. This is the kind of problem that can be solved by using higher level peak picking and sinusoidal tracking.

Further Reading

A few classic papers were pivotal in establishing the phase vocoder as an effective tool for audio signal processing, such as [94–96], and two quite different review articles appeared in 2013 [98,99]. New phase vocoder-based techniques for pitch shifting and time scaling continue to be proposed, though there is also new work on identifying when and how much these techniques have been applied [100]. However, Auto-tune's popularity and the drastic impact that pitch correction technologies have had on pop music have resulted in a large amount of discussion on their impact and its implications. The subject is nicely reviewed in the book Rise of Auto-Tune [101].

Problems

1. Write pseudo-code to perform a short-time-Fourier transform on a sampled signal. You can call another function to perform the FFT.
2. Consider the block diagram of the phase vocoder given in Figure 9.1. If a Hamming window of 1,024 samples is used, what is the maximum hop size that will allow perfect reconstruction of the signal? Why?

3. Suppose the sampling frequency is 32.768 kHz and the FFT size is 1,024 points. What is the width of one frequency bin (i.e. how far apart in frequency are two adjacent bins)?
4. We can perfectly reconstruct any signal under the right conditions, but the FFT has limited frequency resolution. Suppose the input is a sine wave that doesn't exactly align with any bin frequency. Explain how phase information can be used to determine the exact frequency.
5. Explain the operation of the robotization effect, and explain what parameter determines the frequency of the robot voice.
6. Explain the operation of the whisperization effect. What is the restriction on the window length and why is it necessary?
7. Explain the steps involved in a pitch shifter effect (changing pitch without changing speed).

10

Reverberation

Reverberation (*reverb*) is one of the most often used effects in audio production. In this chapter, we will look at the causes, the main characteristics, and the measures of reverb. We then focus on how to simulate reverb. First we describe algorithmic approaches to generating reverb, focusing on two classic designs. Then we look at a popular technique for generating a room impulse response, the image source method. Next, we describe convolutional reverb, which adds reverb to a signal by convolving that signal with a room impulse response, either recorded or simulated (like from using the image source method). This chapter concludes by looking at implementation details and applications.

Theory

In a room, or any acoustic environment, there is a direct path from any sound source to a listener, but sound waves also take longer paths by reflecting off the walls, the ceiling or objects, before they arrive at the listener, as shown in Figure 10.1. These reflected sound waves travel a longer distance than the direct sound and are partly absorbed by the surfaces, so they take longer to arrive and are weaker than the direct sound.

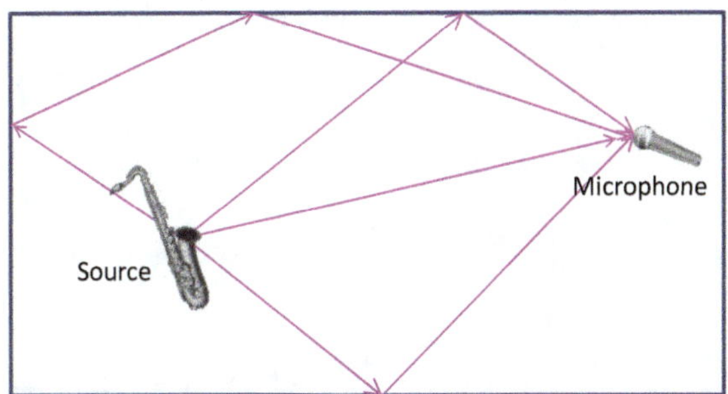

FIGURE 10.1
Reverb is the result of sound waves traveling many different paths from a source to a listener.

DOI: 10.1201/9781003593942-10

These sound waves can also reflect off multiple surfaces before they arrive at the listener. These delayed and attenuated copies of the original sound are what we call *reverberation,* which is essential to the perception of spaciousness in the sounds.

Reverberation is more than just a series of echoes. An echo is the result of a distinct, delayed version of a sound, as could be heard with a delay of at least 40 ms. With reverberation from a typical room, there are many, many reflections, and the early reflections arrive on a much shorter time scale. So these reflections are not perceived as distinct from the sound source. Instead, we perceive the effect of the combination of all the reflections.

Reverberation is also more than a simple delay device with feedback. With reverb, the rate at which the reflections arrive will change over time, as opposed to just simulating reflections that have a fixed time interval between them. In reverberation, there are a set of reflections that occur shortly after the direct sound. These *early reflections* are related to the position of the source and listener in the room, as well as the room's shape, size, and material composition. The later reflections arrive much more frequently, appear more random, usually decay exponentially, and are difficult to directly relate to the physical characteristics of the room. These *late reflections* give rise to *diffuse reverberation.* An example impulse response for a room is depicted in Figure 10.2. Each vertical line marks when the original sound is repeated, and the height of each of these lines is the amplitude of the sound at that time.

A measure that is often used to characterize the reverb in an acoustic space is the *reverberation time,* often denoted RT_{60}. This is the time that it takes for sound pressure level or intensity to decay by 60 dB, i.e., 1/1,000,000th of its original intensity, or 1/1,000th of its original amplitude (see Chapter 1). A long reverberation time implies that the reflections remain strong for a long time before their energy is absorbed. The reverberation time is

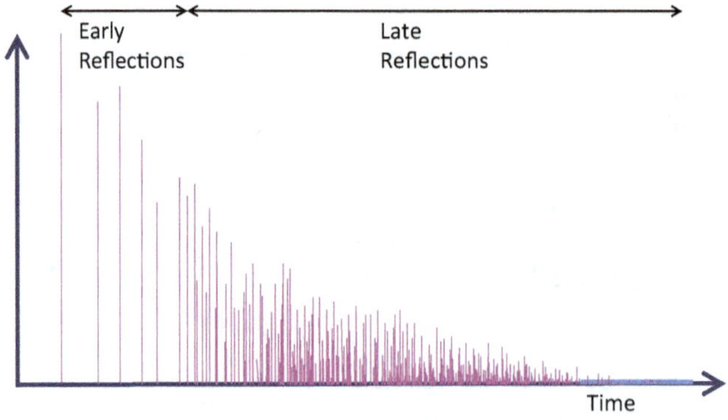

FIGURE 10.2
Impulse response of a room.

associated with room size. Small rooms tend to have reverb times on the order of hundreds of milliseconds, though this can vary greatly depending on the acoustic treatment and other factors. Larger rooms will usually have longer reverberation times since, on average, the sound waves will travel a larger distance between reflections. Concert halls typically have reverberation times of around 1.5 to 2 s. Cathedrals and other highly reverberant environments may have reverb times of more than 3 s.

It is possible to construct a large room with short reverberation time, and vice versa. The reverberation time is dictated primarily by the size of the room and the surfaces in the room. The surfaces determine how much energy is lost or absorbed each time the sound is reflected. Highly reflective materials, such as a concrete or tile floor, will result in a long reverb time. Absorptive materials, such as curtains, cushions or heavy carpet, will reduce the reverberation time. People and their clothing absorb a lot of sound. This explains why a room may sound 'bigger' during a soundcheck prior to a performance, but smaller and more intimate once the audience has arrived. The absorptivity of most materials usually varies with frequency, which is one reason why the reverb time is dependent on the spectral content of the source signal (formal measurement of reverb time is performed using an impulse or turning off a noise generator). The air in the room will also attenuate the sound waves, reducing the reverberation time. This attenuation is a function of temperature and humidity and is most significant for high frequencies. Because of this, many audio effect implementations of reverb will include some form of low pass filtering, as will be discussed later in this chapter.

Another important measure of reverberation is the *echo density*, defined as the frequency of peaks (number of echoes per second) at a certain time t during the impulse response. The more tightly packed together the reflections are, the higher the echo density. If the echo density is larger than 20–30 echoes per second, the ear no longer hears the echoes as separate events, but fuses them into a sensation of continuous decay. In other words, the early reflections become a late reverberation.

Sabine and Norris-Eyring Equations

The *mean free path* of a room gives the average distance a sound wave will travel before it hits a surface. Assuming a rectangular box room, this is given by $4V/S$ where V is the volume and S is the surface area of the room. Dividing this by the speed of sound c, the mean time τ until a sound source hits a wall is $\tau = 4V/(cS)$. So the mean number of reflections over a time t is,

$$n(t) = t/\tau = t\frac{cS}{4V}. \tag{10.1}$$

At each reflection off a surface S_i, α_i is the proportion of the energy absorbed. So $1-\alpha_i$ is the proportion of the energy that is reflected back into the room. Now assume all surfaces have the same absorption coefficient, α. So after n reflections, the sound has been reduced by a factor $(1-\alpha)^n$. Thus, after a time t, the sound energy has been reduced to,

$$E(t)=(1-\alpha)^{tcS/(4V)}E(0). \tag{10.2}$$

Taking the log base $(1-\alpha)$ of both sides, we have

$$\log_{(1-\alpha)}(E(t)/E(0))=tcS/(4V). \tag{10.3}$$

Now recall that $\log_A(x)=\ln(x)/\ln(A)$. So, putting the time t on one side,

$$t=\frac{4V\log_{(1-\alpha)}(E(t)/E(0))}{cS}=\frac{4V\ln(E(t)/E(0))}{cS\ln(1-\alpha)}. \tag{10.4}$$

Recall that the reverberation time, RT_{60}, is the time for the sound energy to decrease by 60 dB. Thus,

$$RT_{60}=\frac{4V\ln(10^{-6})}{cS\ln(1-\alpha)}\approx\frac{-0.161V}{S\ln(1-\alpha)}. \tag{10.5}$$

This is known as the Norris-Eyring formula. However, it has several approximations. The most important is that it assumes a single absorption coefficient, α. A more accurate form of Eq. (10.5) is,

$$RT_{60}=\frac{-0.161V}{S\ln\left(1-\frac{1}{S}\sum_i S_i\alpha_i\right)}. \tag{10.6}$$

Alternatively, it can be simplified even further. For small α, $\ln(1-\alpha)\sim-\alpha$. So Eq. (10.5) becomes,

$$RT_{60}\approx\frac{0.161V}{S\alpha}. \tag{10.7}$$

This is the Sabine equation, which was first found empirically by Wallace Clement Sabine in the late 1890s.

The absorption coefficient of a material ranges from 0 to 1 and indicates the proportion of sound that is not reflected by (so either absorbed by or transmitted through) the surface. A thick, smooth painted concrete ceiling would have an absorption coefficient very close to 0. Analogous to a mirror reflecting light, almost all sound would be reflected with very little attenuation. Conversely, a large, fully open window would have an absorption coefficient of 1, since any sound reaching it would pass straight through and not be reflected. In which case, Sabine's formula would be a very poor approximation, since it could still generate significant reverberation time.

As mentioned, absorption, and hence the reverberation time, is a function of frequency. Usually, less sound energy is absorbed in the lower frequency ranges, resulting in longer reverb times at lower frequencies. Nor do the equations above take into account room shape or losses from the sound travelling through the air, which is important in larger spaces. More detailed discussion of the reverberation time formulae (with differing interpretations!) are available in [102,103].

Direct and Reverberant Sound Fields

The reverberation due to sound reflection off of surfaces is extremely important. Reverberation keeps sound energy contained within a room, raising the sound pressure level and distributing the sound throughout. Outside in the open, there are far fewer reflective surfaces, and hence, much of the sound energy is lost.

For music, reverberation helps ensure that one hears all the instruments, even though they may be at different distances from the listeners. Also, many acoustic instruments will not radiate all frequencies equally in all directions. For example, without reverberation, the sound of a violin may change considerably as the listener moves with respect to the violin. The reverberation in the room helps to diffuse the energy a sound wave makes so that it can appear more uniform when it reaches the listener. But when the reverberation time becomes very large, it can affect speech intelligibility and make it difficult to follow intricate music.

A distinction is often made between the direct and reverberant sound fields in a room. The sound heard by a listener will be a combination of the direct sound and the early and late reflections due to reverberation. When the sound pressure due to the direct sound is greater than that due to the reflections, the listener is in the *direct field*. Otherwise, the listener is in the *reverberant field*.

The *critical distance* is defined as the distance away from a source at which the sound pressure level of the direct and the reverberant sound fields are equal. This distance depends on the shape, size and absorption of the space, as well as the characteristics of the sound source. A highly reverberant room generates a short critical distance, and a non-reverberant or anechoic room generates a longer critical distance.

For an omnidirectional source, the critical distance may be approximately given by Eq. (10.8) [104],

$$d_c \sim \sqrt{\frac{V}{100\pi RT_{60}}} \sim 0.141\sqrt{S\alpha}, \tag{10.8}$$

where critical distance d_c is measured in meters, volume V is measured in m^3, and reverberation time RT_{60} is measured in seconds. More accurate approximations are also available [105].

THE AVANT-GARDE ANECHOIC CHAMBER

An acoustic anechoic chamber is a room designed to be free of reverberation (hence non-echoing or echo-free). The walls, ceiling, and floor are usually lined with a sound-absorbent material to minimize reflections and insulate the room from exterior sources of noise. All sound energy will travel away from the source with almost none reflected back. Thus, a listener within an anechoic chamber will only hear the direct sound, with no reverberation.

The anechoic chamber effectively simulates a quiet open space of infinite dimension. Thus, they are used to conduct acoustics experiments in 'free field' conditions. They are often used to measure the radiation pattern of a microphone or of a noise source, or the transfer function of a loudspeaker.

An anechoic chamber is very quiet, with noise levels typically close to the threshold of hearing in the 10–20 dBA range (the quietest anechoic chamber has a decibel level of –9.4 dBA, well below hearing). Without the usual sound cues, people find the experience of being in an anechoic chamber very disorienting and often lose their balance. They also sometimes detect sounds they would not normally perceive, such as the beating of their own heart.

One of the earliest anechoic chambers was designed and built by Leo Beranek and Harvey Sleeper in 1943. Their design is the one upon which most modern anechoic chambers are based. In a lecture titled 'Indeterminacy,' the avant-garde composer John Cage described his experience when he visited Beranek's chamber.

> in that silent room, I heard two sounds, one high and one low. Afterward I asked the engineer in charge why, if the room was so silent, I had heard two sounds… He said, 'The high one was your nervous system in operation. The low one was your blood in circulation.'

After that visit, he composed his famous work entitled *4′33″*, consisting solely of silence and intended to encourage the audience to focus on the ambient sounds in the listening environment.

In his 1961 book 'Silence,' Cage expanded on the implications of his experience in the anechoic chamber. "Try as we might to make silence, we cannot… Until I die there will be sounds. And they will continue following my death. One need not fear about the future of music."

Implementation

Algorithmic Reverb

Early digital reverberators that tried to emulate a room's reverberation primarily consisted of two types of infinite impulse response (IIR) filters, allpass and comb filters, to produce a gradually decaying series of reflections.

Schroeder's Reverberator

Perhaps the first important artificial reverberation was devised in the early 1960s by Manfred Schroeder of Bell Telephone Laboratories. Early Schroeder reverberators [106,107] consisted of three main components: comb filters, allpass filters, and a mixing matrix. The first two are still used in many algorithmic reverbs of today, but the mixing matrix, designed for multichannel listening, is either not used or replaced with more sophisticated spatialization methods.

Figure 10.3 is a block diagram of the Schroeder Reverberator. This design does not create the increasing arrival rate of reflections, and modern algorithmic approaches are more realistic. Nevertheless, it provides an important basic framework for more advanced approaches.

There is a parallel bank of four feedback comb filters. The comb filter shown in Eq. (10.9) is a special case of an IIR digital filter, because there is feedback from the delayed output to the input. The comb filter effectively simulates a single room mode. It represents sound reflecting between two parallel walls and produces a series of echoes. The echoes are exponentially decaying and uniformly spaced in time. The advantage is that the decay time can be used to define the gain of the feedback loop.

The comb filters are connected in parallel. The comb filters have an irregular magnitude response and can be considered a simulation of

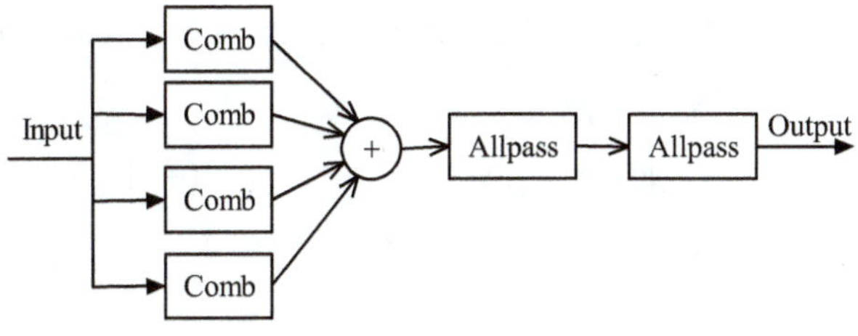

FIGURE 10.3
Schroeder's reverberator.

four specific echo sequences. The delay lengths in these comb filters may be used to adjust the illusion of 'room size', although if they are shortened, there should be more of them in parallel according to Schroeder. They also serve to reduce the spectral anomalies. Each comb filter has different parameters in order to attenuate frequencies that pass through the other comb filters. By controlling the delay time, which is also called loop time, of each filter, the sound waves will become reverberated.

The time domain and z domain form of Schroeder's comb filter is given in Eq. (10.9)

$$\begin{gathered} y[n] = x[n-d] + gy[n-d] \\ H_{\mathrm{Comb}}(z) = z^{-d} / \left(1 - gz^{-d}\right)' \end{gathered} \tag{10.9}$$

where d is a delay in samples, and g is the filter's feedback coefficient. The block diagram is shown in Figure 10.4a. The parallel comb-filter bank is intended to give an appropriate fluctuation in the reverberator frequency response. A feedback comb filter can simulate a pair of parallel walls, so one could choose the delay-line length in each comb filter to be the number of samples it takes for a wave to propagate from one wall to the opposite wall and back. Schroeder based his approach on trying to get the same number of impulse response peaks per second as would be found in a typical room.

The comb-filter delay-line lengths can be more or less arbitrary, as long as enough of them are used in parallel (with mutually prime delay-line lengths) to achieve a perceptually adequate fluctuation density in the frequency-response magnitude. In Schroeder's paper, four such delays are chosen between 30 and 45 ms, and the corresponding feedback coefficients g_i are set to give the desired overall decay time.

One problem with a single comb filter is that the distances between adjacent echoes are decided by a single delay parameter. Thus, the output of a comb filter has very distinct periodicity, producing a metallic sound, sometimes known as 'flutter echoes.' This rapid echo is often found in the acoustics of long narrow spaces with parallel walls. To counteract this, Schroeder's main design connects two allpass filters in series.

The allpass filter is given below in (10.10), and a block diagram is depicted in Figure 10.4b.

$$\begin{gathered} y[n] = -gx[n] + x[n-d] + gy[n-d] \\ H_{\mathrm{AP}}(z) = \frac{z^{-d} - g}{1 - gz^{-d}} \end{gathered} \tag{10.10}$$

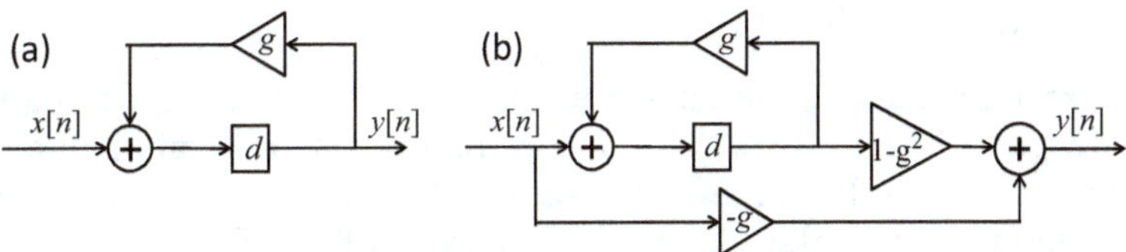

FIGURE 10.4
(a) Schroeder's comb filter. (b) A Schroeder allpass section. A typical value for g is 0.7.

The allpass filters provide 'colorless' high-density echoes in the late impulse response of the reverberator. Essentially, these filters transform each input sample from the previous stage into an entire IIR, which results in a higher echo density. For this reason, Schroeder allpass sections are sometimes called *impulse diffusers*. Unlike the comb filters, these allpass filters give the same gain at every frequency. But though they do not provide an accurate physical model of diffuse reflections, they succeed in expanding the single reflections into many reflections, which produces a similar qualitative result.

Schroeder recognized the need to separate the coloration of reverberation (changing the frequency content) from its duration and density aspects. In Schroeder's original work, and in much work which followed, allpass filters are arranged in *series* as shown in Figure 10.3, thus maintaining the uniform magnitude response. Since all of the filters are linear and time-invariant, the series allpass chain can go either before or after the parallel comb-filter bank.

Schroeder suggests a progression of allpass delay-line lengths close to the following;

$$d_i \sim 100 \text{ ms}/3^i, \quad i = 0,1,2,3,4. \tag{10.11}$$

The delay-line lengths *di* are typically mutually prime and spanning successive orders of magnitude. The 100 ms value was chosen so that when $g = 0.708$ in Eq (10.10), the time to decay 60 dB (T_{60}) would be 2s. Thus, for $i = 0$, T_{60}~2, and each successive allpass has an impulse-response duration that is about a third of the previous one. Using five series allpasses in this way yields an impulse-response echo density of about 810 echoes per second, which is close to the desired thousand per second.

A system using one of these reverberators simply adds its output, suitably scaled, to the reverberator input sample at the current time.

Moorer's Reverberator

The next important advance in algorithmic reverberators is attributable to James Moorer in 1979 [108]. Moorer's reverberator is the combination of an FIR filter that simulates the early impulse response activity of the room in cascade with a bank of low pass and comb filters. The FIR filter coefficients were based on the simulation of a concert hall.

Although comb filters can be used for modeling the decaying impulse response, as in Schroeder's reverberator, they do not simulate the tendency of the reverb to be attenuated at high frequencies, corresponding to absorption by the air and walls. To solve this problem, Moorer embedded a low pass filter in the feedback loop of each comb filter. Now reverberation time is a function of frequency. This helps simulate a more natural sounding reverberation, avoiding the unnatural metallic sound that can be observed without the presence of the filter. A simple form of low pass filter was sufficient to get satisfactory results. An all pass filter then followed the low pass comb filters, in order to increase the reflection intensity.

Moorer's reverberator is depicted in Figure 10.5. The top block takes care of early reflections, and the bottom block, consisting of six parallel comb filters with different delay lengths, takes care of late reflections.

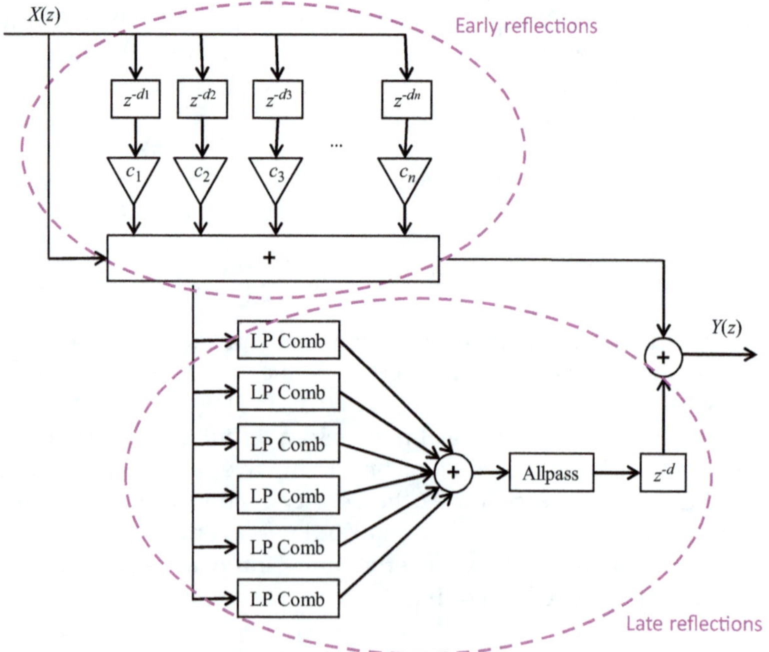

FIGURE 10.5
Moorer's reverberator used an FIR filter to simulate the early reflections and parallel comb filters with low pass filtering to simulate the late reflections.

Feedback Delay Networks

Schroeder's and Moorer's reverberators are both examples of Feedback Delay Networks (FDNs). Feedback delay networks (FDNs) are recursive filters and are widely used for artificial reverberation. They typically consist of delay lines, a set of gains, and a feedback matrix through which the delay outputs are coupled to the delay inputs. It can be seen as a general architecture, or design methodology, for implementing many forms of artificial reverberation (as well as other applications), and can be easily modified or extended to multichannel or time-varying reverb. FDNs are the most common means by which algorithmic reverb is implemented. For more on this, the reader is referred to some classic texts on the subject [109,110].

Generating Reverberation with the Image-Source Method

Background

The reverberation heard when a source is recorded in a room is characterized by the *room impulse response* (RIR), which gives the impulse response corresponding to a sound source and a listener in a room.

The *image-source model* is a popular method for generating simulated room impulse responses. Once an RIR is available, reverberation can be applied to an audio signal by convolving it with the RIR. This approach generates a sound that appears as if the source signal had been recorded by a microphone in the room. The technique gives a relatively simple way to generate a number of RIRs with differing characteristics, such as their reverberation times.

The Image-Source Model

The image-source method was originally presented for rectangular enclosures in a seminal paper by Allen and Berkley from 1979 [111]. To explain this method, we will first show how the individual echoes that together produce reverberation can be visualized as virtual sources. Then we will find a unit impulse response for each echo with the proper time delay, given the listener's position. After this, we calculate the magnitude of each echo's unit impulse response. So we have then combined the times and magnitudes of each echo to generate the room impulse response. A discrete time implementation of the impulse response can be achieved with a finite impulse response (FIR) filter.

Modeling Reflections as Virtual Sources

Figure 10.6 shows a rectangular room (often referred to as a 'shoebox') on the left. Within it are a sound source and a microphone. We are trying to calculate the impulse response at the location of the microphone. The direct sound path is the line between the source and the microphone. The sound is also reflected off a wall before arriving at the microphone. The listener perceives this echo as coming from a location past that wall. So we create a mirror image of the room and place it on the other side of the wall. The mirror image of the source, known as a *virtual source*, resides in this virtual room at the location from which we perceive this reflected sound to be located.

The solid line in Figure 10.6 is the actual path taken by the sound wave, and the dotted line is the perceived path. We treat the virtual sources as if they were individual sound sources and ignore each virtual source's echo. This process is repeated by making mirror images of the room's mirror image, each with another virtual source. We can extend this to two dimensions, as in Figure 10.7, or three dimensions.

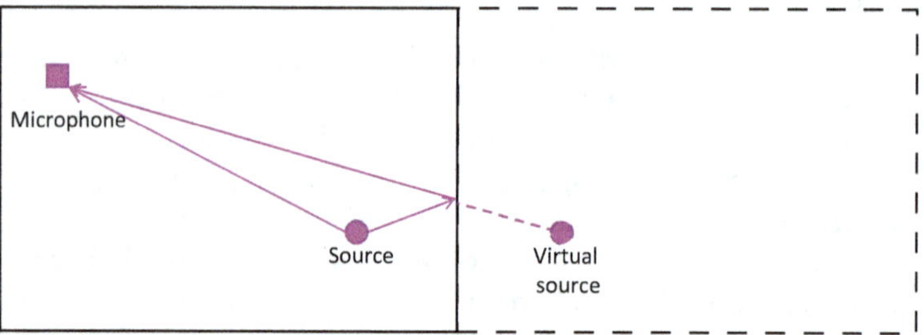

FIGURE 10.6
A room with a source sound and microphone, and its mirror image containing a virtual source.

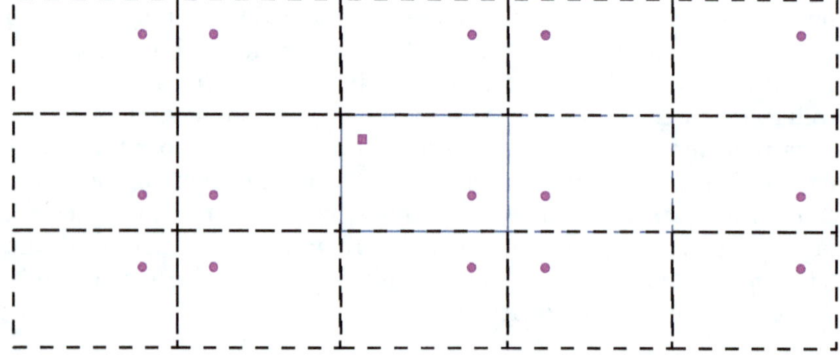

FIGURE 10.7
A room with multiple mirror images in two dimensions, corresponding to reflections of the sound source off the walls.

Locating the Virtual Sources

Let us first consider virtual sources along just one dimension. We set the origin to be the position of the microphone, and x_r is the length of the room along the x-axis. The ith virtual source is located at x_i. If $i=0$, then the virtual source is actually the real sound source. If i is negative, then the virtual source is located on the negative x-axis. The distance between the microphone and the ith virtual sound source along the x-axis is given by;

$$x_i = \begin{cases} (i+1)x_r - x_0 & i \text{ odd} \\ ix_r + x_0 & i \text{ even} \end{cases}. \tag{10.12}$$

Similarly, we can find the positions of virtual sources along the y and z axes;

$$y_j = \begin{cases} (j+1)y_r - y_0 & j \text{ odd} \\ jy_r + y_0 & j \text{ even} \end{cases}. \tag{10.13}$$

$$z_k = \begin{cases} (k+1)z_r - z_0 & k \text{ odd} \\ kz_r + z_0 & k \text{ even} \end{cases}. \tag{10.14}$$

The three-dimensional distance from the microphone to a virtual source is then given by,

$$d_{ijk} = \sqrt{x_i^2 + y_j^2 + z_k^2}. \tag{10.15}$$

The Impulse Response for a Virtual Source

We now define the following delta function,

$$\delta_{ijk} = \begin{cases} 1 & d_{ijk} = tc \\ 0 & \text{otherwise} \end{cases}, \tag{10.16}$$

where t is the time, d_{ijk} is the distance to a virtual source, and c is the speed of sound. So $d_{i,j,k}/c$ is the effective time delay of each echo. Equation (10.16) gives the unit impulse response for a virtual source, with unity magnitude when the sound from that virtual source reaches the microphone.

Now note that the magnitude is inversely proportional to the distance it travels to get from the source to the microphone.

$$m_{ijk} \propto 1/d_{ijk}. \tag{10.17}$$

Also, the magnitude is affected by the number of reflections that the sound wave makes before arriving at the microphone. Assuming that all walls have the same absorption coefficients, we can take the coefficient α and raise it to the exponent $n = |i| + |j| + |k|$, which represents the total number of reflections the sound has made.

$$\alpha_{ijk} = \alpha^{|i|+|j|+|k|}. \tag{10.18}$$

This can be extended to the case where each wall could have a different reflection coefficient. Let $\alpha_{x=0}$ be the reflection coefficient for the wall perpendicular to the x-axis that is closest to the origin, and let $\alpha_{x=xr}$ be the reflection coefficient for the wall opposite that, and use similar notation for the walls opposite the y and z axes. The reflection coefficients for all the reflections made are given by the following equation.

$$\alpha_{x_i} = \begin{cases} \alpha_{x=0}{}^{|(i-1)/2|}\alpha_{x=x_r}{}^{|(i+1)/2|} & i \text{ odd} \\ \alpha_{x=0}{}^{|i/2|}\alpha_{x=x_r}{}^{|i/2|} & i \text{ even} \end{cases}$$

$$\alpha_{y_j} = \begin{cases} \alpha_{y=0}{}^{|(j-1)/2|}\alpha_{y=y_r}{}^{|(j+1)/2|} & i \text{ odd} \\ \alpha_{y=0}{}^{|j/2|}\alpha_{y=y_r}{}^{|j/2|} & i \text{ even} \end{cases}. \tag{10.19}$$

$$\alpha_{z_k} = \begin{cases} \alpha_{z=0}{}^{|(k-1)/2|}\alpha_{z=z_r}{}^{|(k+1)/2|} & i \text{ odd} \\ \alpha_{z=0}{}^{|k/2|}\alpha_{z=z_r}{}^{|k/2|} & i \text{ even} \end{cases}$$

To find the total reflection coefficient of a virtual source with the indices i, j, and k we multiply *these* components together;

$$\alpha_{ijk} = \alpha_{x_i}\alpha_{y_j}\alpha_{z_k}. \tag{10.20}$$

Now we can generate the impulse response by multiplying the delta function from Eq. (10.16) with the magnitude of each echo, magnitude m_{ijk}, and the reflection coefficient α_{ijk}, and then sum over all three indices. This can be thought of as the summation of all the sounds arriving from all the virtual sources.

$$h(t) = \sum_{i=-n}^{n}\sum_{j=-n}^{n}\sum_{k=-n}^{n} \delta_{ijk} m_{ijk} \alpha_{ijk}. \tag{10.21}$$

In Figure 10.8, a circle has been shown, with the microphone at its center and with its edge touching the nearest wall. The portion of the impulse response that comes from virtual sources within this sphere represents a truncated room impulse response. When extended into three dimensions,

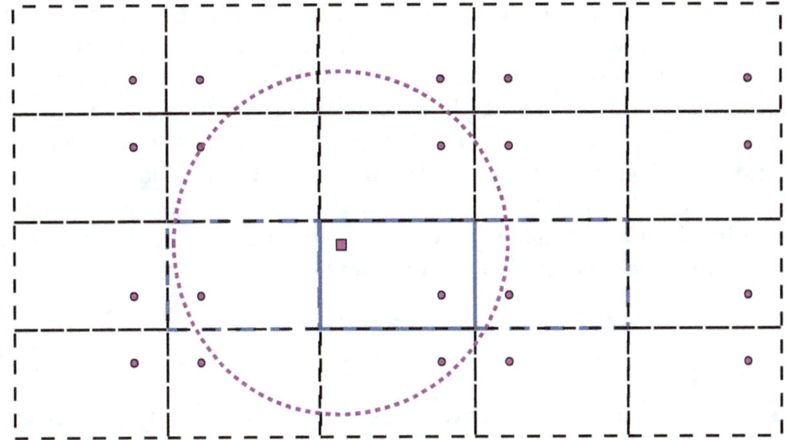

FIGURE 10.8
Image source representation, with a circle drawn to indicate virtual sources used for early reflections.

Figure 10.8 would form a partitioned cuboid with and the truncated RIR would become a sphere.

This approach makes several assumptions that are known to be inaccurate. It ignores any attenuation of the sound that may result from traveling through the air. Furthermore, it assumes that no change in phase takes place upon reflection, and that the reflection coefficients are independent of angle and frequency.

Many image source model variants and extensions based on this standard technique have been proposed in the literature. For instance, [112] extended the method to the calculation of late reflections, as well as speeding up computation, and [113] proposed an improvement where each virtual source is given using fractional delays.

Convolutional Reverb

Convolution and Fast Convolution

Convolution is an important mathematical technique for combining two signals. By convolving a signal with a room impulse response, we can create the reverberated signal as it would be heard in a room. Given an input signal s, the filtered output r is the result of the convolution of s by the FIR $h[0] \ldots h[n-1]$;

$$r[n] = (s * h)[n] = \sum_{k=0}^{N-1} s[n-k]h[k]. \tag{10.22}$$

However, though convolution is very useful, it is excessively computationally expensive. For realistic simulation of a natural environment, the required impulse response can last more than 2 s. For processing one sample with convolutional reverb, addition and accumulation operations must be performed on the entire length of the impulse response. Thus, assuming a 44.1 kHz sampling frequency and an impulse response of just 1 s, roughly 2×10^9 adds and multiplies are required each second. Fast convolution provides a means to address this complexity issue.

The *fast convolution* uses the well-known convolution theorem: performing multiplication in the Fourier domain is equivalent to performing convolution in the time domain (and vice-versa).

$$\begin{aligned} r = s * h \leftrightarrow R = S \cdot H \\ \rightarrow r = F^{-1}\left\{F(s) \cdot F(h)\right\} \end{aligned} \quad (10.23)$$

That is, rather than convolving two signals together directly, one can compute their Fourier transforms, multiply them together, and then take the inverse Fourier transform. This may seem like more steps, but the Fast Fourier Transform (FFT) can be used, which offers vast efficiency savings over convolution operations. Whereas convolving two signals of length N requires N^2 operations, the FFT requires on the order of $N\log(N)$ operations. Again assuming a sampling frequency of 44.1 kHz and an impulse response of 1 s, only about 10^6 operations are needed each second if this Fast Convolution is used.

Block-Based Convolution

There are at least two main approaches to performing fast convolution on a real-time signal: the overlap and add method and the overlap and save method. They both rely on cutting the input signal into blocks. Here, we will describe the overlap and add method, which is slightly simpler to understand and is discussed in more detail in Chapter 9. The input signal s is cut into blocks of length N. The kth block $s_{kN:(k+1)N-1} = s[kN], \ldots s[(k+1)N-1]$ is convolved with the impulse response h of length N as shown in Figure 10.9.

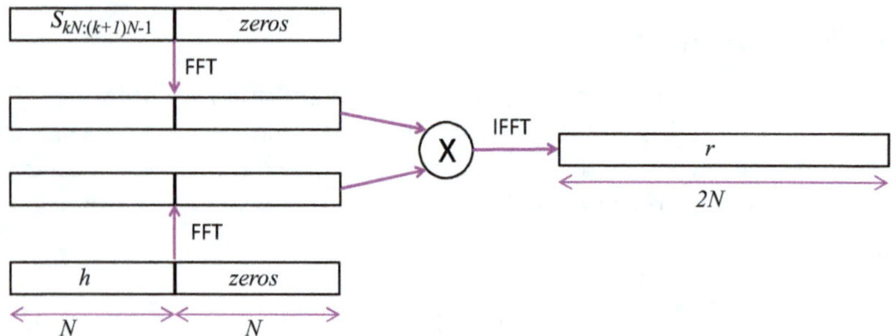

FIGURE 10.9
One block convolution, as implemented in block-based convolutional reverb. The kth block is convolved with the impulse response h of length N.

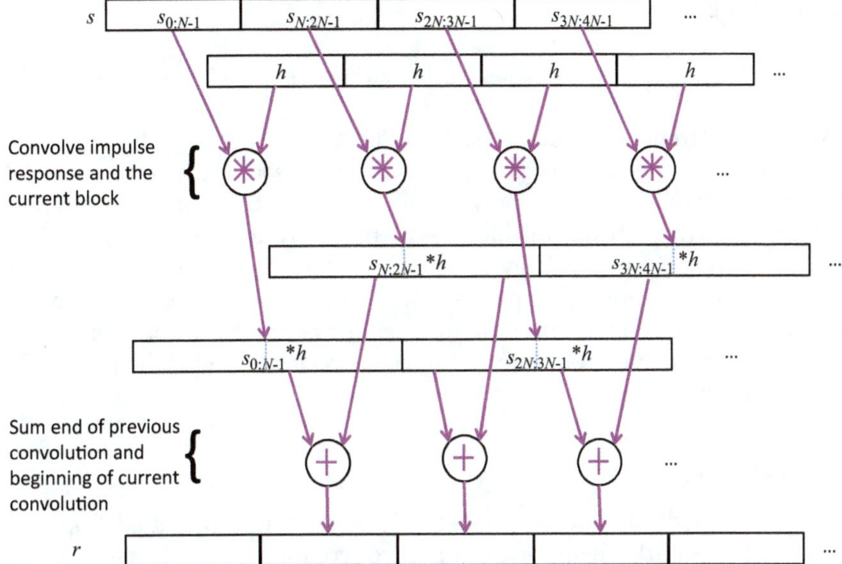

FIGURE 10.10
Partitioned convolution for real-time artificial reverberation.

This convolution is performed by zero padding the input signal block of length N in order to realize a $2N$ long Fourier transform. As a result, the fast convolution of one block with h produces a filtered signal r of length two blocks. The second of these blocks is summed with the first resulting block to create a complete filtered response, as shown in Figure 10.10. To further improve performance, the Fourier transform of h can be pre-computed and reused until h is changed.

From the definition of convolution and a change of variables,

$$u = l - j \rightarrow r[kN + l] = \sum_{j=0}^{N-1} s[kN + l - j]h[j]$$

$$= \sum_{u=l-N+1}^{l} s[kN + u]h[l - u]. \tag{10.24}$$

This summation can be separated into positive and negative terms for u.

$$r[kN + l] = \sum_{u=0}^{l} s[kN + u]h[l - u]$$

$$+ \sum_{u=l-N+1}^{-1} s[kN + u]h[l - u], \tag{10.25}$$

which, again from the definition of convolution, becomes:

$$r[kN + l] = s_{kN:kN+l}{}^{*}h + s_{kN+l-N+1:kN-1}{}^{*}h. \tag{10.26}$$

Thus, we obtain:

$$r_{kN:(k+1)N-1} = (s_{(k-1)N:kN-1} * h)_{N:2N-1} + (s_{kN:(k+1)N-1} * h)_{0:N-1}. \quad (10.27)$$

That is, the kth output block is obtained by adding together the tail of the convolution of the $(k-1)$th block and h and the head of the convolution of the kth block and h. This is shown for the first few blocks in Figure 10.10.

Note that the kth output block will be ready at the earliest at time $(k+1)N$. This minimum delay of one block size can be problematic. For instance, if the impulse response lasts 1 s, a device that uses this overlap and add fast convolution method will have at least 1 s of delay. Also, blocks are convolved with the full impulse response, which can be quite computational.

Physical Meaning

The partitioned convolution has a simple physical interpretation. Suppose h is a room impulse response and that convolutional reverb is applied to the input signal. Assuming a source produces a sound frame at time t, the direct path brings this frame to the listener. The other paths have reflections from the walls and ceiling, resulting in modified versions of the original frame with different attenuation and delay. The delays depend on the length of the path, including the number of reflections.

If sounds are produced continuously, then at any point in time, a listener will hear the direct sound and delayed and transformed versions of previous sounds. The partitioned convolution is just an expression of this phenomenon. That is, at any point in time, the sum is made over the length of the filter h plus the length of the last block. Since input blocks are convolved with h, we effectively obtain a delayed and transformed version of sounds that were emitted in the past.

Other Approaches

As mentioned, one issue with the partitioned convolution is the partition size. Since it is based on convolving with the whole impulse response, there is still a lot of processing and latency. Gardner [114] developed a solution to the high delay issue that uses the same idea as fast convolution. Since the input signal is partitioned into blocks, the impulse response can also be partitioned. Block convolution is then performed on the appropriate combination of input blocks and filter blocks and summed in order to produce the desired output. This efficient approach relies on a fast convolution that is performed many times, but on smaller blocks.

Today, many of the fastest convolutional reverbs are improvements on this approach. Another technique used to speed up calculation and maintain low latency is to use a low latency, partitioned convolutional reverb for the early reflections and then an algorithmic reverb for the late reflections. In which case, a FIR filter is typically used to generate the early reflections, and then IIR filters may be applied to create the diffuse reverberation. Low pass filters may also be used to account for air absorption.

More advanced algorithms can be developed to model specific room sizes, whether for generating a room impulse response or for designing an algorithmic reverb. Ray tracing techniques can also be used to derive the reverberation for a given room geometry, source, and listener location.

ACOUSTIC REVERBERATORS

Many recording studios have used special rooms known as *reverberation chambers* to add reverb to a performance. Elevator shafts and stairwells (as in New York City's Avatar Recording Studio) work well as highly reverberant rooms. The reverb can also be controlled by adding absorptive materials like curtains and rugs.

Spring reverbs are found in many guitar amplifiers and have been used in Hammond organs. The audio signal is coupled to one end of the spring by a *transducer* that creates waves traveling through the spring. At the far end of the spring, another transducer converts the motion of the string into an electrical signal, which is then added to the original sound. When a wave arrives at the end of the spring, part of the wave's energy is reflected. However, these reflections have different delays and attenuations from what would be found in a natural acoustic environment, and there may be some interaction between the waves in a spring, thus this results in a slightly unusual (though not unpleasant) reverb sound.

Often several springs with different lengths and tensions are enclosed in a metal box, known as the *reverb pan*, and used together. This avoids uniform behavior and creates a more realistic, pseudo-random series of echoes. In most reverb units, though, the spring lengths and tensions are fixed in the design process, and not left to the user to control.

The *plate reverb* is similar to a spring reverb, but instead of springs, the transducers are attached at several locations on a metal plate. These transducers send vibrations through the plate, and reflections are produced whenever a wave reaches the plate's edge. The location of the transducers and the damping of the plate can be adjusted to control the reverb. However, plate reverbs are expensive and bulky, and hence not widely used.

Water tank reverberators have also been used. Here, the audio signal is modulated with an ultrasonic signal and transmitted through a tank of water. The output is then demodulated, resulting in the reverberant output sound. Other reverberators include pipes with microphones placed at various points.

These acoustic and analogue reverberators can be interesting to create and use, but they lack the simplicity and ease of use of digital reverberators. Ultimately, the choice of implementation is a matter of taste.

Applications

Why Use Reverb?

We usually inhabit the reverberant field, with many sources of reverberation already around us. Yet it is still useful to add reverb to recordings. We often listen to music in environments with very little or poor reverb. A dry signal may sound unnatural, so the addition of reverb to recordings is used to compensate for the fact that we cannot always listen to music in well-designed acoustic environments. The reverberation in a car may not adequately recreate the majestic sound of a symphony orchestra. And when listening over headphones, there is no reverberation added to the music.

Stereo Reverb

Another important aspect of reverb is the correlation of the signals that reach the listener's ears. In order for a listener to perceive the 'spaciousness' of a large room, the sounds at each ear should be slightly offset. This is one reason why concert halls have such high ceilings. With a low ceiling, the first reflections to reach the listener usually are reflections from the ceiling, and reach both ears at the same time. But if the ceiling is high, the first reflections come from the walls of the concert hall. And since the walls are generally different distances away from a listener, the sound arriving at each ear is different. Ensuring some slight differences in the reverberation applied to left and right channels is important for stereo reverb design.

Gated Reverb

A gated reverb is created by truncating the impulse response of a reverberator, thus changing the IIR filters to FIR. The amount of time before the response is cut off is known as the *gate time*, as labeled in Figure 10.11.

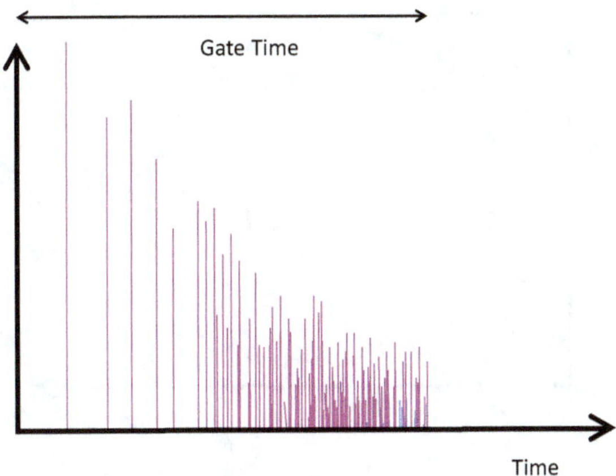

FIGURE 10.11
Impulse response of a gated reverb.

Some reverberation implementations provide for a more gradual decay of the sound, rather than a sharp cut off that produces abrupt silence. Gated reverbs are often used on percussive instruments.

Reverse Reverb

Rather than generating reflections that become quieter and gradually fade away, the impulse response can be reversed. This will generate reflections that get louder over time and then abruptly cut off. This sounds a bit like slapback delay because it ends suddenly, but analogous to the difference between echo and reverb, it has more complicated and less uniform behavior than slapback delay.

The length of time it takes for the reflections to build up is known as the *reverse time*, or the gate time, since it works like a gated reverb that has been reversed in time. Figure 10.12 depicts a possible reverse reverb impulse response.

An interesting, related technique is to reverse the signal, apply the reverb, and reverse again.

Common Parameters

The available parameters to control a reverberator can vary widely. The following parameters apply mainly to algorithmic reverb. For convolutional reverb, the choice of the room impulse response is of primary importance, and a commercial audio effect will often come with a wide range of pre-computed room impulse responses that can then be fine-tuned to fit the needs of the user. Alternatively, some convolutional reverbs will allow

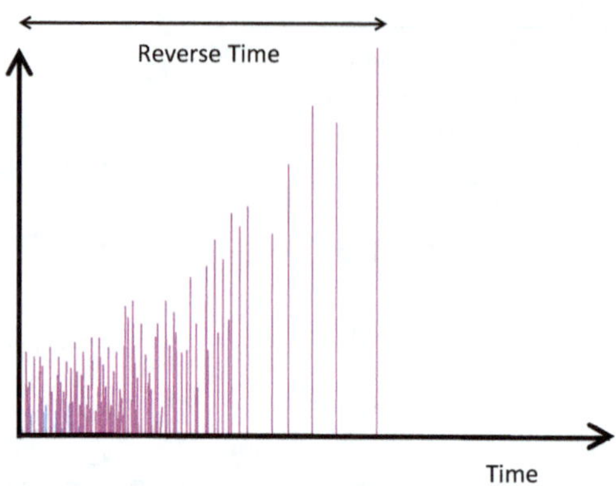

FIGURE 10.12
Impulse response for reverse reverb.

the user to specify the room size. In which case, the image source method or a similar approach is used to generate the impulse response.

Reverb time: This is also known as the Decay Time, or the Reverb Decay, and usually represents the reverberation time as described in the beginning of this chapter. It indicates how long the reverb can be heard after the input stops, usually for a 1 kHz input. The actual measure of what can be 'heard' can vary among manufacturers. This parameter is typically in terms of milliseconds, which can be thought of as something like the reverb time.

Long reverb times applied to tracks often work well in a sparse mix where there are few sources, but can produce clutter when there are lots of sources. Short but distinct reverbs can be applied to each source in a busy ensemble mix so that each one will have a unique ambience and they won't mask each other in the final mix.

Diffusion/density: This parameter is usually related to the echo density, also discussed at the beginning of the chapter. A highly diffuse reverb will sound smooth but can also result in noticeable filtering of the signal, or coloration. Low diffusion values often sound nice on vocals, but may result in reverb with a coarse or grainy sound reverberation.

Note that some commercial reverbs will have both a diffusion and a density parameter. In which case, diffusion is often specific to just the early reflections, whereas density refers just to the late reflections.

Direct to reverberant ratio: This is essentially the dry/wet mix parameter, as found on many other effects. It determines how much of the original sound is used. That is, if the reverb is simulating an acoustic environment, it determines to what extent the source travels directly to the listener, without reflecting off any surfaces.

Predelay: The predelay is usually defined as the amount of time before the first reflection in the impulse response, i.e., the time until the first reverberations are heard.

More advanced reverberation units may allow the user to set a second predelay for the amount of time before the first late reflection, as shown in Figure 10.13. Of course, for simulation of a realistic environment, the predelay for the early reflections should always be less than the predelay for the late reflections.

Filtering/room damping: Typically, high-frequency content in the reflections will be attenuated more than low-frequency content, both due to absorption of sound while traveling through the air and absorption when reflecting off surfaces. Thus, most reverberators will provide some form of filtering to simulate this damping of high frequencies. This is often in the form of a single frequency value that defines the cut-off frequency of a low pass filter. Reverb designs that give more control to the user may also specify a crossover point for both low and high shelving filters along with the gain for each shelf.

Having minimal attenuation of the high frequency content may give a bright, airy sound to the reverb, and is often used on vocals. Strong attenuation of this content is more typical of emulation of classic reverberators, since they often couldn't reproduce the high frequencies.

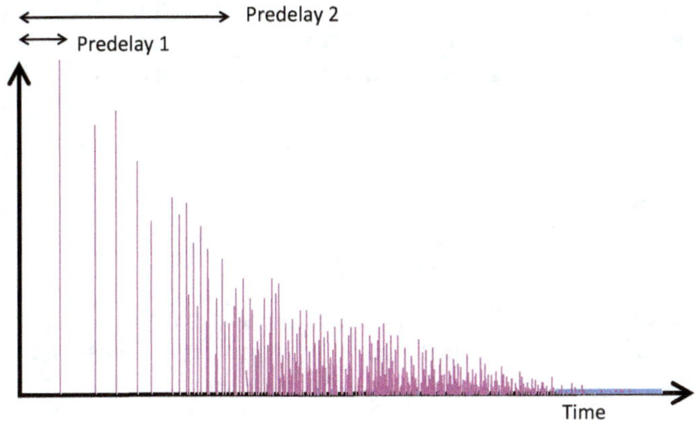

FIGURE 10.13
A room impulse response with the predelay parameters labeled.

Gate time: This parameter applies to gated reverbs. The gate time is just the amount of time, usually in milliseconds, for which the reverb is allowed to sound. This may also refer to the duration of a reverse reverb.

Gate decay time: Some gated reverberators will also provide a decay time, which controls the behavior of the gate as it is closed. A short gate decay time means that the reverb will attenuate rapidly, as is the case with the gated reverb depicted in Figure 10.11. A long decay time implies that the reverb will gradually fade away.

Gate threshold: Instead of applying gated reverb to an entire signal, the gating of the reverb can be made level dependent. Typically, the gate on a reverb will be kept open (the impulse response is not truncated) when signals are above the threshold. But if the signal level drops below the threshold, the gate closes and the reverb is truncated. If the signal level rises back above the threshold, the gate will then reopen.

Further Reading

An excellent overview of artificial reverberation was provided in [115] in 2012, and followed up 4 years later with [116]. Since then, FDNs have become more popular and usable, see [117,118], with similar advances in convolutional reverb [119]. Other advances have been concerned more with the perception of reverb [18,120], or attempts to automate the application of reverberation [121]. And as with almost all audio effects, significant recent research has focused on implementing reverb in a differentiable form [122], so that it can be embedded in a neural network architecture.

Problems

1. Explain why *impulse responses* are useful for modeling the properties of an acoustic space. How is an impulse measured with a static sound source and a static mono microphone?

2.
   i. Sketch an impulse response plot for a typical room. Label the early and late reflections.

   ii. How would the plot in part (i) be changed for a larger room, and for the same size room with higher reflectivity off all surfaces?

3. Define the main difference between *echo* and *reverberation*.
4. Why are FIR filters often used to generate early reflections and IIR filters used to generate late reflections?
5. Show that Schroeder's allpass section has the required properties of an allpass filter. That is, show that the magnitude response is always equal to 1.
6. Give the impulse response of a single Schroeder comb filter.
7. Explain the function of a *gated reverb* as compared to a standard reverb. What is the effect of the *gate time* control?
8. Explain the difference between convolutional and algorithmic reverb. What are the advantages and disadvantages of each?
9. Use both Sabine's formula and the Norris-Eyring formula to estimate the reverberation time RT_{60} for a room with floor area 12 m×10 m and 5 m high ceilings. The floor is carpeted (α=0.15) ceiling is acoustic tiles (α=0.6) and the walls are covered with thick drapes (α=0.5). Now estimate reverberation time for the same room but without drapes and all surfaces made of brick (α=0.03).

11

Spatial Audio

Previously, we described a digital audio signal as a discrete series of values, sampled uniformly in time. But the listener has two ears, and hears the sound differently in each ear depending on the location of the source. Stereo audio files encode two signals, or *channels*, for listening over headphones or loudspeakers, so that sources can be localized. Furthermore, spatial audio reproduction systems will often use a large number of loudspeakers, recreating an entire sound scene, thus requiring more channels.

For reproduction of spatial audio via multiple loudspeakers, we should first consider how we localize sound sources. Consider a listener hearing the same content coming from two different locations, and at different times and levels. Under the right conditions, this will be perceived as a single sound source, but emanating from a location between the two original locations. This fundamental aspect of sound perception is a key element in many spatial audio rendering techniques.

In this chapter, we describe some of the main techniques used to spatialize audio signals. We concentrate on the placement of sound sources using level difference (*panorama*) and time difference (*precedence*), and variations and advances on these approaches. We also discuss some advanced multichannel spatial audio reproduction methods, such as *ambisonics* and *wave field synthesis*. Finally, it should be noted that spatial audio continues to be a very active research area. New approaches, such as Directional Audio Coding [123], have gained popularity and may offer some advantages over the methods mentioned here.

THE BEGINNING OF STEREO

The sound reproduction systems for the early 'talkie' movies often had only a single loudspeaker. Because of this, the actors all sounded like they were in the same place, regardless of their position on screen.

In 1931, the electronics and sound engineer Alan Blumlein and his wife Doreen went to see a movie where this monaural sound reproduction occurred. According to Doreen, as they were leaving the cinema, Alan said to her, 'Do you realise the sound only comes

DOI: 10.1201/9781003593942-11

from one person?' And she replied, 'Oh does it?' 'Yes,' he said, 'And I've got a way to make it follow the person.'

The genesis of these ideas is uncertain (though it might have been while watching the movie), but he described them to Isaac Shoenberg, managing director at EMI and Alan's mentor, in the late summer of 1931. Blumlein detailed his stereo technology in the British patent "Improvements in and relating to Sound-transmission, Sound-recording and Sound-reproducing systems," which was accepted June 14, 1933.

Theory

Panorama

Suppose we have two loudspeakers in different locations. Then the apparent position of a source can be changed just by giving the same source signal to both loudspeakers, but at different relative levels. When a camera is rotated to depict a *panorama*, or wide angle view, this is known as *panning*. Panning in audio is derived from this, and describes the use of level adjustment to move a virtual sound source. During mixing, this panning is often accomplished separately for each sound source, giving a panorama of virtual source positions in the space spanned by the loudspeakers.

Consider a standard stereo layout. The listener is placed in a central position as depicted in Figure 11.1. In this figure, there is a 60° angle between loudspeakers, which is typical, but not a requirement. ϕ is the angle of the apparent source position, known as the *azimuth* angle, and θ is the angle (in this case, 30°) formed by each loudspeaker with the frontal direction. $\mathbf{p}$ defines the unit length vector pointing towards the source, and $\mathbf{l}_1$ and $\mathbf{l}_2$ are the unit vectors in the directions of the two loudspeakers.

The unit vectors $\mathbf{p}$, $\mathbf{l}_1$, and $\mathbf{l}_2$ can be written in terms of the angles ϕ and θ,

$$
\begin{aligned}
p_1 &= \cos\phi, p_2 = \sin\phi \\
l_{11} &= \cos\theta, l_{12} = \sin\theta \quad , \\
l_{21} &= \cos\theta, l_{22} = -\sin\theta
\end{aligned}
\tag{11.1}
$$

The unit vectors in the loudspeaker directions can be defined as a matrix.

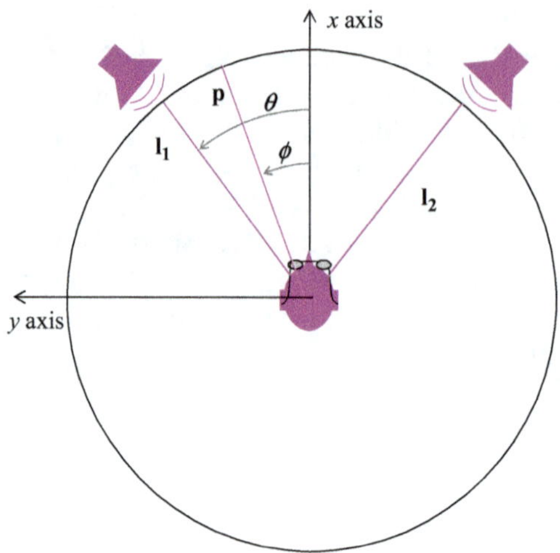

FIGURE 11.1
Listener and loudspeakers configuration for placing a sound source using level difference.

$$\mathbf{L} = \begin{bmatrix} \mathbf{l}_1 \\ \mathbf{l}_2 \end{bmatrix} = \begin{bmatrix} \cos\theta & \sin\theta \\ \cos\theta & -\sin\theta \end{bmatrix}. \tag{11.2}$$

Importantly, the vector pointing toward the source can be constructed by applying gains, g_1 and g_2, to the vectors pointing to the loudspeakers.

$$\mathbf{p} = \begin{bmatrix} p_1 & p_2 \end{bmatrix} = \mathbf{gL}. \tag{11.3}$$

So we can invert this to find $\mathbf{g} = \mathbf{pL}^{-1}$,

$$\begin{aligned} \begin{bmatrix} g_1 & g_2 \end{bmatrix} &= \frac{1}{l_{11}l_{22} - l_{21}l_{12}} \begin{bmatrix} p_1 & p_2 \end{bmatrix} \begin{bmatrix} l_{22} & -l_{12} \\ -l_{21} & l_{11} \end{bmatrix} \\ &= \frac{1}{-2\cos\theta\sin\theta} \begin{bmatrix} \cos\phi & \sin\phi \end{bmatrix} \begin{bmatrix} -\sin\theta & -\sin\theta \\ -\cos\theta & \cos\theta \end{bmatrix} \\ &= \begin{bmatrix} \dfrac{\cos\phi\sin\theta + \sin\phi\cos\theta}{2\cos\theta\sin\theta} & \dfrac{\cos\phi\sin\theta - \sin\phi\cos\theta}{2\cos\theta\sin\theta} \end{bmatrix}. \end{aligned} \tag{11.4}$$

However, as is, these gains may change the power of the signal, causing the signal level to change depending on the direction of the source. So the gains must be scaled.

$$\mathbf{g}_{\text{scaled}} = \frac{\mathbf{g}}{\sqrt{g_1^2 + g_2^2}}. \tag{11.5}$$

Thus, we can find the gains that need to be applied for a given azimuth angle for the source. This also gives a simple relationship for the azimuth angle in terms of the gains, known as the *tangent law.*

$$\frac{g_1 - g_2}{g_1 + g_2} = \frac{2\sin\phi\cos\theta}{2\cos\phi\sin\theta} = \frac{\tan\phi}{\tan\theta} \rightarrow \tan\phi = \frac{g_1 - g_2}{g_1 + g_2}\tan\theta. \tag{11.6}$$

This is known as the tangent law, and has been found to yield accurate perceived direction in listening tests under anechoic conditions. A similar law is often used, known as the *sine law* or *Blumlein law.*

$$\sin\phi = \frac{g_1 - g_2}{g_1 + g_2}\sin\theta. \tag{11.7}$$

Figure 11.2 depicts the gain and power for constant power panning as a function of the azimuth angle ϕ. Here the total power is constant, but total gain varies as a function of azimuth angle, reaching its maximum when the source is positioned equidistant from both loudspeakers.

Figure 11.3 shows the perceived azimuth angle as a function of the level difference between the applied gains, using the tangent law, for a typical loudspeaker placement.

If θ is 45°, then stereo panning can be put in a more compact form using a simple rotation matrix. From (11.4) and (11.5), it can be seen that

$$\mathbf{g} = \mathbf{g}^{\text{scaled}} = \begin{bmatrix} \cos\phi & \sin\phi \\ -\sin\phi & \cos\phi \end{bmatrix} \begin{bmatrix} 1/\sqrt{2} \\ 1/\sqrt{2} \end{bmatrix}. \tag{11.8}$$

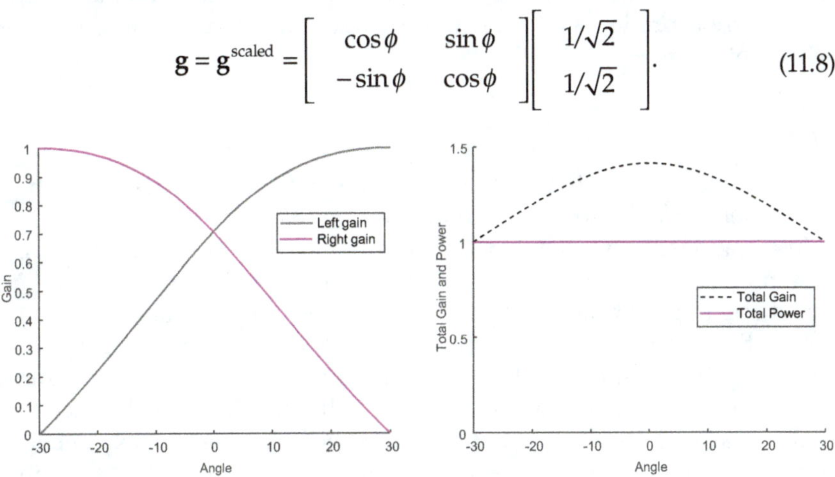

FIGURE 11.2
Constant power panning for two channels. On the left is the gain for each channel, and on the right is the total power and total gain.

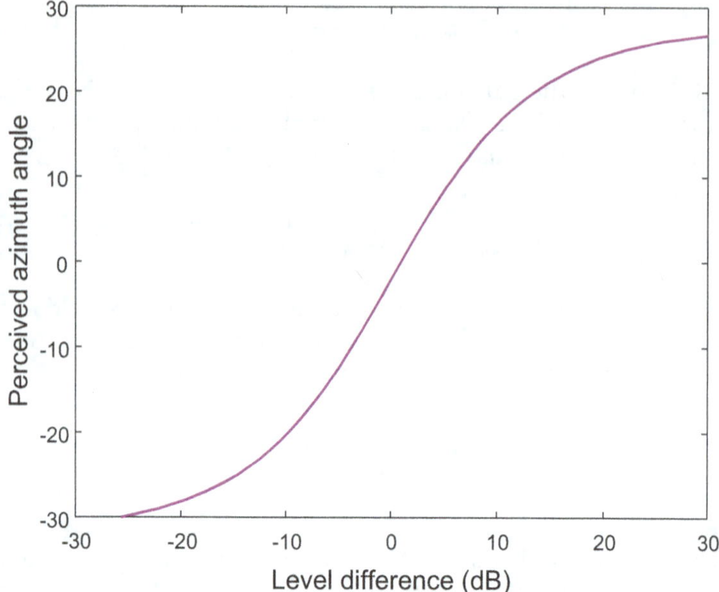

FIGURE 11.3
Perceived azimuth angle as a function of level difference.

Note that this approach is much preferred over crossfade panning, where the apparent position of the sound source is shifted by linearly interpolating the amplitude between the two extremes of hard left and hard right. This method produces a "hole" in the middle due to the reduced power when not scaling.

Precedence

The precedence effect is a well-known phenomenon that plays a large part in our perception of source location.

In a standard stereo loudspeaker layout, the perceived azimuth angle of a monophonic source that is fed to both loudspeakers can be changed by having a small time difference between the two channels. Figure 11.4 shows the qualitative dependency of the apparent azimuth on the relative delay between channels. If the time difference between the signals fed to the two loudspeakers is below the echo threshold, the listener will perceive a single auditory event. This threshold can vary widely depending on the signal, but it ranges from about 5 to 40 ms. When this time difference is below about 1 ms (so much lower than the echo threshold), the source angle can be perceived as being between the two loudspeakers. In between 1 ms and the echo threshold, the sound appears to come

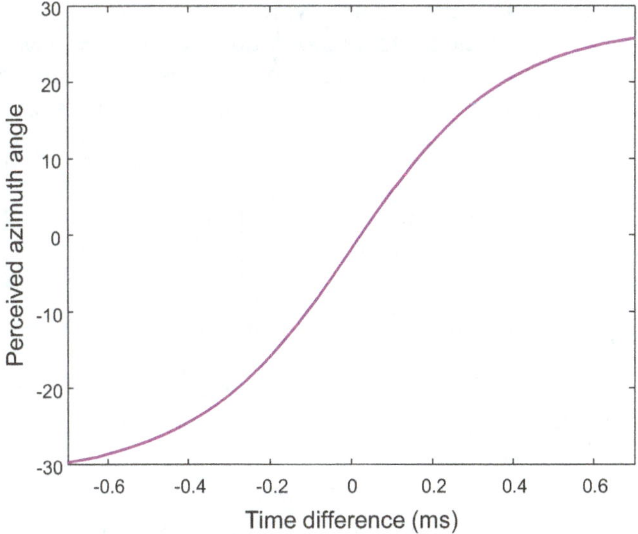

FIGURE 11.4
Perceived azimuth angle as a function of time difference.

just from whichever loudspeaker has the least delay. More generally, the fact that, for appropriate choice of delay, we perceive the two sounds as a single entity coming from the location of the first to arrive is known as the *precedence* effect.

This effect has several implications. In a stereo loudspeaker layout, if the listener is far enough away from the central position, he or she will locate the source as being located at the closest loudspeaker, and this apparent position does not change even if the other channel is significantly louder.

Subjective testing has shown that an almost equivalent relationship can exist between the effect of time difference and the effect of level difference on perceived azimuth angle [124]. However, the actual relationship can depend on the characteristics of the sounds that are played, the exact placement of the loudspeakers, and the listener [125].

Vector Base Amplitude Panning

Vector base amplitude panning (VBAP) is an extension of stereo panning to arbitrary placement of loudspeakers in three dimensions [126]. A unit vector,

$$\mathbf{l}_n = \begin{bmatrix} l_{n1} \\ l_{n2} \\ l_{n3} \end{bmatrix}, \tag{11.9}$$

points from the listening position toward loudspeaker n. Thus, unit vectors $\mathbf{l}_n$, $\mathbf{l}_m$, and $\mathbf{l}_k$ give the directions of loudspeakers n, m, and k. Then $\mathbf{p}$ gives a panning direction for a virtual source in terms of gains g_n, g_m, and g_k applied to the unit vectors in the directions of each loudspeaker.

$$\mathbf{p} = \begin{bmatrix} p_n \\ p_m \\ p_k \end{bmatrix} = \begin{bmatrix} \mathbf{l}_n & \mathbf{l}_m & \mathbf{l}_k \end{bmatrix} \begin{bmatrix} g_n \\ g_m \\ g_k \end{bmatrix} = \begin{bmatrix} l_{n,x} & l_{m,x} & l_{k,x} \\ l_{n,y} & l_{m,y} & l_{k,y} \\ l_{n,z} & l_{m,z} & l_{k,z} \end{bmatrix} \begin{bmatrix} g_n \\ g_m \\ g_k \end{bmatrix}$$

$$= \begin{bmatrix} l_{n,x}g_n + l_{m,x}g_m + l_{k,x}g_k \\ l_{n,y}g_n + l_{m,y}g_m + l_{k,y}g_k \\ l_{n,z}g_n + l_{m,z}g_m + l_{k,z}g_k \end{bmatrix}. \tag{11.10}$$

This can now be solved to find the gains that need to be applied to each loudspeaker in order to place the source in a given position.

$$\mathbf{g} = \begin{bmatrix} g_n \\ g_m \\ g_k \end{bmatrix} = \begin{bmatrix} l_{n,x} & l_{m,x} & l_{k,x} \\ l_{n,y} & l_{m,y} & l_{k,y} \\ l_{n,z} & l_{m,z} & l_{k,z} \end{bmatrix}^{-1} \begin{bmatrix} p_x \\ p_y \\ p_z \end{bmatrix}. \tag{11.11}$$

And finally, the applied gain should be normalized so that

$$\sqrt{g_n^2 + g_m^2 + g_k^2} = 1. \tag{11.12}$$

Ambisonics

Ambisonics comprises methods for both spatial audio recording and reproduction. By encoding and decoding sound signals for a number of channels, a two- or three-dimensional sound field can be captured and rendered. However, it can also be used to give virtual positions to sound sources that were not recorded using ambisonic techniques [127].

Unlike some other spatial audio formats, the transmission channels for ambisonics do not carry loudspeaker signals. Instead, they contain a speaker-independent representation of the sound field, which is then decoded to the listener's setup. This allows the producer to think in terms of source directions rather than loudspeaker positions, and offers the listener flexibility as to the layout and number of speakers used for playback.

Ambisonics can be used for full three-dimensional sound reproduction. However, we will focus on the two-dimensional case, and follow the approach taken in [128]. We consider an incoming reference wave and attempt to create an outgoing wave that provides a faithful representation of the original using a finite number of loudspeakers. Although we refer to incoming and outgoing waves for the original plane wave and the attempt at recreating it, both waves are converging on the centre of the listening area. These terms are used since the original reference plane wave is incoming on the loudspeakers, but the recreation attempt is outgoing from the speakers.

We will assume that both the incoming reference sound and the outgoing sound from the loudspeakers are plane waves, which is roughly valid if the listener is far from the source and the loudspeakers. Note, though, that these assumptions are not general requirements for ambisonics [129].

Now assume that we have the configuration shown in Figure 11.5. The x-axis points toward the front of the room, and the y-axis points toward the left of the room (this is typical in representing ambisonics). The listener is at a radial distance r at an angle ϕ, represented by the vector $\mathbf{r}$, and a plane wave comes from an angle ψ with respect to the x-axis.

So the plane wave is given by $P_\psi e^{j\mathbf{k}\cdot\mathbf{r}}$. P_ψ is the pressure of the plane wave, and $\mathbf{k}$ represents a wave at an angle ψ and with wave number $k = 2\pi f$, where f is the frequency (in acoustics, signals are often represented by wave number, which gives the number of wavelengths per 2π units of distance). The plane wave arriving at the listener may be expressed as:

$$S_\psi = P_\psi e^{jkr\cos(\phi-\psi)}. \tag{11.13}$$

Ambisonics attempts to reproduce this plane wave in the center of the listening area. If this wave can be expressed with a series expansion, then it should be possible to reproduce the wave if one were able to add an

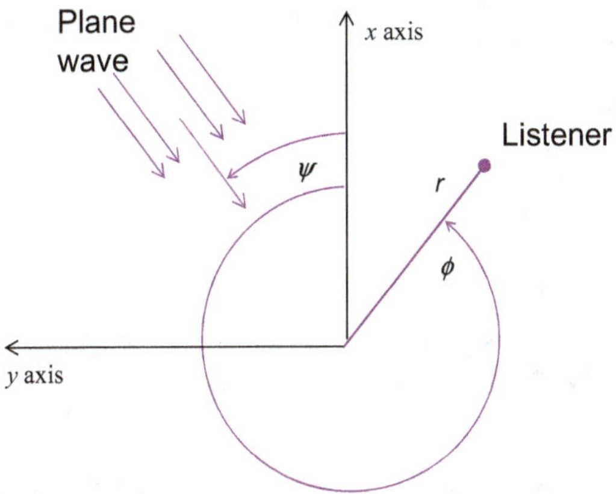

FIGURE 11.5
Standard depiction of the coordinate system in two-dimensional ambisonics.

infinite number of signals or channels to represent each term in the series. And the lower terms in the series can be used to give an approximation to that wave, and used to drive a system of loudspeakers. This is the essence of ambisonics.

From wave propagation theory [130,131], a plane wave can be expanded as a series of cosines and Cylindrical Bessel Functions,

$$S_{\psi} = P_{\psi} J_0(kr) + P_{\psi} \sum_{m=1}^{\infty} 2i^m J_m(kr)[\cos(m\psi)\cos(m\phi) + \sin(m\psi)\cos(m\phi)]. \tag{11.14}$$

The subscript ψ is used to denote a quantity with respect to the original plane wave and the subscript n is used to denote a quantity with respect to the nth speaker.

An ambisonic loudspeaker array should consist of N loudspeakers, each at the same distance from the centre point in a regular polygonal array, as shown in Figure 11.6. If each loudspeaker produces plane wave output, then the output of the nth speaker will be given by;

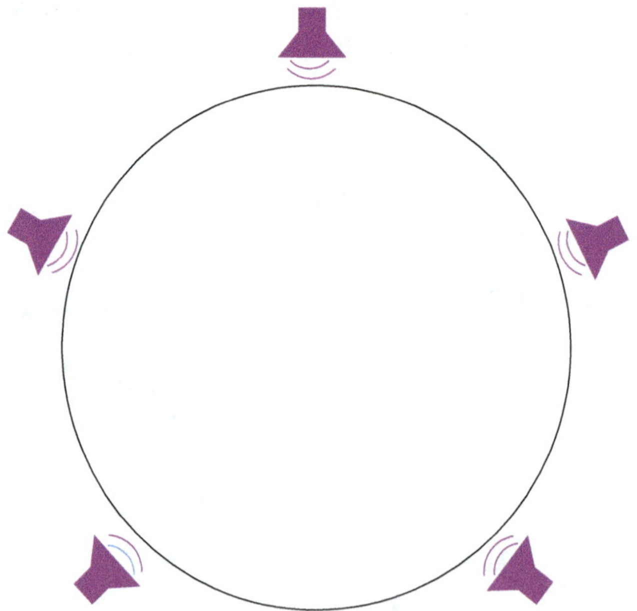

FIGURE 11.6
A typical ambisonics layout. Five loudspeakers arranged in a regular layout, suitable for second-order, two-dimensional ambisonics.

$$S_n = P_n J_0(kr) + P_n \sum_{m=1}^{\infty} \begin{matrix} 2i^m J_m(kr)[\cos(m\psi)\cos(m\phi_n) \\ + \sin(m\psi)\cos(m\phi_n)], \end{matrix} \tag{11.15}$$

where ϕ_n is the angle of the nth speaker. And summing this over the N loudspeakers gives the total reproduced wave;

$$S_n = \sum_{n=1}^{N} P_n J_0(kr) + \sum_{m=1}^{\infty} 2i^m J_m(kr) \begin{bmatrix} \sum_{n=1}^{N} P_n \cos(m\phi_n)\cos(m\phi) + \\ \sum_{n=1}^{N} P_n \sin(m\phi_n)\sin(m\phi) \end{bmatrix}. \tag{11.16}$$

This reproduced plane wave must match the original plane wave (1.1). So,

$$\begin{aligned} P_\psi &= \sum_{n=1}^{N} P_n \\ P_\psi \cos(m\psi) &= \sum_{n=1}^{N} P_n \cos(m\phi_n) . \\ P_\psi \sin(m\psi) &= \sum_{n=1}^{N} P_n \sin(m\phi_n) \end{aligned} \tag{11.17}$$

Equation (11.17) represents a set of criteria, known as *matching conditions*, that must be met in order for the original plane wave to match the reproduced plane wave. Each value m gives the mth order spherical harmonic that needs to be matched. Since we truncate the series, the higher spherical harmonics will not be reproduced.

For two-dimensional ambisonics, $2m+1$ channels are needed to reproduce the signal, where m is the order of the system. If a three-dimensional case were to be considered, the number of channels would be $(m+1)^2$ [128]. Thus, it's clear that three-dimensional ambisonics requires many more loudspeakers (and more complicated maths!).

For a zeroth-order match to a plane wave, only the lowest order in (11.17) will need to be matched, i.e., $m = 0$. This would match just the pressure of the waves, so that the original and reproduced plane waves have the same amplitude. The first part of Eq. (11.17) gives the pressure of the plane wave, denoted by W in ambisonic notation.

This first-order match, $m = 1$, provides the velocity of the plane wave. That is, the velocities of the original and reproduced plane waves will be identical, and hence both waves will be in the same direction. We can rewrite Eq. (11.17) for just zeroth- and first-order terms,

$$\begin{aligned} W &\equiv P_\psi = \sum_{n=1}^{N} P_n \\ X &\equiv P_\psi \cos\psi = \sum_{n=1}^{N} P_n \cos\phi_n \\ Y &\equiv P_\psi \sin\psi = \sum_{n=1}^{N} P_n \sin\phi_n. \end{aligned} \tag{11.18}$$

At this point, the feed or gain applied to the nth loudspeaker, P_n, will need to be considered. Here, there are multiple approaches, so we will use one similar to that provided in Michael Gerzon's original work [128,132]. The feed for a first-order N loudspeaker ambisonic system is

$$P_n = \left(W + 2X\cos\phi_n + 2Y\sin\phi_n\right)/N, \tag{11.19}$$

which will satisfy Eq. (11.17). As an example, let's suppose we have four loudspeakers, placed at 0°, 90°, 180°, and 270°. So (11.19) gives

$$\begin{aligned} P_1 &= (W+2X)/4 \\ P_2 &= (W+2Y)/4 \\ P_3 &= (W-2X)/4 \\ P_4 &= (W-2Y)/4 \end{aligned} \tag{11.20}$$

We can see that the matching conditions of Eq. (11.17) hold.

$$\begin{aligned} \sum_{n=1}^{N} P_n &= W \\ \sum_{n=1}^{N} P_n \cos\phi_n &= (W+2X)/4-(W-2X)/4 = X. \\ \sum_{n=1}^{N} P_n \sin\phi_n &= (W+2Y)/4-(W-2Y)/4 = Y \end{aligned} \tag{11.21}$$

We can now plug in Eq. (11.18) into Eq. (11.20) to arrive at the gain applied to each loudspeaker in order to render a plane wave source arriving from an angle ψ.

$$\begin{aligned} P_1 &= P_\psi\left(1+2\cos\psi\right)/4 \\ P_2 &= P_\psi\left(1+2\sin\psi\right)/4 \\ P_3 &= P_\psi\left(1-2\cos\psi\right)/4 \\ P_4 &= P_\psi\left(1-2\sin\psi\right)/4 \end{aligned} \tag{11.22}$$

By matching up to second order, we obtain two additional terms, $U = P_\psi \cos(2\psi)$ and $V = P_\psi \sin(2\psi)$. A second-order system comprises the five signals: *W, X, Y, U,* and *V*. These terms are then added to the feeds such that the above matching conditions still hold;

$$P_n = \left(W + 2X\cos\phi_n + 2Y\sin\phi_n + 2U\cos(2\phi_n) + 2V\sin(2\phi_n)\right)/N. \qquad (11.23)$$

If a system were to contain those five signals with enough loudspeakers, a second-order approximation of the plane wave would be reproduced in the center of the listening area.

Although ambisonics places a minimum requirement on the number of channels for a given order reproduction, it does not explicitly place a requirement on the number of loudspeakers. However, the matching conditions of Eq. (11.17) suggest that the number of loudspeakers be at least as many as the number of channels. For instance, if four loudspeakers are used with second-order ambisonics, then not all the channels will be used. In which case, it makes sense to just use a lower order reproduction.

Recall also that ambisonics typically requires that the angles between the loudspeakers must be equal. We can relax this constraint by adding parameters in front of the channels in equations (11.19) and (11.23) for the second-order case. The parameters are then determined for a particular layout by solving the system of equations given by (11.17).

Three-dimensional ambisonics is a straightforward extension of the two-dimensional case, for use with three-dimensional loudspeaker setups. But now all the maths and derivations become more challenging. Directions are expressed using either an Angle/Elevation representation or Cartesian coordinates, related by,

$$\begin{aligned} x &= \cos\alpha\cos\varepsilon \\ y &= \sin\alpha\cos\varepsilon \\ z &= \sin\varepsilon, \end{aligned} \qquad (11.24)$$

where α corresponds to the azimuth angle and ε to the elevation angle.

Wave Field Synthesis

Wave field synthesis (WFS) is another spatial audio reproduction method, used to create a virtual acoustic environment. It uses a large number of individually driven speakers to produce an approximation to the wave fronts produced by sound sources. The wave fronts that it produces are perceived as originating from virtual sources. Unlike some other spatial audio techniques, such as stereo or surround sound, the localization of virtual sources in wave field synthesis is not dependent on the listener's position.

Fundamental to understanding of wave field synthesis is Huygens' Principle. This states that any wave front can be represented as a superposition of elementary spherical waves. So we can attempt to simulate the wave fronts from a finite number of these elementary waves. To implement this, a large number of loudspeakers are controlled by software that, for each loudspeaker, plays a processed version of the source sound at the same time that the wave front would pass through the loudspeaker location.

The math behind WFS (and to some extent, behind ambisonics) is based on the Kirchhoff-Helmholtz integral, which states that when the sound pressure and velocity are known on the entire surface of a volume, then the sound pressure is completely determined within that volume. So we can position a large number of loudspeakers on that surface to approximately reconstruct the sound field. As with ambisonics, we'll focus on the two-dimensional space. In this case, we will consider a horizontal line of loudspeakers used to synthesize a wave field on the horizontal plane beyond that line.

Now look at this in more detail. If a wave emitted by a point source P_S with a frequency f is considered at any time t, all the points on the wave front can be taken as point sources for the production of a new set of waves with the same frequency and phase. The pressure amplitude of these secondary sources is proportional to the pressure due to the original source at these points. This principle of wave diffraction was discovered by Huygens in 1690, and is depicted in Figure 11.7.

Wave field synthesis is a direct application of this concept. Each loudspeaker in a loudspeaker array acts as a source of a secondary wave, and the whole system will recreate the sound field generated by sounds positioned behind the array. The challenge then is to determine the driving signals applied to each loudspeaker due to the original sources, their locations, and the locations of the loudspeakers.

Consider a source-free volume V enclosed by a surface S. The Kirchhoff-Helmholtz integral gives the Fourier transform of the sound pressure at a listening position L inside V,

$$P(\mathbf{r},\omega)=\frac{1}{4\pi}\int_S [P(\mathbf{r}_S,\omega)\frac{\partial}{\partial n}(\frac{e^{-j\omega|\mathbf{r}-\mathbf{r}_S|/c}}{|\mathbf{r}-\mathbf{r}_S|})-\frac{\partial P(\mathbf{r}_S,\omega)}{\partial n}\frac{e^{-j\omega|\mathbf{r}-\mathbf{r}_S|/c}}{|\mathbf{r}-\mathbf{r}_S|}]dS. \quad (11.25)$$

ω is the angular frequency of the wave, c is the speed of sound, $\mathbf{r}$ defines the position of the listening point inside V and $P(\mathbf{r}_S, \omega)$ is the Fourier transform of the pressure distribution on S.

This integral is incredibly complicated, and it is not necessary to delve into its derivation or fine details here. But it implies that by setting the correct pressure distribution $P(\mathbf{r}_S, \omega)$ and its gradient on a surface S, a sound field in the volume enclosed within this surface can be created. In our

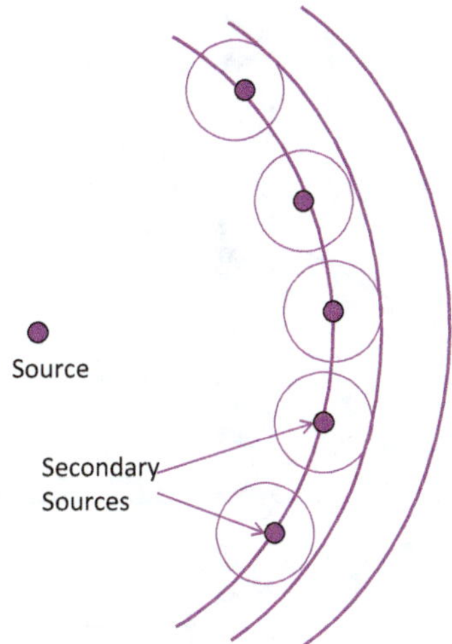

FIGURE 11.7
Illustration of Huygens' principle.

case, we have a finite set of loudspeakers, so we can set the pressure at discrete locations. So we transform the integral into a summation.

The pressure field produced at a distance d from a source with a spectrum $S(\omega)$ is:

$$P(d,\omega) = S(\omega)\frac{e^{-j\omega d/c}}{d}. \tag{11.26}$$

So we know the sound field that should be produced by a virtual source. Now consider the set-up shown in Figure 11.8. We won't go into the steps here, but we can find a driving function that determines the signal to send to each loudspeaker [133]. The driving functions for the loudspeakers are derived from the synthesized sound field produced by the notional sources.

$$Q_n(\mathbf{r},\omega) = S(\omega)\cos\theta_n\sqrt{\frac{j\omega}{2\pi c}}\sqrt{\frac{a}{b}}\frac{e^{-j\omega r/c}}{\sqrt{r}}. \tag{11.27}$$

This driving function $Q_n(\mathbf{r}, \omega)$ is the sound pressure at the nth loudspeaker due to the virtual source. Each loudspeaker is fed with a filtered version of the source signal.

The driving function's terms can be interpreted as follows:

- e^{-jkr} describes the delay due to sound traveling from the virtual source location to the nth loudspeaker.

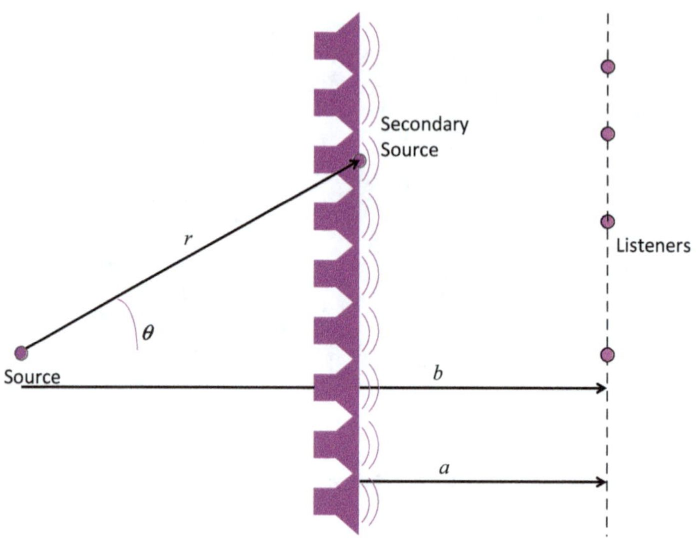

FIGURE 11.8
Geometry for the calculation of the driving functions, given a virtual source behind a line of loudspeakers.

- The amplitude factor $1/\sqrt{r}$ is the dispersion factor for a cylindrical wave.
- The factor $\sqrt{a/b}\cos\theta_n$ is an attenuation based on the distance between the virtual source, the loudspeaker array and the listener position. It does not change much within the listening area.
- $\sqrt{j\omega/(2\pi c)}$ gives high pass filtering of the signal. Essentially, for high frequencies, the line of loudspeakers corresponds to less samples per period, and so this filter ensures equal weighting regardless of frequency.

So, given the distance from a virtual source to the loudspeaker array, the distance from the loudspeaker array to the listener position, the number and distance between loudspeakers, we have all the necessary information to synthesize a wave field for the line of listener positions.

Note that the driving function is given in the frequency domain. So most implementations will split the signal into overlapping segments (overlap of 50%), and form the columns of a matrix with the N-point discrete Fourier transforms of these segments. This is repeated for as many loudspeakers as there are on the array. These matrices are then transformed into the time domain with an inverse Fourier transform.

The Head-Related Transfer Function

When spatializing sound for listening on headphones, a quite different approach is taken. We use both ears and other cues to determine the location of sound sources. A listener is able to determine the distance, elevation and azimuth (horizontal) angle of sound sources by analyzing and comparing the sounds received at each ear. Before entering the ear canal, a sound is modified by, among other things, the acoustics of the room, the shape of the listener's body, head, and outer ear. However, this is all missing if the sound is simply played back over headphones. So *binaural rendering* is applied to artificially recreate the sort of processing that would naturally be applied when sounds reach our ears.

There are two primary cues for azimuth: the Interaural Level Difference (ILD) and Interaural Time Difference (ITD). They relate to the different signal levels and delays received by each ear due to a source arriving from a given location. Much of the differences, especially level differences, are due to *head shadowing*, which describes sound traveling either through or around the head in order to arrive at an ear. The head provides filtering and significant attenuation of the source signal.

ILD and ITD are what we exploit for panorama and precedence. At low frequencies below about 1.6 kHz, the auditory system analyses the interaural time shifts between the signal's fine structure. But at high frequencies, the Inter Level Difference (ILD) is used, along with the timing between the envelopes of the signals [134]. The maximum naturally occurring ITD is about 0.65 ms [135], and the maximum ILD is about 20 dB, and both can be estimated based on the shape of the human head. For sources at the side, both ILD and ITD are at their maximum. And for sources directly in front or behind the listener, they are near zero.

However, the ILD and ITD do not provide much information about the sound source elevation. This information is mostly given by the filtering done by the head, body and outer ear before sounds reach the inner ear. This effect is referred to as the *Head-Related Transfer Function* (HRTF), though of course it relates to much more than just the shape of the head.

An HRTF is a transfer function that characterizes how sound from a particular point in space is received by the ear. A pair of HRTFs, one for each ear, can be used to synthesize binaural sound. HRTFs are very useful for accurate localization of sounds, especially those not coming just from the horizontal plane. HRTFs for the left and right ear give the filtering of a sound source (represented here in the time domain) $x[n]$ before it is perceived by the ears as $x_L[n]$ and $x_R[n]$, respectively. The HRTF may be found by measuring a head-related impulse response (HRIR).

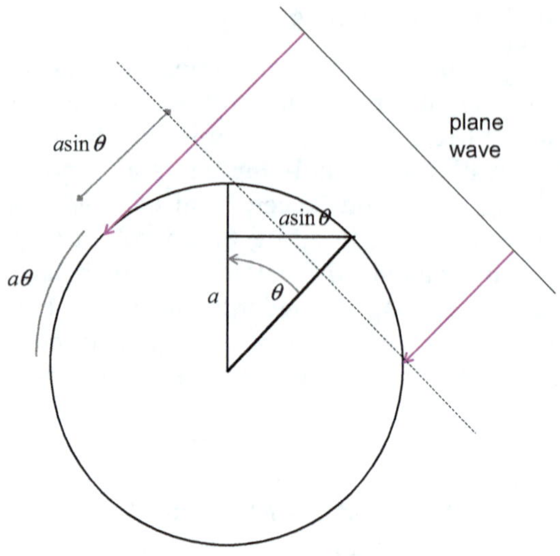

FIGURE 11.9
A sound wave from a distant source acts like a plane wave, and approaches a spherical head of radius a from a direction specified by the azimuth angle θ.

ITD Model

The physicist Lord Rayleigh (1842–1919) was perhaps the first to explain the ITD [136], in his description of sound localization, known as *duplex theory*. Assume the listener's head is completely spherical (this is a ludicrous assumption, but a common starting point for modeling binaural sound), with radius a. A sound source is located far from the listener, at an angle θ, and so the sound is approximately a plane wave. This is depicted in Figure 11.9.

Unless $\theta = 0$, the sound will arrive at one ear first. In the figure, this is the right ear. The sound must first travel an additional distance $a\sin(\theta)$ to reach the left-hand side of the head, and then a distance $a\theta$ to reach the left ear. From this, we have the following estimate of the interaural time difference,

$$\text{ITD} = a(\theta + \sin\theta)/c. \tag{11.28}$$

From Eq. (11.28), the maximum ITD occurs when the sound arrives from one side, along the line that intersects both ears, and is given by $a(\pi/2+1)/c$. This is approximately 0.65 ms for a typical human head, and agrees with the measured ITD maximum mentioned above. As with the interchannel time difference that can be exploited in a stereo loudspeaker set-up, we can use this effect to change the perceived azimuth angle of a source when listening over either loudspeakers or headphones.

Equation (11.29) provides an implementation of the ITD, where the azimuth is offset to the positions of the ears [137]. It introduces two azimuth-dependent delays, one for each ear.

$$T_d(\theta) = \begin{cases} a(1-\cos\theta)/c & |\theta| < \pi/2 \\ a(|\theta|+1-\pi/2)/c & \pi/2 < |\theta| < \pi \end{cases} \quad . \tag{11.29}$$
$$\text{ITD} = T_{d,L}(\theta+\pi/2) - T_{d,R}(\theta-\pi/2) = a(\theta+\sin\theta)/c$$

This simple model generates a sound that moves smoothly from the left ear to the right ear as the azimuth goes from –90° to +90. But it is not effective in creating the impression that the sound came from outside the head, or in assisting with front/back discrimination. Furthermore, though the ITD cue suggests that the source is displaced, the energy at the two ears is the same, and thus the ILD cue indicates that the source is in the center. So the listener may sometimes get the impression of two sounds, one displaced and one at the center of the head. This problem can be addressed by adding head shadow, as discussed next.

ILD Model

Based on Lord Rayleigh's duplex theory, the magnitude response for sound at a frequency ω and azimuth angle θ can be approximated by a simple transfer function [138,139].

$$H(\omega,\theta) = \frac{\alpha(\theta)\omega+\beta}{\omega+\beta} \quad \text{where } \alpha(\theta) = 1+\cos\theta, \beta = 2c/a. \tag{11.30}$$

This transfer function will give high-frequency attenuation for azimuth angle $\theta=\pi$, and will boost the high-frequency content when $\theta=0$, as is expected for head shadowing. As in Eq. (11.29), if we again offset the azimuth to the ear positions, we get the following model for the ILD.

$$H_L(\omega,\theta) = \frac{\alpha(\theta+\pi/2)\omega+\beta}{\omega+\beta}$$
$$H_R(\omega,\theta) = \frac{\alpha(\theta-\pi/2)\omega+\beta}{\omega+\beta} \quad \text{where } \alpha(\theta) = 1+\cos\theta, \beta = 2c/a \tag{11.31}$$

As with the ITD model, the ILD model does not provide good front/back discrimination or externalization. However, as the azimuth angle is changed, it can give a smooth perceived motion of for the virtual sound source.

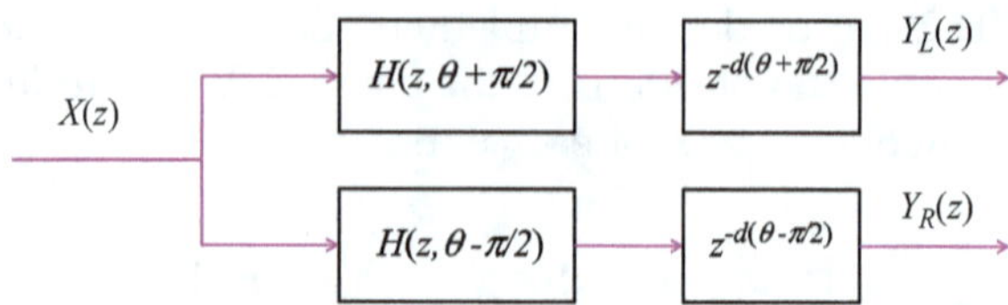

FIGURE 11.10
HRTF modeling ITD and ILD as a filter based on the input signal and azimuth angle.

At high frequencies, this transfer function will not produce a significant delay between the response for left and right ears. The ITD and the ILD models need to be combined so that both time and level information give the same spatial cues.

The Spherical Head

By cascading the ITD and the ILD, as in Figure 11.10, we obtain a spherical-head model of the HRTF.

The HRTF that we have developed here only deals with the azimuth angle. It could be extended to incorporate both azimuth and elevation angles. However, the elevation cues are much harder to notice and pick up on than the azimuth cues. One reason for this is that the HRTF parameters often need to be fine-tuned to a particular user's ears.

Monaural Pinna Model

The pinnae are the ridges on the outer part of the ear. The way that they reflect incoming sound depends on the angle of the source, and so they provide significant information used for sound localization. Pinna reflections are used in the perception of elevation. And notably, the way that sound reflects off of the ridges on the pinnae is used to help distinguish whether a sound source is arriving from in front or in back of the listener. Sounds from behind the listener will have very few of these reflections. So, without this information (like when listening over headphones without binaural processing), the sound will usually appear to have come from behind the listener, or even from inside his head.

The pinnae are also involved in resolving the Cone of Confusion [134], which are virtual source locations that all produce the same ITD and same ILD values.

Simulation of the pinnae reflections can be achieved by tapping a delay line. The pinnae reflections are short echoes that result in notches in the spectrum whose positions depend on the elevation angle. A typical simulation of pinnae reflections is shown in Figure 11.11. This can be applied independently to the left and right signals, $Y_L(z)$ and $Y_R(z)$ that result from filtering the source.

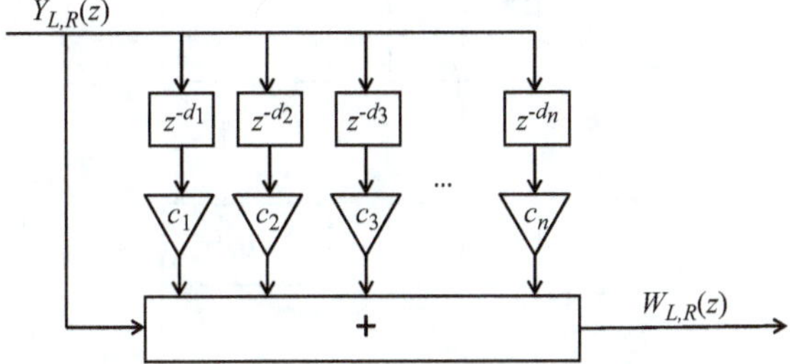

FIGURE 11.11
The pinna echoes implemented as an FIR filter.

The gains and time delays are a function of azimuth and elevation. This dependence is non-trivial and may vary from person to person. However, even a simple model helps provide more effective spatialization of the sound.

Implementation

Joint Panorama and Precedence

As mentioned, sound sources can be "placed" in an apparent location using two factors: delay and amplitude. A delay of 1 ms is enough to give the impression of coming almost completely from one side. A difference in amplitude of 30 dB or more between two equally spaced speakers will have a similar effect.

A simple block diagram that provides control of both panorama and precedence is shown in Figure 11.12. The range of gains and delay values should be set to cover the values in Figures 11.3 and 11.4.

Ambisonics, and Its Relationship to VBAP

One popular first order, 3D ambisonics encoding format is known as B-Format. It uses the following channels:

$$\begin{aligned} W &\equiv P_{\alpha,\varepsilon}/\sqrt{2} \\ X &\equiv P_{\psi} \cos\alpha \cos\varepsilon \\ Y &\equiv P_{\psi} \sin\alpha \cos\varepsilon \\ Z &\equiv P_{\psi} \sin\varepsilon \end{aligned} \tag{11.32}$$

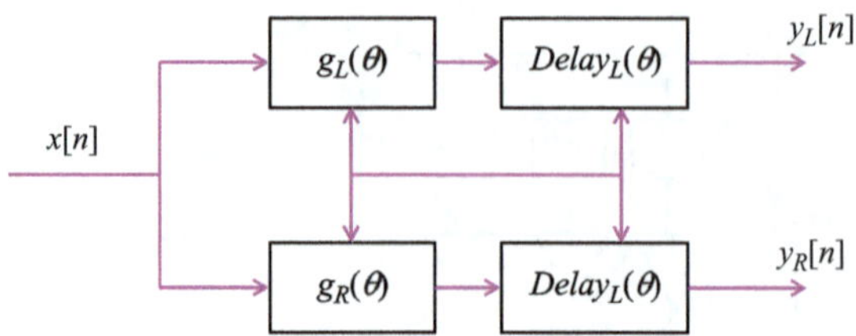

FIGURE 11.12
Joint control of panorama and precedence.

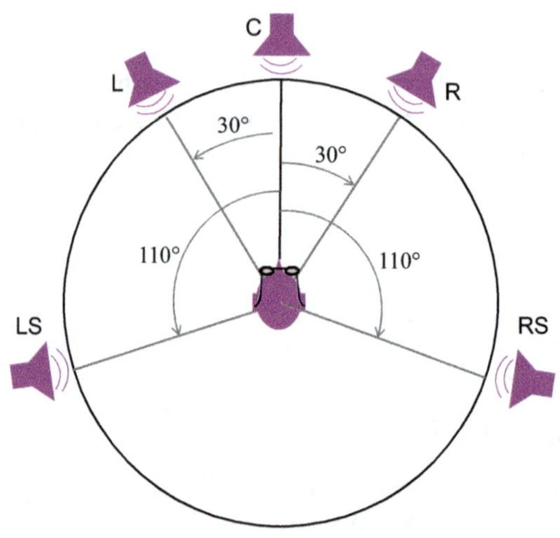

FIGURE 11.13
A five-channel surround sound system.

The loudspeakers are arranged on the surface of a virtual sphere, and the gains applied to each loudspeaker are found by using a linear combination of these four channels, where each gain depends on the angle of the line between that loudspeaker and the center of the sphere. A popular decoding function is

$$P_n = \left(W/\sqrt{2} + X\cos\alpha_n\cos\varepsilon_n + Y\sin\alpha_n\cos\varepsilon_n + Z\sin\varepsilon_n\right)/N. \quad (11.33)$$

Note that this format is written slightly differently from the one we presented for two-dimensional ambisonics, since B-Format has a root two term in Eq. (11.32).

In other approaches to decoding ambisonics, spatial equalization is applied to the signals to account for the differences in the methods we use

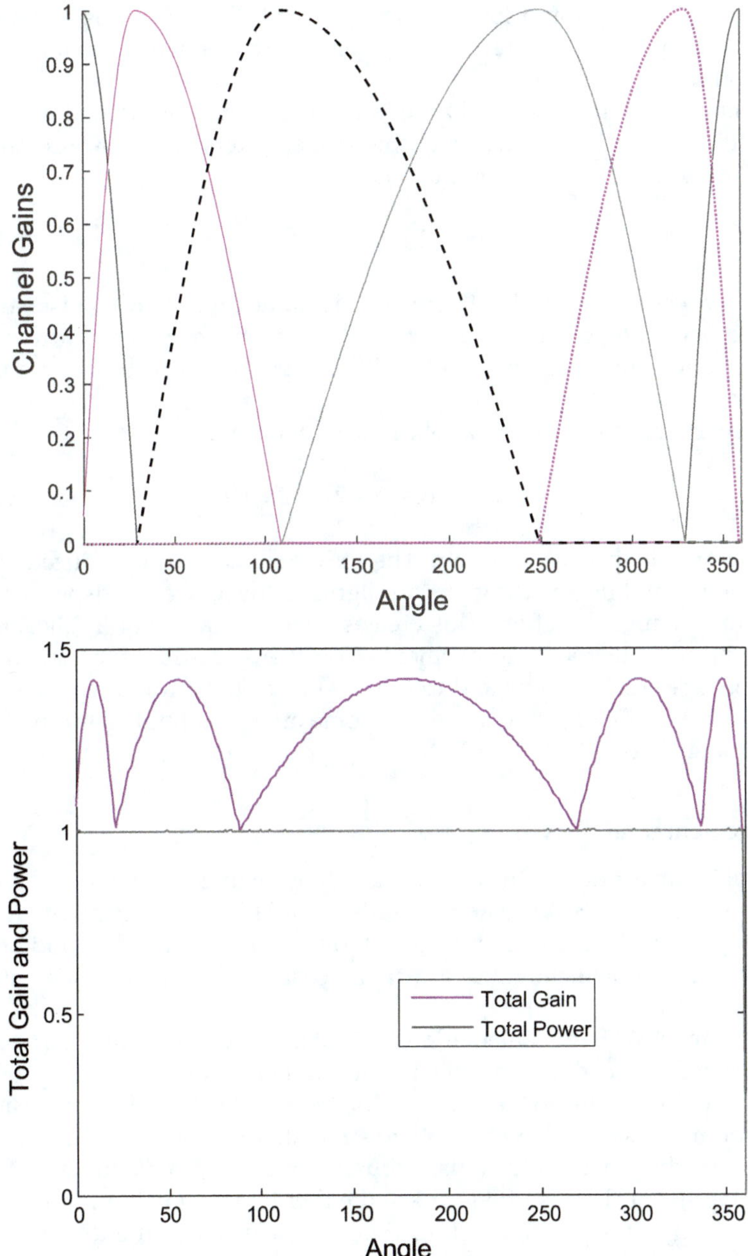

FIGURE 11.14
Constant power panning for five channels: (top) channel gains, (bottom) total power and total gain.

for localizing high and low frequency sound. Further refinements take into account the distance from the listener to the loudspeakers and to the virtual source location.

Ambisonics can be viewed as an amplitude panning method in which the source signal is applied to all the loudspeakers. For first-order ambisonics, the gains are set as in Eq. (11.34) [140],

$$g_i = (1 + 2\cos\alpha_i)/N. \tag{11.34}$$

where g_i is the gain of the ith speaker, N is the number of loudspeakers, and α is the angle between the loudspeaker and the panning direction. The sound signal is generated by all loudspeakers, which may result in spatial artifacts.

Second-order ambisonics applies the gain factors

$$g_i = (1 + 2\cos\alpha_i + 2\cos 2\alpha_i)/N. \tag{11.35}$$

to a similar loudspeaker system. The source signal is still produced by all loudspeakers, but now the gains are significantly less for loudspeakers far from the source direction. This creates fewer artifacts, and higher order designs offer further improvement. However, we still have constraints on loudspeaker positions and requirements for the minimum number of loudspeakers that are not needed for some other amplitude panning techniques, such as VBAP.

Implementation of WFS

Though wave field synthesis has incredibly high potential, it has so far been mostly limited to custom installations. This is at least partly due to the high cost. A very large number of loudspeakers is needed, and hence most WFS implementations are more costly than other spatial audio alternatives.

With most WFS implementations, one of the most perceptible difference between the intended and rendered sound field is error in the reproduction of ambience due to the reduction of the sound field to two dimensions. Furthermore, since WFS attempts to simulate the acoustic characteristics of a recording space, the actual acoustics of the reproduction area must be suppressed. This is often achieved either by minimizing room reflections, or by limiting playback to the near field where the direct sound dominates.

There are also undesirable spatial distortions due to spatial aliasing. That is, the discretization to a finite number of loudspeakers results in an inability to effectively render narrow frequency bands. Their frequency depends on the angle of the virtual source and on the angle of the listener to the loudspeaker arrangement:

Since the reproduced wavefront is a composite of elementary waves, a sharp pressure change can occur at the last loudspeaker in the array. This is known as the *truncation effect* [133,141]. It is essentially spectral leakage, but occurring in the spatial domain, and can be viewed as a rectangular window function applied to an infinite array of speakers.

HRTF Calculation

The most common method used to estimate an HRTF from a given source location is to measure the head-related impulse response (HRIR), $h[n]$. This is done by generating an impulse at the source location and then measuring the received signal at the ear. The HRTF $H(z)$ is then the Fourier transform of the HRIR. Alternatively, HRTFs can also be calculated in the frequency domain using a sine wave whose frequency increases over time, otherwise known as a chirp signal [142].

HRTFs are complicated functions of both spatial location and frequency of the source. But for distances greater than one meter from the head, the HRTF magnitude is roughly inversely proportional to distance. So at least in most cases, the dependence on distance can be estimated, and it is for this far field case that the HRTF, $H(f, \theta, \varphi)$, that has most often been measured.

An anechoic chamber (see Chapter 10) is often used to measure HRTFs in order to prevent reverberation from causing inaccuracies in the measured response. The HRTFs are measured with varying azimuth and elevation angle, and thus HRTFs at angles that were not measured can be estimated by interpolating between nearby measurements. But even with small increments in the direction of arrival for measured HRTFs, these interpolated values can contain significant errors.

Applications

Transparent Amplification

For a live performance, it is often desirable to place the virtual sound sources all in front of the listener, so that they are perceived as coming from the sound stage. Furthermore, it's important that the time at which the sound is heard agrees with the time it would take for the sound to travel to the listener from the performer seen creating the sound on stage. However, if loudspeakers are placed near the back of the venue, if no delay is applied, they won't be aligned with the front speakers or with the performance. And if only the front loudspeakers are used, listeners near the back of the venue would hear the performance at reduced levels, and without much stereo width. So the audience would not be provided with a well-balanced sound.

One solution, known as *transparent amplification*, is to delay the signal sent to the rear loudspeakers. Ideally, it could be made to match the delay introduced by sound traveling from the front loudspeakers to the rear. However, as long as the delay is at least this long, the precedence effect will help ensure that the perceived location of the sound source is near the front speakers, which are placed close to the stage. So for a room of length L, the delay should be set to $D = L/c$ seconds, where c is the speed of sound.

Surround Sound

Constant power stereo panning can be extended easily to the case of an arbitrary number of loudspeaker channels. This is fundamental to surround sound systems, which aim to reproduce spatial audio to an audience, especially in front of the listeners. In Figure 11.13, angles are given to the right (R), center (C), left (L), surround left (LS), and surround right (RS) channels in a possible surround sound system. Note that most specifications for surround sound will allow some flexibility in the placement of the loudspeakers.

The gain of the two loudspeakers that are on either side of the desired source angle is the same as for the case of two channel stereo panning, and all other gains are set to zero. Figure 11.14 shows the channel gains, total gain, and total power for five channel surround sound, using a constant power panning law.

The most common surround sound system is known as 5.1. The 'point one' is due to the addition of a subwoofer, which delivers the low-frequency content. The location of this subwoofer is not considered important, since we do not have a good perception of the localization of (very) low frequencies.

Sound Reproduction Using HRTFs

Similar to convolutional reverb (Chapter 10), convolution of a sound signal with the HRIRs for each ear will produce the sound that would have been heard by a listener due to the source being played at the appropriate angle and distance. Using this approach, HRIRs have been employed to create virtual surround sound. When played over headphones, recordings that have been convolved with an HRTF will be perceived as if they comprise sounds coming from various locations around the listener, rather than directly into the ears without any binaural cues. The perceived localization accuracy from this approach will depend on how well the used HRTF matches the characteristics of the listener's actual HRTF.

Some consumer products that are intended to reproduce spatial sound when listened to over headphones will use HRTFs. Variations on HRTF processing are also sometimes used to give the appearance of surround sound playback from computer loudspeakers.

Using an overlap and add method (described in Chapter 9), it is possible to create the illusion of moving, rotating a sound source about the listener's head. The procedure consists of performing HRTF processing on a small segment of the audio, and then overlap-adding this segment together with another at a slightly different azimuth or elevation angle. The overlap-add does not attempt to perform any processing on the input as one might do in the case of most phase vocoder applications. It simply attempts to smooth different frames of output.

Further Reading

Seminal papers in the field of spatial audio include Michael Gerzon's original paper on ambisonics [143], Ville Pulkki's original paper on Vector-Based Amplitude Panning [126], and what might be the first paper to apply measured HRTFs for binaural rendering [144]. Although VBAP is still described and implemented in a very similar manner to Pulkki's original work, ambisonics as described and implemented today bears only a slight resemblance to Gerzon's original description. For those interested in the historical and personal perspective, Robert Charles Alexander has written interesting biographies of the inventor of stereo, Alan Blumlein [145], and of the inventor of Ambisonics, Michael Gerzon [146]. Of course, Blumlein and Gerzon are both significant figures for many more innovations.

Though HRTFs are still frequently measured, especially for personalized binaural rendering based on measuring the listener's HRTFs, there also exist a large number of widely used databases of HRTFs, such as CIPIC [147]. These collections, as well as databases of room impulse responses, have led to the Spatially Oriented Format for Acoustics (SOFA) standard [148].

Spatial audio formats continue to be proposed and implemented, such as Directional Audio Coding (DirAC) [149] and Spatial Audio Object Coding (SAOC) [150]. Many recent innovations in spatial audio focus on binaural rendering, with [151] providing a review of the field.

Problems

1. Using the formula for ITD, estimate the ITD for a spherical head of radius 8.5 cm, a source in the far field arriving from an angle of 45°, and where the speed of sound is assumed to be 340 m/s.
2. Consider a set of loudspeakers that could be positioned at arbitrary angles on the perimeter of a circle. Given these conditions, explain how ambisonics, stereo panning, multichannel surround sound, vector-based amplitude panning, and wave-field synthesis could be used for positioning sources. Comment on the number of loudspeakers, their positions, and the possible positions of sources in each case.
3. Which of the spatial audio techniques that we have discussed, and under which conditions, can represent a source at a given azimuth angle, at a given elevation angle, and/or with a given depth?
4. Assume loudspeakers are at 45° from the frontal position, and a virtual source is placed at a 30° angle. Estimate the gain applied to each loudspeaker for stereo panning using both the tangent law and the sine law.
5. Consider two-dimensional first-order ambisonics with only three loudspeakers at the angles 0°, 120°, and 240°. What would be the driving equations for each loudspeaker? Derive the gain applied to each loudspeaker, given a plane wave source at an angle ψ.
6. Assume four loudspeakers, placed at angles $\theta = 45°$, 135°, 225°, and 315° (so they are equidistant, each 90° apart). Sketch/generate figures for amplitude as a function of virtual azimuth angle under first-order and second-order two-dimensional ambisonics.
7. Compare and contrast ambisonics, wave field synthesis, and vector-based panning. What do they have in common, and what are the advantages and disadvantages of each one?

12

Immersive Audio

Immersive audio is a general term used to describe audio technologies and sound design practices that aim to create a three-dimensional soundscape, enveloping the listener in a realistic and engaging sonic experience. There is a strong relationship between immersive and spatial audio. Though both spatial audio and immersive audio are intended to create a sense of space and depth in the audio experience, audio may be considered immersive without specifying the spatialization technique. Similarly, audio may be spatialized without being immersive.

But what is meant by *immersion*? Immersion is often confused with terms like envelopment and engulfment, and [152] suggests that immersion has become a "vague and diluted concept." Immersion is sometimes treated as a psychological state, and sometimes as a perceptual phenomenon having to do with feeling surrounded. Here, we are interested in the perceptual aspect, that sense of being 'in it'.

Here, we will focus on immersion as an audio effect, i.e., an approach to processing audio to make it immersive. So, for example, the listener may move around, and their orientation is taken into account. Spatial audio effects, such as panning and ambisonics, are typically applied to place sound sources anywhere in the two- or three-dimensional space around the listener. Immersive audio processing often goes much further, allowing both the listener and all sources to be placed in arbitrary positions, with arbitrary orientations, as well as to define directional properties of the sources, how sound levels decay with distance and how the spatial sound is finally rendered.

The potential of this immersive audio is vast and exciting. It places listeners in a virtual auditory world. There is an element of responsivity to immersion, i.e., a listener could move around in the virtual environment, and the presented sound should change accordingly. Sounds are not just presented to the listener; immersive audio aims to provide a sonic experience as if the listener were really there.

This chapter uses a step-by-step approach to explain how immersive audio can be achieved. First, the coordinate system for all spatial audio operations is described. Then the virtual listener's properties and use are given.

DOI: 10.1201/9781003593942-12

Properties for positioning and orientation of a sound source are then given so that one can fully understand the geometry. Further properties for establishing a source's directivity, the distance model, and the panning model are all explained, including the relationships between parameter settings and their associated processing of the audio stream. This is all put together to show the full workings and use of an immersive audio renderer. A simplified code example is then given for a listener in a 2D space, with a specified distance model and panning model.

Finally, we discuss and implement a strong perceptual cue that takes into account the listener's position in a changing sound field, the Doppler effect.

The approach taken here is similar to that of the PannerNode in the Web Audio API [13], or Microsoft's Direct 3D. Similar immersive rendering systems are also part of game audio and virtual reality (VR) simulation engines, such as the Unreal Engine or Unity.

Spatial Audio in an Immersive Environment

Audio sources can be positioned and moved around in a three-dimensional space relative to a listener. Such spatialized audio can greatly increase the immersiveness of a game or VR experience, or help bring out the rich detail in a piece of music. Typically, parameters are available for controlling a source's position, orientation, directivity (the sound cone), distance rendering, and panning model. Furthermore, it takes as input the listener's location and forward and up vectors representing the direction in which the listener is facing.

Our immersive audio effect will spatialize an incoming audio stream in three-dimensional space and render it appropriately for a given listener.

The input audio is assumed to be mono (1 channel), and the output is multichannel when the audio effect is active. Inputs with more channels can be downmixed appropriately.

Essentially, the immersive audio effect applies a series of processing steps on its input to arrive at a spatialized output.

1. Distance attenuation, based on a distance model and the distance between source and listener
2. Directivity attenuation, based on sound cone parameters for the source and the angle between the source's orientation vector and the vector from listener to source

3. Panning of the source, based on a panning model and the azimuth and elevation angles between source and listener.

To fully explain the effect, we will first explain the geometry for a source and listener, and then deal with each of these aspects in turn, before finally giving a simplified 2D code example. Be sure to use headphones (or at least stereo speakers) when listening to the example, so that you can appreciate how the left and right channels are transformed by the spatialization approach.

The Coordinate System

Both the listener and the spatialized sources have a position in a three-dimensional space using a right-handed Cartesian coordinate system.[1] The units used in the coordinate system are not defined since the spatialization depends on orientation angles and relative, not absolute, distances.

The right-hand coordinate system is depicted in Figure 12.1, with X, Y, and Z dimensions. Typically, X represents horizontal (left to right), Y vertical (down to up), and Z longitudinal (front to back). The orientation is maintained if one considers X pointing to the right, Y pointing forward, and Z pointing up.

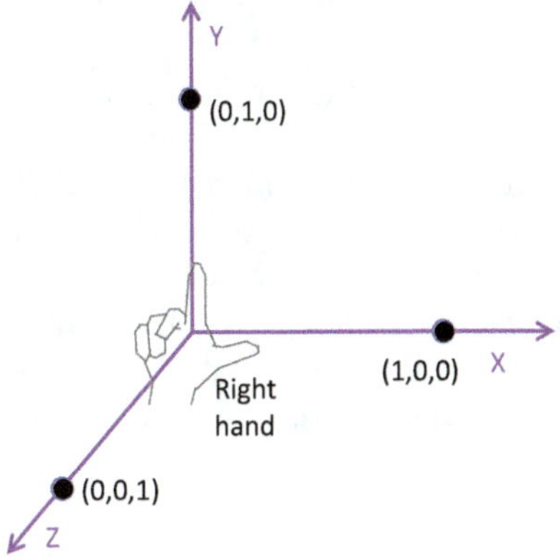

FIGURE 12.1
Right-hand coordinate system. When holding out your right hand in front of you as shown, the thumb points in the X direction, the index finger in the Y direction, and the middle finger in the Z direction.

The Listener

There is a single listener whose properties can be set so that the actual listener will hear the sounds as if they are at that position and orientation in space.

By defining the listener's position, forward direction and listener's head position, we have nine properties representing three-dimensional vectors;

- `Position` (X,Y,Z): horizontal/vertical/longitudinal position of the listener. Default is to position the listener at the origin, (0,0,0).
- `Forward` (X,Y,Z): horizontal/vertical/longitudinal position of the listener's forward direction. Default is (0,0,–1).
- `Upward` (X,Y,Z): horizontal/vertical/longitudinal position of the top of the listener's head. Default is (0,1,0).

This represents the location and orientation of the person listening to the audio scene. So the default situation is that the listener is at the center of the coordinate system facing forward, with the top of their head pointing upwards.

The `position` vector represents the location of the listener in 3D Cartesian coordinate space. Note that the sound source (the input audio) may also be at any given location in that space.

The `forward` vector represents the direction the person's nose points toward. The up vector represents the direction the top of a person's head is pointing. The forward and up vectors should be normalized and orthogonal to each other. Together these two vectors determine the orientation of the listener.

The Listener is not a separate audio effect, but the same listener could be used with many different immersive audio effects. That is, a single listener can receive many different sound sources with different spatial properties. The listener's properties can also be time varying or automated.

Typically, there is only one listener per renderer. This is not an issue because you can imagine a situation where many people are playing a multiplayer game, each one has their own renderer with different settings of their listener's parameters, and so each will hear sounds as if they were in a different location.

The Sound Sources

Though there is only one listener, there can be many different sound sources placed at many different positions. The position and orientation vectors for each sound source are similar to the listener's position,

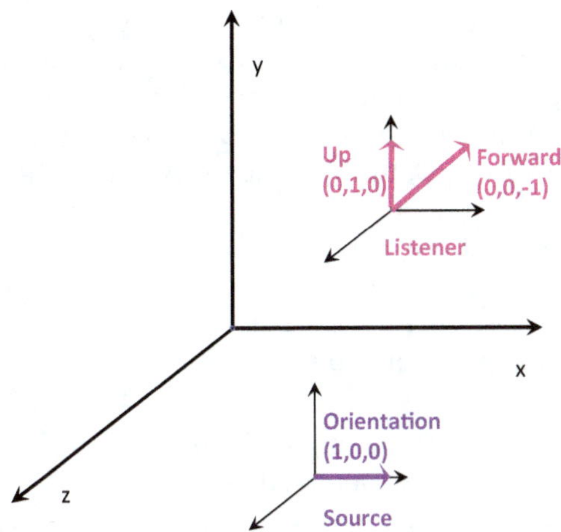

FIGURE 12.2
Listener and source in space.

forward and up properties vectors, and define the source's location and the direction that it points toward.

- Position (X,Y,Z): horizontal, vertical, and longitudinal position of the audio source.
- Orientation (X,Y,Z): horizontal, vertical, and longitudinal direction in which the audio source is facing. The sound is projecting toward this direction.

So far, so good. We can now position and orient the listener and any sources, as shown in Figure 12.2. The coordinate system for spatialization is shown with typical values. The locations for the listener and sound source are moved from the default positions so we can see things better.

Directivity

Each sound source's sound directivity characteristics are given by an inner and outer *sound cone* that gives sound intensity as a function of the source/listener angle from the source's orientation vector. Thus, a sound source pointing directly at the listener will be louder than if it is pointed off-axis. Sound sources can also be omnidirectional, in which case they are heard equally regardless of orientation.

Note that the physical basis of audio directivity is far more detailed than this. In particular, higher frequencies typically experience stronger attenuation off-axis. This frequency-dependence can be simulated using the game audio middleware Wwise, but is lacking in the sound cone renderers for MS Direct3D, the Web Audio API, and the implementation described herein.

The sound cone that we will implement is defined by the following three properties;

- `coneInnerAngle`: Cone angle (degrees) where there is no volume reduction.
- `coneOuterAngle`: Cone angle (degrees) outside which the volume is reduced by coneOuterGain.
- `coneOuterGain`: Attenuation applied outside cone defined by coneOuterAngle attribute. This is in the range 0 to 1, with 1 meaning there is no attenuation and 0 meaning that no sound is heard.

Figure 12.3 illustrates the relationship between the source's sound cone with respect to the listener. In the diagram, `coneInnerAngle` = 60 and `coneOuterAngle` = 120. That is, the inner cone extends 30° on each side of the direction vector, and the outer cone is 60° on each side.

Once you have specified an inner and outer cone, you end up with a separation of space into three parts:

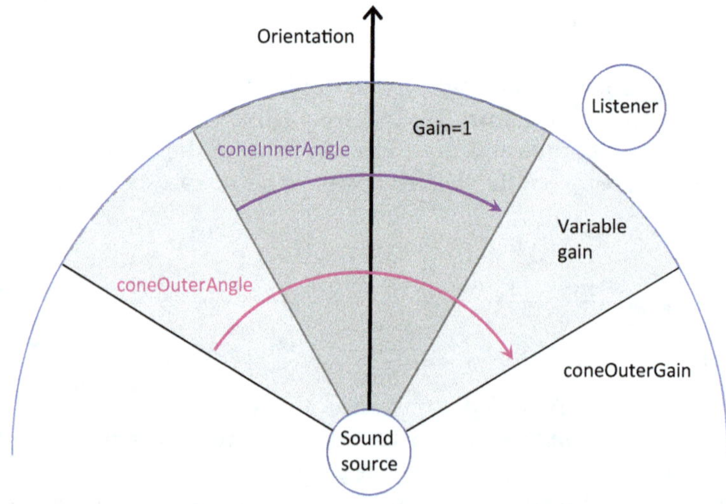

FIGURE 12.3
Cone angles for a source in relation to the source orientation and the listener's position and orientation.

1. Inside the inner cone
2. Between inner and outer cone
3. Outside the outer cone.

Each of these sub-spaces can have a gain multiplier associated with it, calculated as follows.

Assume **S** is the source position vector and **L** is the listener's position. First, we define the normalized vector from listener to source,

$$\mathbf{V} = (\mathbf{S} - \mathbf{L}) / |\mathbf{S} - \mathbf{L}|. \tag{12.1}$$

This translates the problem to a coordinate system where the listener is at the origin, and we can just use normalized vectors to calculate the angles.

Let $\mathbf{S_O}$ be the source orientation vector, C_I the cone's inner angle, C_O the cone's outer angle and G_O the cone outer gain. If $\mathbf{S_O} = \mathbf{0}$, or $C_I = C_O = 360$, then no sound cone is specified. So the source is omnidirectional and the gain is set to 1. Otherwise, the angle between source orientation vector and source-listener vector is given by

$$A = 2 \cdot 180 \left| \operatorname{acos}(\mathbf{V} \cdot \mathbf{S}_0) \right| / \pi. \tag{12.2}$$

Here, the angle was multiplied by 2 since the inner and outer angles are given in terms of the entire range from 0° to 360°.

The gain G due to the sound cone is now determined by

$$G = \begin{cases} 1 & A \le |C_I| \\ 1 + (G_O - 1)\dfrac{A - |C_I|}{|C_O| - |C_I|} & |C_I| < A < |C_O| \\ G_O & |C_O| \le A \end{cases}. \tag{12.3}$$

So there is no attenuation when the source falls inside the inner cone, maximal attenuation when it is outside the outer cone, and a linear transition between the two values when it is between the inner and outer cones.

For example, to produce a highly directional sound, we can set $C_I = 10$, $C_O = 30$, and $G_O = 0.1$, giving a very narrow range where the sound is heard without attenuation, and a quick transition to heavily attenuated sound. On the other hand, setting $C_I = 180$; $C_O = 270$; $G_O = 0.4$ gives no attenuation for any sound in front of the listener, and only mild attenuation as the sound transitions to being completely behind the listener.

Distance Models

So now we have dealt with how the *angle* between a source and listener affects the perceived source level, but we haven't yet dealt with how the *distance* between a source and listener affects the perceived source level. For that, we introduce four more properties;

- **distanceModel**: Algorithm to reduce source volume as it moves away from listener. 'linear', 'inverse' and 'exponential'.
- **refDistance**: For distances between source and listener greater than this, the volume will be reduced based on rolloffFactor and distanceModel.
- **rolloffFactor**: How quickly volume reduced as source moves away from listener.
- **maxDistance**: Maximum distance between source and listener, after which volume is not reduced more.

The refDistance, rolloffFactor, and maxDistance properties are only used as input parameters for calculating the gain attenuation with distance. So they can be explained just by giving the distance models.

Let $d = |\mathbf{S} - \mathbf{L}|$ be the distance between source and listener, d_{ref} the reference distance and $d_{\max}$ the maximum distance, all in some arbitrary distance units, and R the rolloff factor. The `distanceModel` property determines how attenuation is applied as the source moves away from the listener. The options are;

- `'linear'`: $G = \begin{cases} 1 & d \le d_{ref} \\ 1 - R\dfrac{d - d_{ref}}{d_{\max} - d_{ref}} & d_{ref} < d < d_{\max} \\ 1 - R & d_{\max} \le d \end{cases}$.
- `'inverse'`: $G = \begin{cases} 1 & d \le d_{\text{ref}} \\ \dfrac{d_{\text{ref}}}{d_{\text{ref}} + R(d - d_{\text{ref}})} & d_{\text{ref}} \le d \end{cases}$.
- `'exponential'`: $G = \begin{cases} 1 & d \le d_{\text{ref}} \\ \left[\dfrac{\text{d}}{d_{\text{ref}}}\right]^{-R} & d_{\text{ref}} \le d \end{cases}$,

Notice that $d_{\max}$ is only used in the linear model. And for a rolloffFactor of 1, the inverse and exponential models are equivalent.

Panning Model

Azimuth and Elevation

The *azimuth* and *elevation* angles determine the angle, on horizontal and vertical planes respectively, between the listener and the sound source. Suppose the normalized vector from source to listener is **V**, as described for the section describing the sound cone, and **F** and **U** are the listener's normalized forward direction and normalized up direction.

The situation is visualized in Figure 12.4. Azimuth angle α is the angle between **F** and **V** when traveling to the right. This can be found straight from the geometry, $\cos(\alpha) = \mathbf{F}{\cdot}\mathbf{V}$.

A similar relationship is used for elevation angle ε, which is the angle between the horizontal plane (orthogonal to the upwards direction) and **V** when traveling upwards, or 90° minus the angle between **U** and **V**, $\cos(90° - \varepsilon) = \mathbf{U}{\cdot}\mathbf{V}$. For both azimuth and elevation, care must be taken for the cases where the source is behind or below the listener, in which case $|\alpha| > 90°$ or $|\varepsilon| > 90°$.

Panning Algorithm

Finally, we need to set a `panningModel` property, which is the spatialization algorithm that positions the sound in a 3D space relative to the listener. This could, for instance, be any of the spatial audio algorithms from Chapter 11: surround sound, VBAP, Ambisonics, Wavefield Synthesis, depending on possible loudspeaker layouts. It could also be

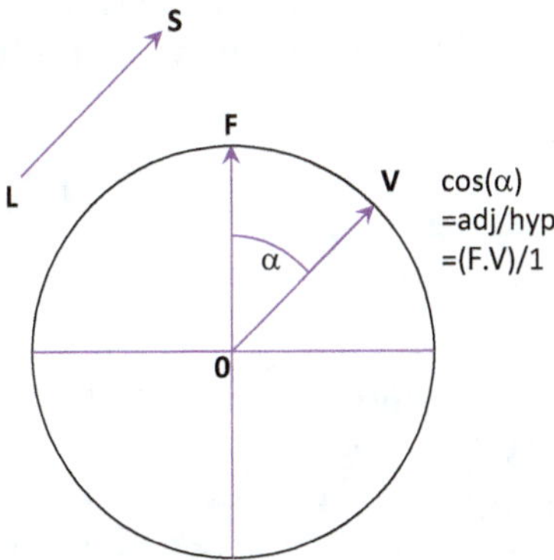

FIGURE 12.4
Calculation of azimuth angle α from source position S, listener position P, and listener forward direction F. V is the normalized S-L vector, and α is calculated by projecting V onto the forward direction, e.g., adjacent (F·V) over hypotenuse (1).

a binaural mapping of one of those techniques, or a binaural rendering using measured Head Response Transfer Functions (HRTFs) from human subjects.

For simplicity, and to ensure compatibility with almost any playback system, we will use standard level-based stereo panning in the code example that follows. This is based only on the azimuth angle, and the elevation angle is ignored. It does not distinguish between sources in front or in back of the listener.

Putting It All Together

Table 12.1 gives descriptions of all the parameters of both an immersive audio effect and the Listener, along with what they are used for and their default values for the Web Audio API implementation. Changing these parameters will change how a source is spatialized.

The spatialization procedure is shown in Figure 12.5. It is actually a simple combination of what has been described. First, the distance vector between source and listener is found using their positions. Then a gain is applied to the source based on the distance, distance model, reference distance, rolloff and maximum distance. Then, the angle between source and listener is found using the distance vector and source orientation. A gain is applied using this angle, coneInnerAngle, coneOuterAngle, and coneOuterGain.

Finally, the azimuth and elevation angles of the source are found using the distance vector, the listener's forward direction and the listener's up direction. These angles are used to apply the panning model.

Of course, the user doesn't have to calculate any of this. The parameters just need to be specified, and the immersive audio effect does the work.

Code Example

This code example shows a simple use of the immersive audio effect where we make the following assumptions. We assume that the spatialization model is amplitude panning, as introduced in Chapter 11, that the distance model is *inverse*, and that the source and listener are on the horizontal plane. So the elevation angle is 0°, and this becomes a two-dimensional situation. We do this just to present a straightforward example.

We first introduce some simple functions to handle the vector geometry.

TABLE 12.1

List of All the Parameters That Can be Changed to Pan a Source with Respect to a Listener

Used in	Used for	Parameter	Description	Default
Immersion	All	position	(x,y,z)-coordinates of the audio source	(0, 0, 0)
	Distance gain	distanceModel	Specifies the distance model that is used	'inverse'
		refDistance	The distance beyond which attenuation is applied as source moves away from listener	1
		maxDistance	maximum distance between source and listener, after which distance gain is not reduced further	10,000
		rolloffFactor	How quickly gain reduced as source moves away from listener	1
	Directivity gain	coneInnerAngle	Angle inside of which there will be no gain reduction	360
		coneOuterAngle	Angle outside of which coneOuterGain is applied	360
		coneOuterGain	gain outside of the coneOuterAngle.	0
		orientation	(x,y,z) components of source's direction vector	(1, 0, 0)
	Panning	Panning model	Specifies the panning model used	'equalpower'
Listener	All	position	(x,y,z)-coordinates of the listener	(0, 0, 0)
	Panning	Forward	(x,y,z)-coordinates of the listener's forward direction	(0, 0, –1)
		up	(x,y,z)-coordinates of direction of top of listener's head	(0, 1, 0)

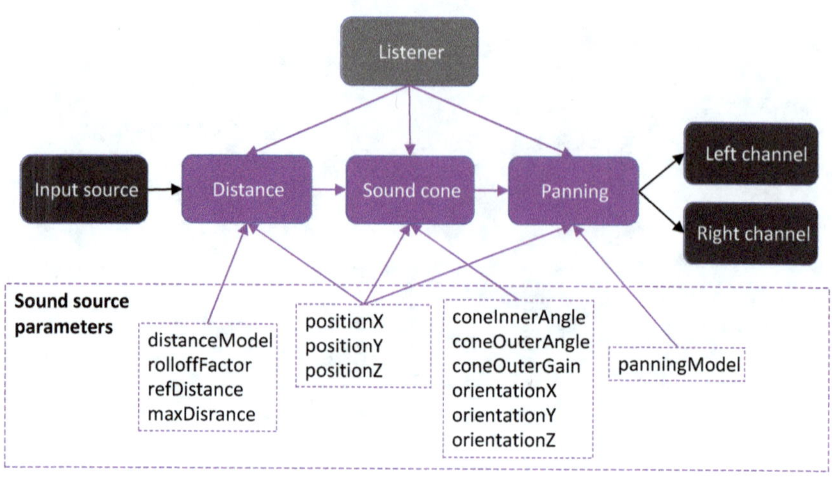

FIGURE 12.5
The spatialization procedure used in the immersive audio effect.

```
// Dot product with another vector.
float dotProduct(const std::vector<float>& x, const std::vector<float>& y) {
  float dot = 0;
  for (int i = 0; i < x.size(); i++)
  {
   dot += x[i] * y[i];
  }
  return dot;
}
// Get the magnitude of this vector.
float magnitude(const std::vector<float>& x)
{
 return sqrt(dotProduct(x, x));
}
// Difference with another vector.
std::vector<float> difference(const std::vector<float>& x, const
   std::vector<float>& y)
{
  int i, s = x.size();
  std::vector<float> diff(s);
  for (i = 0; i < s; i++) diff[i] = x[i] - y[i];
  return diff;
}
// Get a copy of this vector multiplied by a scalar.
const std::vector<float> scale(const std::vector<float>& x, float s)
{
  int lengthX = x.size();
  std::vector<float> scaledX(lengthX);
  for (int i = 0; i < lengthX; i++) scaledX[i] = s * x[i];
  return scaledX;
}
// Get normalized copy of this vector
const std::vector<float> normalize(const std::vector<float>& x)
{
  float m = magnitude(x);
  if (m == 0) return x;
  return scale(x, 1 / m);
}
```

```
// Compute the angle between two vectors. If direction matters, then negative
// angles are those that are counterclockwise from first to second vector
float AngleBetweenVectors(const std::vector<float>& vector1, const
  std::vector<float>& vector2, bool direction)
{
  const std::vector<float>& v1norm = normalize(vector1);
  const std::vector<float>& v2norm = normalize(vector2);
  float dot = dotProduct(v1norm, v2norm);
  float angle = 180 * acos(dot) / juce::MathConstants<float>::pi;
  if (direction)
  if (v1norm[0] * v2norm[1] - v1norm[1] * v2norm[0] > 0) angle *= -1;
  return angle;
}
```

These functions are needed to perform operations such as finding the azimuth angle. Once these vector operations are defined, sources can be immersed in the space around the listener as follows;

```
// Assume we have these values from the interface
std::vector<float> sourcePosition { 1, 1 };
std::vector<float> listenerPosition { 0, 0 };
// Directivity parameters
std::vector<float> sourceOrientation { 0, -1 };
float coneInnerAngle = 0;
float coneOuterAngle = 180;
float coneOuterGain = 0.5;
// Distance parameters
float rollOff = 1;
float refDistance = 1;
float* channelDataL = buffer.getWritePointer(0);
float* channelDataR = buffer.getWritePointer(1);
// Panning parameters
std::vector<float> listenerForward{ 0, 1 };
std::vector<float> listenerToSource = difference(sourcePosition,
  listenerPosition);
/* GAIN FROM SOUND CONE */
// Find angle between source orientation vector and source-listener
  vector
std::vector<float> sourceToListener = difference(listenerPosition,
  sourcePosition);
float angle = AngleBetweenVectors(sourceToListener, sourceOrientation,
  0);
// Compute gain due to sound cone
float coneGain;
if (2 * angle < coneInnerAngle) coneGain = 1.0;
else if (2 * angle > coneOuterAngle) coneGain = coneOuterGain;
else coneGain = 1.0 + (coneOuterGain - 1.0) * (2 * angle -
  coneInnerAngle) / (coneOuterAngle - coneInnerAngle);

/* GAIN FROM DISTANCE */
// Find distance between listener and source
float distance = magnitude(listenerToSource);
// Compute gain due to distance
float distanceGain;
if (distance < refDistance) distanceGain = 1;
else distanceGain = refDistance / (refDistance + rollOff * (distance
  - refDistance));
```

```
/* PANNING */
// Compute azimuth angle
float azimuth = AngleBetweenVectors(listenerForward, listenerToSource, 1);
float gainLeft = cos((azimuth / 2.0 + 45) *
  juce::MathConstants<float>::pi / 180.0);
float gainRight = sin((azimuth / 2.0 + 45) *
  juce::MathConstants<float>::pi / 180.0);

for (int sample = 0; sample < numSamples; ++sample) {
channelDataL[sample] = channelDataL[sample] * gainLeft * distanceGain *
  coneGain;
channelDataR[sample] = channelDataR[sample] * gainRight * distanceGain *
  coneGain;
}
```

The Doppler Effect

So far, we have given the geometry to describe a space with a listener and sources, and used this to render sources as heard by the listener. This mainly involves gain changes. But if either source(s) or listener is moving, then there is also a pitch change. This is the Doppler effect, named after Austrian mathematician and physicist, Christian Andreas Doppler (1803–1853). Formally, it is the apparent change in frequency of a wave that is perceived due to either motion of the source or motion of the observer.

The best way to describe this effect is with a familiar example. Most people have heard an ambulance go by. The pitch of the siren first becomes higher, then lower. This change of pitch as the vehicle moves toward the listener, then moves away, is basically the Doppler effect in action. The perceived change of pitch is due to a shift in the frequency of the sound wave.

As the ambulance moves toward the listener, as shown in Figure 12.6a, the sound waves from its siren appear condensed, relative to the listener. Thus, intervals between the waves are reduced, which results in an increase in frequency or pitch. Then once the ambulance has passed and is now moving away from the listener, as in Figure 12.6b, the sound waves are stretched relative to the listener, causing a decrease in the siren's pitch. The change in pitch of the siren allows one to determine whether the ambulance is approaching nearer or speeding away.

Consider a stationary sound source that is producing a constant frequency f_0. The wavefronts emanating from this source propagate in all directions away from the source at the speed of sound c. The distance between wavefronts with the same phase is the wavelength. Without movement of the source, this wavelength remains constant. So all stationary listeners will hear the frequency of the source.

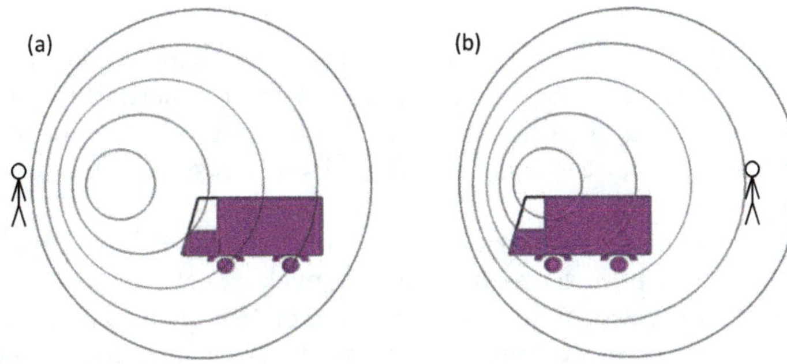

FIGURE 12.6
An ambulance siren produces the Doppler effect. A higher pitch is perceived when the ambulance moves toward the listener (a), and a lower pitch is perceived when the ambulance moves away from the listener (b).

Now suppose this sound source is moving toward the listener. The wavefronts are still created with the same frequency f_0. But because the source is moving, the wavefronts will be compressed in front of the moving source, and will be spread further apart behind the source. Thus, if the source moves toward the listener, he or she will hear a higher frequency $f > f_0$, and a listener placed behind the moving source will hear a lower frequency $f' < f_0$.

BREAKING THE SOUND BARRIER

Consider a source moving at the speed of sound (Mach 1). The sounds it produces will travel at the same speed as the source, so that in front of the source, each new wavefront is compressed to occur at the same point. A listener placed in front of the source will not detect anything until the source arrives. All the wavefronts add together, creating a wall of pressure. This shock wave will not be perceived as a pitch but as a 'thump' as the pressure front passes the listener.

Pilots who have flown at Mach 1 have described a noticeable "barrier" that must be penetrated before achieving supersonic speeds. Traveling within the pressure front results in a bouncy, turbulent flight.

Now consider a sound source moving at supersonic speed, i.e., faster than the speed of sound. In this case, the source will be in advance of the wavefront. So a stationary listener will hear the sound after the source has passed by. The shock wave forms a Mach cone, which is a conical pressure front with the plane at the tip. This cone creates the *sonic boom* shock wave as a supersonic aircraft passes by. This shock wave travels at the speed of sound, and since it is the combination of all the wavefronts, the listener will hear a quite intense sound. However, supersonic aircraft actually produce two sonic booms in quick succession. One boom comes from the aircraft's nose and the other one from its tail, resulting in a double thump.

The speed of sound varies with temperature and humidity, but not directly with pressure. In air at sea level, it is about 343 m/s. But in water, the speed of sound is far quicker (about 1,484 m/s), since molecules in water are more compressed than in air, and sound is produced by the vibrations of the substance. So the sound barrier can be broken at different speeds depending on air conditions, but is far more difficult to break underwater.

Derivation of the Doppler Effect

Simple Derivation of the Basic Doppler Effect

We first start with a simple derivation of the Doppler effect, where we ignore the fact that source and listener velocities may be changing over time. Suppose the listener moves toward the source with velocity v_{ls} and the source moves toward the listener with velocity v_{sl} (these can be negative for source/listener moving away from listener/source). Consider a sound emitted by the source with a frequency f_s. The period is then $T_s = 1/f_s$. Assume that initially the source is at position 0, and the listener is at position x.

At time t_1, the sound produced by the source at time 0 has now traveled to a position $c \cdot t_1$. The listener is at position $x - v_{ls}t_1$. So the sound has reached the listener when these are equal. That is, at time $t_1 = x/(c + v_{ls})$.

At time t_2, the sound produced by the source at time T_s has now traveled to a position $v_{sl}\ T_s + c(t_2 - T_s)$. The listener is at $x - v_{ls}t_2$. So the sound has reached the listener when $v_{sl} \cdot T_s + c(t_2 - T_s) = x - v_{ls}t_2$. That is, at time $t_2 = (x - v_{sl} \cdot T_s + c \cdot T_s)/(c + v_{ls})$

Thus, the Doppler shifted period is $T_l = t_2 - t_1 = T_s\ (c - v_{sl})/(c + v_{ls})$. The Doppler shifted frequency is just one over this Doppler shifted period, and hence we have the one dimensional Doppler formula.

$$\frac{f_l}{f_s} = \frac{c + v_{ls}}{c - v_{sl}} \tag{12.4}$$

For movement in three dimensions, we simply consider vector velocities and their scalar components along the line between source and listener.

The scalar velocity v_{sl} of a source in the direction from the source to the listener is given by

$$v_{sl} = \overline{v}_s \cdot \hat{x}_{sl} = \overline{v}_s \cdot \overline{x}_{sl} / \|\overline{x}_{sl}\|. \tag{12.5}$$

Similarly, the scalar velocity v_{ls} of a listener in the direction from the source to the listener is given by

$$v_{ls} = \overline{v}_l \cdot \hat{x}_{ls} = \overline{v}_l \cdot \overline{x}_{ls} / \|\overline{x}_{ls}\|. \tag{12.6}$$

We use the scalar projection since it prevents us from having to use the vector velocities, and as we shall see, it correctly provides the sign of the velocity.

These values are then plugged directly into Eq. (12.4), giving

$$\frac{f_l}{f_s} = \frac{c + \overline{v}_l \cdot \overline{x}_{ls} / \|\overline{x}_{ls}\|}{c - \overline{v}_s \cdot \overline{x}_{sl} / \|\overline{x}_{sl}\|}. \tag{12.7}$$

General Derivation of the Doppler Effect

Now let's consider the general case, without approximations. A sound is emitted from a source at time τ and heard by the listener at time t. Both source and listener are moving, with position x_l and velocity v_l for the listener, and position x_s and velocity v_s for the listener. This is depicted in Figure 12.7.

Consider a single frequency component, expressed as a complex exponential. At time t, the listener hears the sound $e^{j\omega_s \tau}$, where $\omega_{s=}2\pi f_s$, produced at some previous time τ, so

$$\|\overline{x}_l(t) - \overline{x}_s(\tau)\| = c(t - \tau). \tag{12.8}$$

That is, the distance between the two positions is the time it takes sound to travel between them multiplied by the speed of sound. Note that τ is

FIGURE 12.7
A moving source emits a sound at time τ, which is heard by a moving listener at time t.

dependent on time, since a small change in t must come from the sound being produced at a different time τ. Now let's consider the time derivative of this.

$$\frac{\partial\left\|\bar{x}_l(t)-\bar{x}_s(\tau)\right\|}{\partial t}=c\left(1-\frac{\partial\tau}{\partial t}\right). \tag{12.9}$$

When the sound is emitted at time τ, the frequency component $e^{j\omega_s\tau}$ has instantaneous phase $\phi = \omega_s\tau$ and instantaneous angular frequency $\omega_s = \frac{\partial\phi}{\partial\tau}$. We need to find the new instantaneous angular frequency when this sound is heard by the listener, $\omega_l(t)=\frac{\partial\phi}{\partial t}$. So

$$\frac{\partial\tau}{\partial t}=\frac{\partial\tau}{\partial\phi}\frac{\partial\phi}{\partial t}=\frac{\omega_l(t)}{\omega_s}. \tag{12.10}$$

Now we use a well-known property from vector algebra. The derivative of the magnitude of a vector is the dot product of the unit length direction vector and the vector velocity (i.e., the derivative of each component of the vector):

$$\frac{\partial\|v\|}{\partial t}=\frac{v}{\|v\|}\cdot\dot{v}. \tag{12.11}$$

So,

$$\begin{aligned}c\left(1-\frac{\omega_l(t)}{\omega_s}\right)&=\frac{\left[\bar{x}_l(t)-\bar{x}_s(\tau)\right]\cdot\frac{\partial\left(\bar{x}_l(t)-\bar{x}_s(\tau)\right)}{\partial t}}{\left\|\bar{x}_l(t)-\bar{x}_s(\tau)\right\|}\\&=\frac{\left[\bar{x}_l(t)-\bar{x}_s(\tau)\right]\cdot\left(\bar{v}_l(t)-\bar{v}_s(\tau)\frac{\omega_l(t)}{\omega_s}\right)}{\left\|\bar{x}_l(t)-\bar{x}_s(\tau)\right\|}.\end{aligned} \tag{12.12}$$

And by rearranging terms,

$$\frac{\omega_l(t)}{\omega_s}=\frac{c-\bar{v}_l(t)\cdot\left[\bar{x}_l(t)-\bar{x}_s(\tau)\right]/\left\|\bar{x}_l(t)-\bar{x}_s(\tau)\right\|}{c-\bar{v}_s(\tau)\cdot\left[\bar{x}_l(t)-\bar{x}_s(\tau)\right]/\left\|\bar{x}_l(t)-\bar{x}_s(\tau)\right\|}. \tag{12.13}$$

This, then, is the full Doppler effect, explaining the change in frequency of a moving source as perceived by a moving listener. This derivation of the Doppler is most closely related to the ones in [153] and [130]. But these derivations are restricted to stationary listener, and assume the source moves with constant velocity. It is also similar, at least in notation, to Julius Smith's vector formulation [16,154], but this is not derived and assumes that the time the sound is emitted by the source and received by the listener is equal. Related derivations are also provided in [155,156].

Simplifications and Approximations

If the listener is stationary, $\overline{v}_l(t)=0$ and Eq. (12.13) reduces to

$$\frac{\omega_l}{\omega_s}=\frac{c}{c-\overline{v}_s(\tau)\cdot\left[\overline{x}_l-\overline{x}_s(\tau)\right]/\left\|\overline{x}_l-\overline{x}_s(\tau)\right\|}. \tag{12.14}$$

Whereas if the source is stationary, $\overline{v}_s(\tau)=0$ and the Doppler effect reduces to

$$\frac{\omega_l(t)}{\omega_s}=\frac{c-\overline{v}_l(t)\cdot\left[\overline{x}_l(t)-\overline{x}_s\right]/\left\|\overline{x}_l(t)-\overline{x}\right\|_s}{c}. \tag{12.15}$$

If we assume that the velocities are much less than c, then the distance between source and listener is large compared to the distance a source or listener moves during the time it takes for sound to travel from source to listener. So, we assume the velocities and positions are roughly constant over the time period from t to τ,

$$\frac{\omega_l}{\omega_s}\approx\frac{c-\overline{v}_l\cdot\left[\overline{x}_l-\overline{x}_s\right]/\left\|\overline{x}_l-\overline{x}_s\right\|}{c-\overline{v}_s\cdot\left[\overline{x}_l-\overline{x}_s\right]/\left\|\overline{x}_l-\overline{x}_s\right\|}=\frac{c+v_{ls}}{c-v_{sl}}, \tag{12.16}$$

where all distances and velocities are taken at the same time. This is the simple version of the vector Doppler formula.

And if the motion of both source and listener is in one dimension (they are travelling along a line), then the position and velocity vectors become scalars. Thus,

$$\frac{\omega_l}{\omega_s}=\frac{c-v_l(t)\operatorname{sgn}[x_l(t)-x_s(\tau)]}{c-v_s(\tau)\operatorname{sgn}[x_l(t)-x_s(\tau)]}. \tag{12.17}$$

Implementation

Spatial audio rendering systems typically do not reproduce the Doppler effect [157], so early versions of the Web Audio API [13] also kept track of sound source velocity for implementation of this effect.[2] Doppler could also be determined by keeping track of changing source and listener positions over time. However, here we present an alternative approach.

It is well known that a time-varying delay-line results in a frequency shift. We have seen in Chapter 3 that time-varying delay is often used, for example, to provide *vibrato* and *chorus* effects. We therefore expect a time-varying delay line to be capable of precise Doppler simulation.

The implementation of the Doppler shift can be achieved by directly applying the movement of sources and listeners in a natural environment. The air between the source and listener can be represented as a delay line. A write pointer can be used to represent the source, and a read pointer represents the listener. If both source and listener are stationary, the listener simply receives the source with a delay. The distance between source and listener is represented by the distance on the delay line between the two pointers, and the speed of sound equates to incrementing the pointers from one sample to the next at each time step. If either listener or source is moving, a Doppler shift is observed by the listener, according to the Doppler equation. For now, let us consider the simplified Doppler equation, given in Eqs. (12.4) and (12.16).

Changing the read pointer increment from 1 sample to $1+v_{ls}/c$ samples (thereby requiring interpolated reads) corresponds to listener motion away from the source at speed v_{ls}. Similarly, changing the *write pointer* increment from 1 to $1+v_{ls}/c$ corresponds to *source motion toward the listener* at speed v_{ls}. But when changing the increment of the write pointer, we use *interpolating writes* into the buffer, also known as *de-interpolation*. A review of time-varying, interpolating, delay-line reads and writes, together with a method using a single shared pointer, is given in [158].

Note that it is the distance between listener and source that matters, not the coordinates of either with respect to some absolute frame of reference. So a simple case of moving source and moving listener could be reduced to a single changing distance, which could be implemented with only an interpolated read, i.e., no need for interpolated write.

Time-Varying Delay Line Reads

If $x[n]$ denotes input to time-varying delay, the output can be written as $y[n]=x[n-d(n)]$ where $d(n)$ denotes the time-varying delay in samples, at sample n. But the delay is typically not an integer multiple of the sampling interval. So $x[n-d(n)]$ may be approximated using *bandlimited interpolation* or other techniques for implementation of *fractional delay*, as discussed in Chapter 3.

Let's analyze the frequency shift caused by a time-varying delay by setting a continuous time signal $x(t)$ to a complex sinusoid at frequency ω_s

$$x(t)=e^{j\omega_s t}. \tag{12.18}$$

The output is now

$$y(t) = x(t - D(t)) = e^{j\omega_s(t-D(t))}. \tag{12.19}$$

The exponent gives the instantaneous phase of this signal,

$$\theta(t) = \omega_s(t - D(t)), \tag{12.20}$$

where $D(t)$ is the time-varying delay in seconds. As mentioned in the derivation of the Doppler effect, the instantaneous frequency is just the derivative of the instantaneous phase,

$$\omega_l = \omega_s(1 - dD(t)/dt), \tag{12.21}$$

where ω_l denotes the output frequency. The time derivative of the delay, $dD(t)/dt$, represents the *delay growth-rate*, or the *relative frequency downshift*:

$$\frac{dD(t)}{dt} = 1 - \frac{\omega_l}{\omega_s}, \tag{12.22}$$

So from Eq. (12.4), if the source is stationary and we have a moving listener, we have

$$dD(t)/dt = -v_{ls}/c. \tag{12.23}$$

That is, the delay growth-rate should be set to the normalized speed of the listener *away* from the source. Simulating source motion is also possible, but the relation between delay change and desired frequency shift is more complex. From Eqs. (12.4) and (12.22),

$$dD(t)/dt = -\frac{v_{ls}/c + v_{sl}/c}{1 - v_{sl}/c} = -\frac{v_{ls} + v_{sl}}{c - v_{sl}}. \tag{12.24}$$

A simplified approach is possible using multiple write pointers to move the delay *input* instead of its output.

Multiple Write Pointers

If multiple write pointers are located with a fixed spacing between them, then they represent a set of stationary sources. But for moving sources, where each source produces its own Doppler effect, the write pointers will move independently of each other. Each write pointer will write a

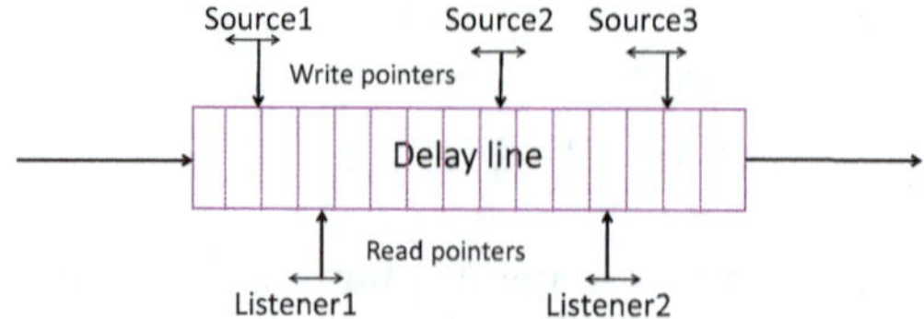

FIGURE 12.8
A delay line with multiple read and write pointers, corresponding to multiple sources and listeners.

different signal, and a unique filter can be applied to that signal. This source-dependent filter can be used to implement the filtering incurred along the propagation path from each source to a single listener.

When a circular buffer is used, the write pointer that is writing farthest ahead in time will *overwrite* memory, instead of summing into it. The write pointers may cross each other without causing issues, since all but the first pointer will just sum into the shared delay line.

So, a single delay line can simulate any number of moving sources and a single listener, or a single source and any number of moving listeners. But is this still the case if we have both moving sources and moving listeners, as in Figure 12.8? The different listeners will not perceive the same Doppler shift for each moving source. So the movement of the read pointer will not accurately represent the movement of multiple sources. So we either need a delay line for every moving source or a delay line for every moving listener. In fact, for M moving sources and N moving listeners, our Doppler simulation approach will require min(M,N) delay lines [154].

Code Example

The time varying delay line was described in Chapter 3. Here, we provide C/C++ pseudocode, based on [16], for a time varying delay line used in simulating the Doppler effect. Note that the code below features interpolated read, but not an interpolated write, which can be implemented in the same manner.

```
// Variables used in this example whose values are set externally:
float *delayData;    // Our own circular delay buffer of audio samples
int delayBufLength; // Length of our delay buffer in samples
int writePointer;    // Write location in the delay buffer
float readPointer;   // Read location in the delay buffer

// User-adjustable effect parameters:
float delta_;        // Derivative of delay time

// Set delay in fractional # samples by finding read pointer relative to
// write pointer
void setDelay(float delayLength)
```

```
{
    readPointer = (float)writePointer - delayLength;
    while(readPointer < 0) readPointer += (float)delayBufLength;
}

// Process group of samples in channelData, put output in same buffer
// input came from
void process(float *channelData, int numSamples)
{
    for (int i = 0; i < numSamples; ++i)
    {
        const float in = channelData[i];
        float out = 0.0;

        int sampleBefore = floor(readPointer);
        int sampleAfter = (sampleBefore + 1) % delayBufLength;
        float fractionalOffset = readPointer - (float)sampleBefore;

        // Calculate output using linear interpolation between samples
        // before and after
        float out = (1.0 - fractionalOffset) * delayData[sampleBefore]
                    + fractionalOffset * delayData[sampleAfter];

        // Store the current sample in the delay buffer
        delayData[writePointer] = in;

        // Increment pointers. Write pointer is integer so will always
        // wrap around
        // strictly to 0; read pointer is fractional so may wrap to
        // fractional value
        if(++writePointer >= delayBufLength) writePointer = 0;
        readPointer += 1.0 - delta_;
        if(readPointer >= delayBufLength) readPointer -= (float)
  delayBufLength;

        // Store the output sample in the buffer, replacing the input
        channelData[i] = out;
    }
}
```

Delta (Δ) is the growth parameter, corresponding to the derivative of the time varying delay, *i.e.*, $dD(t)/dt = \Delta$. When $\Delta = 0$, we have a fixed delay line and there is no relative movement between source and listener. But when $\Delta > 0$, the delay grows by Δ samples per sample, which we may also interpret as seconds per second. By Eq. (12.23), we can see that $\Delta = -v_{ls}/c$ will simulate a listener traveling toward the source at speed v_{ls}.

Note that when the read and write pointers are driven directly from a model of sound propagation in a natural environment, they are always separated by predictable minimum and maximum delay intervals [16]. This implies it should not be necessary to worry about the read pointer *passing* the write pointers or vice versa (though good programming practice suggests that an implementation should always check if this happens).

DOPPLER, LESLIE, AND HAMMOND

Donald Leslie (1913–2004) bought a Hammond organ in 1937 as a substitute for a pipe organ. But at home in a small room, it could not reproduce the grand sound of an organ. Since the pipe organ has different locations for each pipe, he designed a moving loudspeaker.

The Leslie speaker uses an electric motor to move an acoustic horn in a circle around a loudspeaker. Thus, we have a moving sound source and a stationary listener, as shown in Figure 12.9.

It exploits the Doppler effect to produce frequency modulation. The classic Leslie speaker has a crossover that divides the low and high frequencies. It consists of a fixed treble unit with spinning horns, a fixed woofer and spinning rotor. Both the horns (actually, one horn and a dummy used as a counterbalance) and a bass sound baffle rotate, thus creating vibrato due to the changing velocity in the direction of the listener, and tremolo due to the changing distance. The rotating elements can move at varied speeds or stopped completely. Furthermore, the system is partially enclosed, and it uses a rotating speaker port. So the listener hears multiple reflections at different Doppler shifts to produce a chorus-like effect.

The Leslie speaker has been widely used in popular music, especially when the Hammond B-3 organ was played out through a Leslie speaker. This combination can be heard on many classic and progressive rock songs, including hits by Boston, Santana, Steppenwolf, Deep Purple, and The Doors. The Leslie speaker has also found extensive use in modifying guitar and vocal sounds.

Ironically, Donald Leslie had originally tried to license his loudspeaker to the Hammond company and even gave the Hammond company a special demonstration. But at the time, Laurens Hammond (founder of the Hammond organ company) did not like the concept at all.

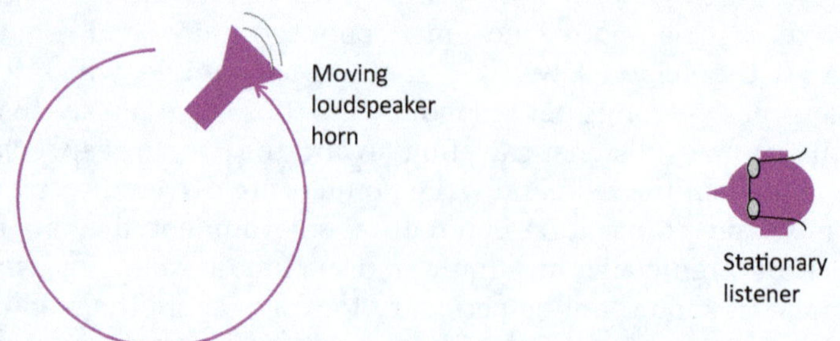

FIGURE 12.9
A moving loudspeaker and stationary listener, as in a Leslie speaker.

Applications

The Doppler effect is used in sound design and composition in order to achieve realistic simulation of moving sound sources, and is one of the main characteristics of the Leslie speaker.

The Doppler effect is used heavily in a lot of game and film special effects, to give a sense of realism or even hyper-realism. Moving sources often have an exaggerated frequency shift to ensure that the listener picks up on auditory cues and focuses attention on the moving source.

SWINGING MICROPHONES AND SLASHING LIGHTSABERS

Sound designers for film and games often use creative methods to generate the appropriate sound from existing sources, rather than through signal processing techniques designed to synthesize or process audio. One well-known technique for generating the Doppler effect is to swing a microphone back and forth in front of a sound source. This was used in the original *Star Wars* to generate the original lightsaber sound. As described by Ben Burtt, the sound designer;

> ... once we had established this tone of the lightsaber of course you had to get the sense of the lightsaber moving because characters would carry it around, they would whip it through the air, they would thrust and slash at each other in fights, and to achieve this additional sense of movement I played the sound over a speaker in a room.
>
> Just the humming sound, the humming and the buzzing combined as an endless sound, and then I took another microphone and waved it in the air next to that speaker so that it would come close to the speaker and go away and you could whip it by. And what happens when you do that by recording with a moving microphone is you get a Doppler's shift, you get a pitch shift in the sound and therefore you can produce a very authentic facsimile of a moving sound. And therefore give the lightsaber a sense of movement...

Further Reading

As has been mentioned, there is often confusion around the term 'immersive' [152], and hence many references to immersive audio actually focus solely on spatial audio aspects. Similarly, the further reading mentioned for Chapter 11 on spatial audio is also relevant here. A 1998 paper on the limits of immersive audio [159], however, clearly distinguished immersive from spatial audio and its discussion of challenges in the field is still relevant today. [160] provides a recent review of the field, and MPEG-I Immersive Audio [161] is the emerging standard for representing and rendering audio for virtual and augmented reality.

The Doppler effect has not just been used for enhanced immersion; it has also been used creatively in sound art [162] and film [163], in an exaggerated form to give a sense of hyperrealism. Finally, it is worth mentioning the field of procedural audio [164], for generating the sounds of a virtual world in which the listener may then be immersed.

Problems

1. Discuss ways in which one might incorporate other aspects of real world sound into an immersive effect, such as reverberation and filtering due to sound traveling through air.
2. Which of the distance models best corresponds to real world attenuation of a sound with distance, and why?
3. Expand the code in the immersive audio effect example to include other types of distance models.
4. From the formulas $\frac{\omega_l}{\omega_s} \approx \frac{c+v_{ls}}{c-v_{sl}}, v_{ls} = \overline{v}_l \cdot \hat{x}_{ls}, v_{sl} = \overline{v}_s \cdot \hat{x}_{sl}$, show that the relative change in frequency for the Doppler effect is given by $\frac{\omega_l - \omega_s}{\omega_s} = \frac{(\overline{v}_l - \overline{v}_s) \cdot \hat{x}_{ls}}{c + \overline{v}_s \cdot \hat{x}_{ls}}$.
5. Find an expression for the Doppler effect for a source moving in a straight line at constant velocity through a stationary listener located at 0. At time 0, the source is at the same location as listener.
6. Find an expression for the Doppler effect for a source and listener moving on x axis with constant velocities. At time 0, both source and listener at 0.
7.
   i. Assume the listener is fixed at 0, and source moves with constant acceleration away from the listener. Find an expression for the Doppler effect.
   ii. Assume this source is emitting a single frequency, $e^{j\omega_s\tau}$ at any time τ. What frequency component(s) does the listener receive at time t.

8. Assume a source is moving with constant angular velocity on a circle (or sphere) of radius r_s, and the listener is fixed at $(r_l,0)$. Find an expression for the Doppler effect.
9. Suppose a source moves in a straight line toward the listener at constant speed $2c$. Show that for each frequency f emitted by the source, the apparent sound heard by the listener has frequency $-f$. In relation to the original sound produced by the source, what does the listener hear?
10. Find an expression for the Doppler effect for a source moving in a straight line on the x-axis past the listener. Specifically, assume the sound source velocity is (10,0), listener position (0,2), initial source position (–10, 0), speed of sound is 340 m/s and frequency emitted by sound source is 1,000 Hz. What frequency is heard by the listener.
11. Simulating Doppler often involves time-varying delay lines. What sort of artifacts and issues might arise? You may wish to refer to discussion of delay lines in previous chapters.

Notes

1 This right-hand coordinate system is used in the Web Audio API. Some spatialization systems, such as Microsoft Direct3D, use left-hand coordinate systems. One must take care to transform data when porting an application between the systems.

2 https://udn.realityripple.com/docs/Web/API/PannerNode/setVelocity

13

Audio Production

In this chapter, we take a look at how audio effects are used in production. We first look at the main devices that host the audio effects that we have covered, and that are used in recording, editing, and mixing audio. We then proceed to discussion of how the effects can be used with these devices, and how they can be ordered or modified to create new effects or to achieve various production goals.

It is important to start with a few basic concepts. A *multitrack recorder* is a device that can record several tracks of audio simultaneously. The term 'track' originates from when recordings were made on tape, and the track referred to the discrete area on the tape where a sequence of audio events was preserved. So multitrack audio is the collection of different audio signals that, when combined, constitute the intended sound, e.g., guitar and vocals can be separate tracks recorded by a singer-songwriter. In contrast, a *mixer* is a device that can process one or more tracks of audio before they are combined.

Most multitrack recording devices will also provide mixing functionality, and vice versa. So various tracks, such as instrument and microphone signals, come into the mixer, where the levels and other attributes are modified. Then the output can be sent to a *sound reinforcement system* for a live performance, or to a multitrack recorder for recording either the processed tracks or a single, processed and combined output track.

The devices for mixing and recording multitrack audio are divided into two main categories: *mixing consoles* (typically dedicated physical devices) and *digital audio workstations* or *DAWs* (typically computers with specialized hardware and software). However, with the growth in functionality and versatility, including the rise of sophisticated control surfaces for DAWs and networking capabilities in mixing consoles, there is now a gray area between the two. Nevertheless, we will provide a formal description of mixing consoles, which tend to follow a standard structure, and then extend the discussion to DAWs, while highlighting some of the more important distinctions. For a more detailed discussion of mixing consoles and their structure, we encourage the reader to read [35].

We begin with a discussion of some of the core components often seen on DAWs and mixing consoles.

DOI: 10.1201/9781003593942-13

The Mixing Console

A mixing console (also known as *audio mixer, sound desk,* or *mixing desk*) may be defined as an electronic device for routing, combining, and changing the characteristics of audio signals. Depending on the type of mixer, the mixing console can mix analogue or digital signals, or both. The modified signals (either discrete digital values or continuous time voltages) are summed to create the output signals.

Mixing consoles are used in many applications, including sound reinforcement systems, public address systems, studio recording, broadcasting, and (television, film, and game) post-production. For example, one application would be to combine multiple microphone signals such that they could be heard simultaneously through one set of loudspeakers.

The mixing console offers four main functionalities: summing, processing, routing, and metering. When summing, various audio signals are combined, and channels are summed to stereo (or surround, or other multichannel formats) via the mix bus. For processing, consoles often have on-board equalizers and sometimes dynamics processors. And to enable the use of effects, processors, and grouping, mixers offer routing functionality via auxiliary sends, insert points, and routing matrices. The various channels and buses may be monitored and metered, to indicate clipping, show estimates of signal levels, and other useful aspects of the signals. Many audio mixers can also provide phantom power as required by some microphones, read and write console automation, create test tones and add external effects.

When used for mixing live performance, the output signal produced by the sound desk will usually be sent to an amplifier. However, the mixing console could also have a built-in amplifier, or be directly connected to powered loudspeakers (i.e., the loudspeakers themselves have built-in amplifiers).

The mixing console also has two distinct sections, the *channel section* and *master section.*

The Channel Section

In a *multitrack recorder,* each input signal to the mixer resides on a track. Each channel is then fed from one of these tracks. The *channel section* in a mixing console is a collection of channels organized in physical strips. Each channel corresponds to a track on the multitrack recorder. Most channels support monaural (usually referred to as just mono) input, though some consoles will have several types of channels (mono, stereo, channels with different EQs).

The channels will have dedicated *channel strips* on the console, where the user can manipulate and route the signals. Figure 13.1 shows a simple

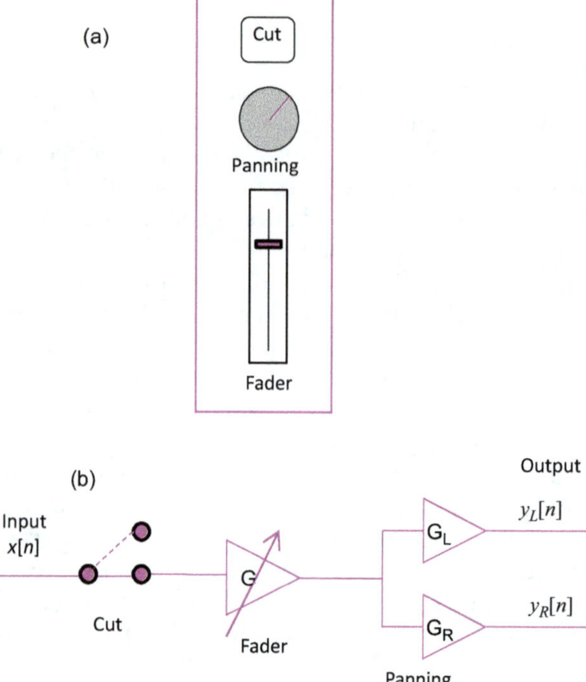

FIGURE 13.1
(a) A simple channel strip having only a cut control, a panning knob, and a level fader. (b) The corresponding signal flow diagram.

channel strip, as might be found on a basic multitrack recorder. The audio signal travels through a *cut switch*, *fader*, and a *pan pot*. The pan pot takes a monaural signal as input and produces a stereo output signal. Note that the layout of controls on the interface need not be the same as the ordering in the signal flow.

The channel strip will often have a lot more than just this basic functionality, and will often have several sections:

- Input socket, for connecting an input jack that provides the audio signal
- Microphone preamplifier that provides a gain to increase the amplitude of an input microphone signal
- Cut or mute switch, used as a gate to allow or prevent the signal from reaching the output. Similarly, a solo switch is often available, which will mute all other channels.
- Faders, or sliding volume controls to manually adjust the level of the output signal

- Equalization, typically parametric and tone controls
- Dynamics processing (e.g., noise gates and dynamic range compressors)
- Panning controls, for positioning the output in the stereo field
- Metering, including clipping indicators and level or volume meters
- Routing, including direct outs, aux-sends, and subgroup assignments, which will be discussed later

A more functional channel strip is shown in Figure 13.2. This also shows the types of audio effects that are most commonly seen on a channel strip in a mixing console: gains and faders, simple filters (especially parametric equalizers), and dynamic range compressors.

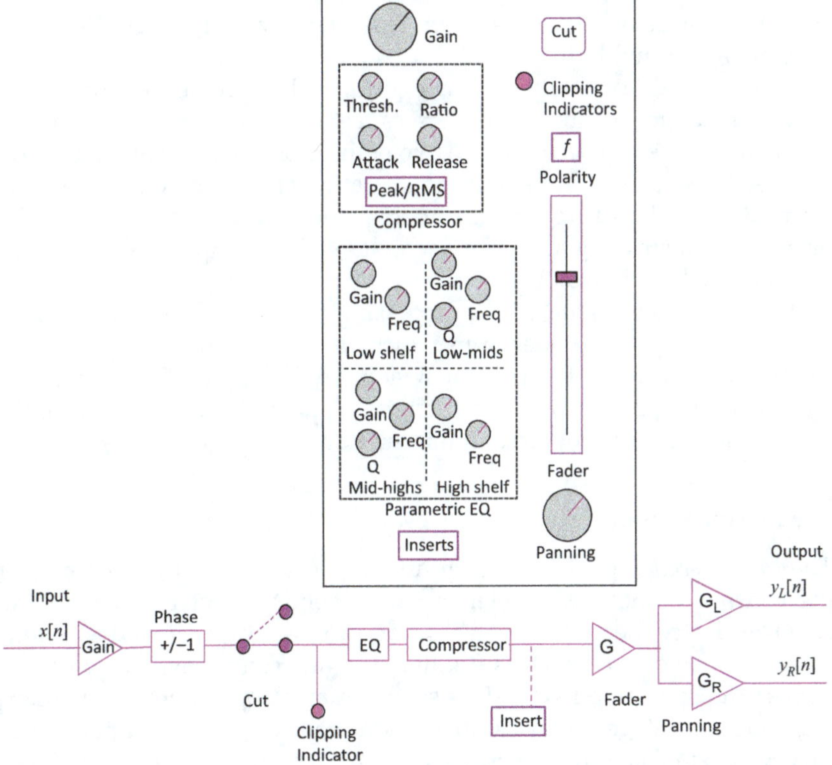

FIGURE 13.2
A channel strip and associated block diagram, with additional controls such as dynamics (a compressor) and frequency (a parametric equalizer) processing.

The line gain is a simple amplifier that boosts or attenuates the signal level before the signal enters the channel. This is used to prevent clipping as well as to get each channel into a reasonable working range. It sets the 0dB point for the channel, used for further operations that depend on relative signal levels. The line-gain control is used to boost or attenuate the level of the signal before it enters the channel path. It lets us optimize the level of the incoming signal. It can also be deliberately used to add distortion.

Audio equipment will often *invert* a signal (positive values become negative, and vice versa). This is particularly problematic when a signal is summed with another, possibly modified, version of itself. Polarity inversion will result in the two signals canceling each other out. This inversion can also occur naturally, such as when two microphone signals from inside and outside a kick drum are combined. When a track is recorded with an inverted sign, a polarity invert or switch may be used to correct the inversion by simply flipping the sign of the signal.

The clipping indicators do not affect or process the signal, but instead act as a simple meter or alert system. They show when the signal level has overshot a threshold.

Processors will typically be provided for each channel, from basic tone controls to a filter section and dynamics section. Dynamic range compression and channel equalization are two of the most common on-board processors. These controls affect the equalization of the signal by separately attenuating or boosting a range of frequencies. Many mixing consoles have a parametric equalizer on each channel. A simple parametric EQ was included in Figure 13.2.

Note also that the sequence of processing depicted on the channel strip isn't necessarily the sequence in which the processing is actually performed. For instance, the fader may actually be applied before filtering and dynamics processing, even though it appears toward the end of the channel strip for ease of use and interface design.

The Master Section

The *Master section* provides global functionality and central control over the console. It includes master auxiliary sends, effect returns, control room level, etc. The master control section almost always includes various fader or level controls, such as auxiliary bus and return level controls, and master and subgroup faders. This section may also include solo monitoring, muting, a stage talk-back microphone control, and an output matrix mixer. On smaller mixing consoles the master controls are often placed on the right of the mixing board, and the inputs are found on the left. In larger mixers, the master controls are typically placed in the center with the inputs on both sides.

Metering and Monitoring

The audio level meters are not considered a separate section since they are often merged into the input and master sections. There are usually VU (Volume Unit) or peak meters that show the levels for each channel and for the master outputs, often in pairs to indicate left and right stereo channel levels. More recently, there has also been a movement toward meters that show more perceptually relevant loudness levels. Clipping indicators are also often provided to show whether the console levels exceed maximum allowable levels.

Since sound is perceived on a logarithmic scale, mixing console displays and control interfaces almost always give level values in decibels. The "professional" nominal level is often given as +4 dBu, where dBu is referenced to 0.775 V RMS. The "consumer grade" level is –10 dBV, where dBV is referenced to 1 V RMS. Thus, 0dB on the mixing console will typically correspond to either +4 dBu or –10 dBV of electrical signal at the output.

When *monitoring* capabilities are provided within the channel strips, this is known as an in-line configuration, and is common to smaller consoles. But many mixers have at least one additional output, besides the main mix. In a split configuration, a separate monitoring section is available to provide a mix heard by the engineer. Mixers may have other outputs as well, including either individual bus outputs or auxiliary outputs. For example, these outputs can be used to provide different mixes to on-stage monitor loudspeakers.

Basic Mixing Console

A basic mixing console will have several channels, each outputting a stereo signal. These stereo signals are then summed to the mix bus. Figure 13.3 depicts such a console. The channel strips could be as simple as the one in Figure 13.1 or more complex than the one depicted in Figure 13.2.

Signal Flow and Routing

As mentioned previously, the mixing console typically includes *summing, processing, routing,* and *metering* functionality. In this section, we focus on the routing. That is, we will be concerned with the signal flow and how signals may be routed or grouped in order to enable external processing and the creation of submixes.

In Figure 13.3, we depicted the most basic aspect of routing, where all channels are routed to a summing amplifier. This may be sufficient in the

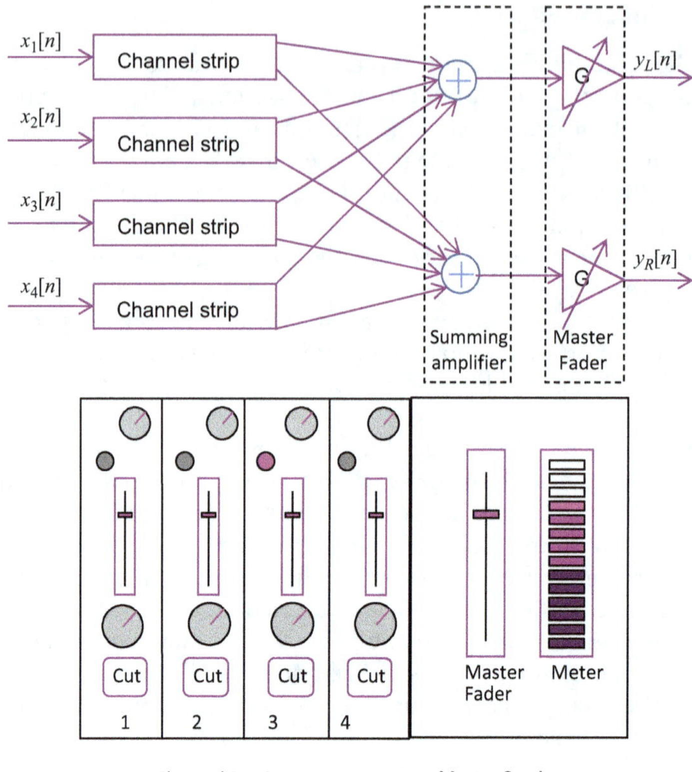

FIGURE 13.3
Signal flow diagram and user interface for a simple mixing console with 4 channels and a master section.

simple situation where no additional mixes are needed and there is no option to perform any processing beyond the default processing in the channel and master sections.

But in most consoles, there are a bewildering number of ways in which signal flow can be modified;

- A processor may be added to the signal flow using an insertion point.
- An effect may be added to the signal flow using an auxiliary send.
- Signals may be grouped together so that a single control can be used on all signals within the group.
- Signals may be routed to a common signal path, known as a bus or buss, to be sent to a new destination.

Inserts for Processors, Auxiliary Sends for Effects

One common point of confusion is the difference between *inserts* and *auxiliary sends*. To understand this difference, we begin with an important distinction between the two types of devices used to treat audio signals. Generally, one refers to audio effects as techniques, usually based on signal processing, to modify audio signals. This is the approach that we take throughout this book. But for the practical aspects of mixing, a different definition is used to distinguish *effects* and *processors*. In this context, effects are usually intended to be added to the original sound, whereas processors are usually intended to replace the input with the processed version at output. Effects typically use delay lines, and almost always have a dry/wet knob, as in Figure 13.4, where a fully dry version is the original sound and a fully wet version is the affected sound. Processors, like EQ, gates, compressors, and panning controls, produce a modified version of the signal and rarely come with a dry/wet mix.

As shown in Figure 13.5, the standard method for connecting effects is using an auxiliary send, whereas processors are normally connected using an insertion point. Having said this, the distinction is often blurred. For instance, it is often possible to add a processed signal to the original, as if it were an effect, and it's possible to connect them in different ways (connecting an effect with an insertion point, for instance).

Insertion points, depicted in Figure 13.6, are used to add an external device into the signal path. They break the signal path, sending the signal

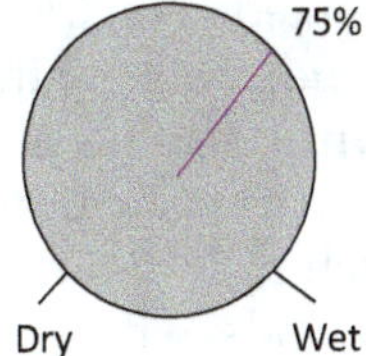

FIGURE 13.4
An implementation of a dry/wet mix.

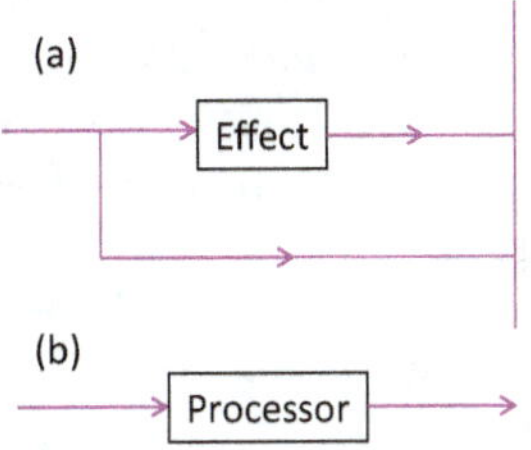

FIGURE 13.5
Standard method to connect an effect (a) and a processor (b). The effect would also typically have some dry/wet mix control to adjust the relative amounts of audio signals in the output.

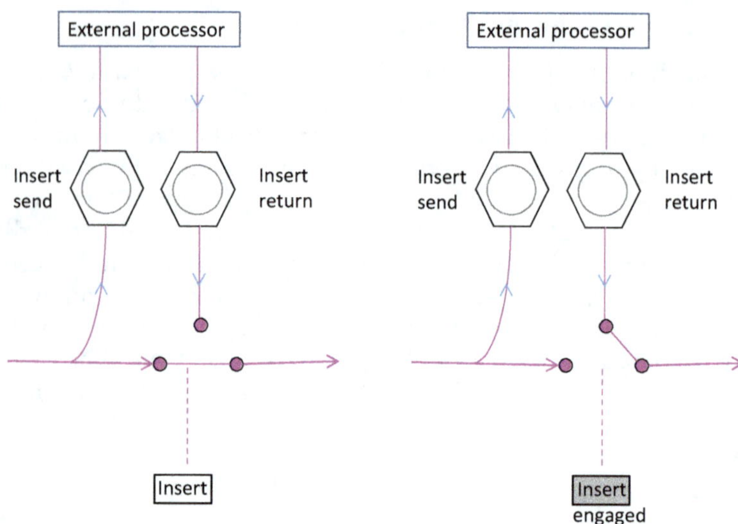

FIGURE 13.6
Signal flow for insertion points.

to an external device, and replace the original with the externally processed signal. If an insert send is not connected to an external device, it can also be used to simply obtain a copy of the signal.

An *auxiliary send* (or just 'aux send') creates a copy of the signal in the channel path, which is then routed to an internal auxiliary bus. As shown in Figure 13.7, the auxiliary send splits the incoming signal, and one path is sent to an auxiliary bus, which can be sent to other effects or devices. Typically, each channel strip on a sophisticated console will have a set of aux send-related controls, such as level and pan controls, and an on/off switch. As opposed to insert sends, where only one channel is sent to an external unit, each channel can be sent to the same auxiliary bus. Thus, aux sends allow effects to be shared between channels.

Since the aux sends are used for effects (reverb, delay), where the output is typically mixed with the original, we would expect the signal to be returned. Thus, there are also *aux returns*, also known as FX returns, which are usually in the master section. However, in most cases, there are unused channels, so the output of effects is often brought back into channels instead of using returns.

Subgroup and Grouping

The signals on selected channels may be combined on a bus as a submix, or just summed to create the main mix [165]. Grouping of all the drum channels is very common. So the tracks representing many microphones

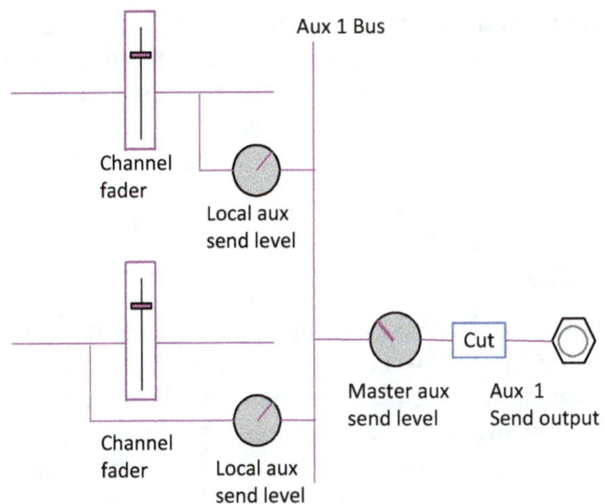

FIGURE 13.7
Signal flow for auxiliary sends.

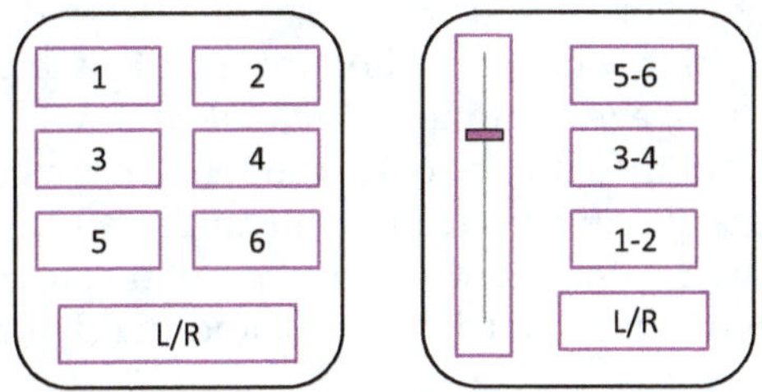

FIGURE 13.8
Routing matrices

around a drum kit can feed channels which are then grouped into a bus, creating a submix, and the level of all the drum signals can be controlled by a single fader.

Grouping is a very common procedure in mixing. While a drum group will often represent a dozen or more individual tracks, groups could consist of just two tracks, e.g., the amplifier microphone and the direct input for a guitar.

Channels are assigned to different groups using routing matrices, depicted in Figure 13.8. Each channel strip may have a routing matrix, which is usually presented on the interface either as a button for each group bus or with a button for each pair of groups. A routing matrix assignment switch may also be provided and used to determine whether the original channel signal is used to feed the mix bus.

Note that this is quite different from *control grouping*. Control grouping is a form of automation that simply allows a control to operate in an identical manner on a group of tracks [35]. So it does not actually process the signals collectively, nor does it change the signal path. Rather, the same processing is applied simultaneously to each signal in the group.

Digital versus Analogue

One of the most important classifications of mixing consoles is whether they operate on digital or analogue signals. This dictates the signal processing approaches and implementations, and is known to affect the resultant sound. Many recording engineers and artists have expressed a strong preference for 'the analogue sound,' but digital mixing consoles have more features and provide more versatility than analogue consoles. Digital circuitry is also more resistant to interference. In addition, digital consoles often include a more extensive range of effects and processors, and may be expandable via third-party software.

Whether analogue or digital systems are preferred for mixing and recording audio is highly dependent on the quality and design choices of the systems under review. Arguments for analogue audio systems include minimal latency and the absence of fundamental errors common in digital systems, such as quantization noise and aliasing. Those in favor of digital approaches often point out the high levels of theoretical performance, including low noise and distortion levels, and highly linear behavior in the audible band.

Latency

Digital mixers have an inherent amount of latency or propagation delay, usually ranging from 1.5 to 10 ms. Every analogue-to-digital conversion, digital sample rate conversion, and digital-to-analogue conversion within a digital device will result in some delay. Hence, audio inserts to external analogue processors may almost double the usual delay. And more delay can be caused by format conversions and normal digital signal processing operations.

The amount of latency in a digital mixer can vary widely depending on the routing and the amount and type of signal processing (for instance, IIR vs. FIR filters) that is performed. And comb filtering can result when a signal is given as input to two parallel paths with differing delays that are later combined.

This small amount of delay is not a serious issue in most situations, but it can still cause problems. For instance, when in-ear monitoring is performed, the performer hears their voice acoustically in their head and may hear a delayed, electronically amplified version in their ears. This can be disorienting and annoying. Furthermore, all the delays in the chain will sum together, making latency more perceptible. Thus, some digital mixers will have built-in techniques for latency correction and avoidance in order to avoid such problems.

Digital User Interface Design

Digital mixing consoles often allow for presets, stored configurations, offline editing of the mix, advanced automation and undo operations. They can also exploit digital interface design techniques to reduce the physical space requirements. Some digital mixers may allow the faders to be used as controls for other inputs or other effects. With most digital mixers, one can also make reassignments so that groupings of related inputs can appear near each other on the interface. However, the design choices needed to present these powerful features in a compact space can present a confusing interface to the operator.

Sound Quality

Microphone signals are often much weaker than what is needed for processing and routing in a mixing console. So both analogue and digital systems use analogue *microphone preamplifiers* to boost the signal. This circuit affects the timbre of the sound and can be the cause of much of the perceived difference in sound quality between consoles. In a digital mixer, an analogue-to-digital converter will occur after the (analogue) microphone preamplifier. Ideally, the quantization and sampling of the converter is carefully designed in order to avoid clipping or overloading, while also producing a highly precise and accurate digital representation of the signal over the whole linear dynamic range. Clipping and requantization need to be avoided, or at least minimized, in any further digital signal processing.

Analogue mixers, too, must also avoid overload at the microphone preamplifier or at the mix busses. However, analogue mixers achieve this with a gradual degradation, rather than the 'all or nothing' approach that is often taken in digital design. Background hiss is also a problem in analogue systems, but gain management is used to minimize its audibility. And low-level gating is used to avoid the addition of further noise from inactive subgroups that may have been left in a mix.

In digital systems, the bandwidth is typically limited by the sample rate used, and the signal-to-noise ratio is limited by the bit depth of the quantization process. In contrast, the bandwidth of an analogue system is restricted by the physical capabilities of the analogue circuits and recording medium. In an analogue system, natural noise sources such as thermal noise and imperfections in the recording medium will lower the signal-to-noise ratio.

Those who favor digital consoles often claim that the 'analogue sound' is more a product of analogue inaccuracies than anything else. This may be true, especially since for high-level signals that overload the system, analogue devices usually produce a fairly smooth limiting behaviour.

Do You Need to Decide?

Many aspects of system design combine to affect the perceived sound quality, which makes the question of analogue versus digital systems difficult to decide. There have been few controlled listening tests, and no conclusive answer can yet be reached. High-performance systems of both types can be built to match most constraints, but it is often more cost-effective to achieve a given level of signal quality with a digital system, except when the requirements are minimal. The distinction is also blurred by digital systems that aim at analogue-like behavior.

Engineers and producers will often mix and match analogue and digital techniques when making a recording. The mixing style and approach have a bigger influence over the recording than the specific choice of audio console, though it is sometimes said that the digital mixer encourages more editing of the content. However, the difference between analogue and digital is only one hardware sound quality issue. Microphones and especially loudspeakers may be considered the bottlenecks in the sound recording, production, and playback process. They both have significant distortion levels and do not easily produce a flat frequency response. In terms of being able to accurately reproduce the sound of a performance, they have a much greater influence over sound quality than the choice of mixer.

Software Mixers

Mixing and editing of audio tracks can be performed on screen, using computer software and appropriate recording and playback hardware. This software-based mixing is an essential part of a digital audio workstation (see below). Space requirements are reduced since the large control surface of the traditional mixing console is not utilized. In a software studio, there is either no physical mixer or there is a small device, possibly touchscreen, with a minimal set of faders.

Digital Audio Workstations

A *digital audio workstation* (DAW) is a term that describes a computer equipped with a high-end sound card and with a software suite for working with audio signals. Digital audio workstations can range from a simple two channel audio editor to a complete, professional recording studio suite. DAWs usually provide specialized software to record, edit, mix and play back multitrack audio content. They have the ability to freely manipulate recorded sounds, akin to how a word processor is used to create, edit, and save text. Although almost any home computer with multitrack and editing software can function as an audio workstation, the DAW is typically a more powerful system with high-quality external analogue to digital (ADC) and digital to analogue (DAC) conversion hardware, as well as extensive audio software. ProTools, Logic Pro, Ableton Live, FL Studio, Cubase, Nuendo, Reaper, and Reason are all popular DAW software platforms.

A common sound card can suffice for many audio applications, but a professional DAC is generally an external and sometimes rackmounted unit and produces wider dynamic range with less noise or jitter.

DAWs can be classified into two categories:

1. *Integrated DAWs* evolved from the traditional mixer, and consist of one device comprised of a mixing console, control surface, and digital interface. Integrated DAWs were more popular when computational power and memory on personal computers were insufficient for many audio production tasks. However, they are still preferred in some markets, and combine the benefits of the robust, physical console with the versatility of a graphical user interface. They have also found a new appeal given the recent rise of touchscreens and embedded devices.
2. *Computer-based DAWs* consist of three components: a computer, an ADC-DAC, and digital audio editor software, such as the one depicted in Figure 13.9. The computer hosts the sound card and software and provides processing power and memory for audio editing. The sound card provides an audio interface, converts analogue audio signals into digital form, and may also assist in processing audio. The software controls the hardware components and provides a user interface to allow for audio recording, editing, and playback.

Common Functionality of Computer-Based DAWs

DAWs are often based around the same concepts as mixing consoles, but with some important distinctions. Most DAWs have a standard layout which includes transport controls (play, pause, rewind, record, etc.), track

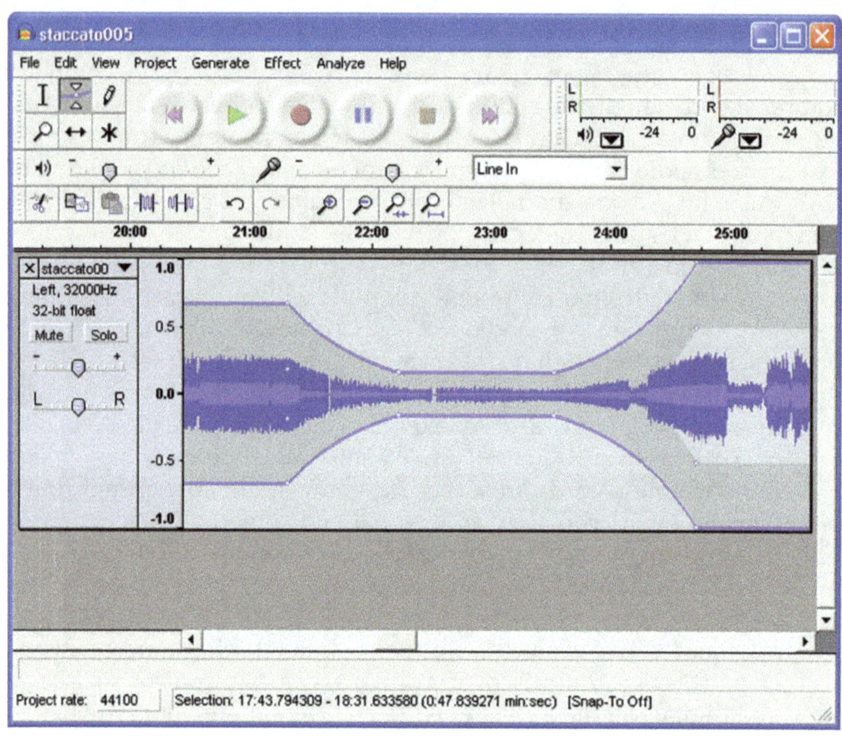

FIGURE 13.9
A screenshot of Audacity, a popular open source audio editor. In addition to editing, a full DAW would also have functionality for routing and summing audio signals.

controls, a waveform display, and a mixer. In multitrack DAWs, each track will have some built-in controls, such as overall level and stereo position adjustment for each waveform in a track. And of course, plug-ins can be placed on the tracks in order to apply the various effects that have been discussed in previous chapters.

Almost all DAWs will also feature automation, often achieved by manipulating *envelope points*. Here, the user can mark up multiple points on a waveform. By adjusting the positions of these markers, the user can shape the sound, or adjust parameters for processing the sound over time, e.g., applying a time-varying panning position.

Unlike the mixing console, in a software mixer, there is generally no separation between the multitracks and the mixer. Instead, the multitrack is represented as a sequence window, also known as arrangement, edit, or project window. It sees tracks and audio regions. So we no longer have channels, and instead we just have tracks and mixer strips. These mixer strips operate similarly to the channel strips on a console. A new track in a sequence window results in a new mixer strip in mixer window.

However, our tracks need not be audio signals. Instead, they could represent MIDI data, which behaves like an audio track but is only converted to a digital audio signal by a virtual instrument at a later stage, as explained below.

MIDI and Sequencers

Musical Instrument Digital Interface (MIDI) is an industry-standard electronic communications protocol that precisely defines each musical note performed on or by a digital musical instrument, thus allowing the instruments and computers to exchange data. The MIDI protocol is not used to transmit audio, but instead to transmit symbolic information about the notes in a music performance.

Today, MIDI is used in recording most electronic and digital music. MIDI is also used to control hardware including live performance equipment, such as effects pedals and stage lighting. MIDI also is the basis of simple ringtones and the music in many games.

One important device for audio production that relies heavily on MIDI information is the *sequencer*. This is a software program for the creation and composition of electronic music. With a MIDI sequencer, one can record and edit a musical performance without requiring audio input. The performance is recorded as a series of events, and is often played on a keyboard instrument.

The MIDI sequencer records events related to the performance, such as what note was played at what time, or how hard a key was pressed, but does not record the actual audio. This MIDI data is played back into software or into a MIDI instrument. So a performer can select a particular instrument for a musical piece, but later choose another one without needing to create a new performance. In fact, the performer can use a single device to record multiple parts, and then modify attributes of each part to give the illusion of a performance by an entire orchestra.

Some music sequencers are intended to be an instrument for live performances, as well as or rather than a tool for composing and arranging. These are often used by DJs for live mixing of tracks.

Although the term "sequencer" is used mainly in reference to software, many hardware synthesizers and almost all music workstations include a built-in MIDI sequencer. Drum machines, for instance, usually will have a built-in sequencer, and there are standalone hardware MIDI sequencers. Furthermore, many sequencers can show the musical score in a piano roll notation, and some also provide traditional musical notation features. Most modern sequencers now have the ability to record audio and feature audio editing and processing capabilities as well, and some well-known DAWs have evolved from MIDI sequencers. Consequently, the terms "music sequencer" and "digital audio workstation" are sometimes used loosely and interchangeably.

PLAYING THE MIXING DESK

King Tubby (1941–1989) was a Jamaican electronics and sound engineer, and his innovative studio work is often cited as one of the most significant steps in the evolution of a mixing engineer from a purely technical role to a very creative one.

In the 50s and 60s, he established himself as an engineer for the emerging sound system scene, and he built sound system amplifiers as well as his own radio transmitter. While producing versions of songs for local deejays, Tubby discovered that the various tracks could be radically reworked through the settings on the mixer and primitive early effects units. He turned his small recording studio into his own compositional tool.

Tuby would overdub the multitracks after passing them through his custom mixing desk, accentuating the drum and bass parts, while reducing other tracks to short snippets. He would splice sounds, shift the emphasis, and add delay-based effects until the original content could hardly be identified.

King Tubby would also rapidly manipulate a tuneable high pass filter, in order to create an impressive narrow sweep of the source until it became inaudible high-frequency content. In effect, he was able to 'play' the mixing desk like a musical instrument, and in his creative overdubbing of vocals, became one of the founders of the 'dub music' genre.

Audio Effect Ordering

Over the past few chapters, we've covered the audio effects that are most often used. So now it is time to see how these effects can be used together to edit a signal in order to achieve a desired result. One essential factor to consider is the order in which the effects are placed. However, the first rule of effect ordering is that *there is no rule to audio effect ordering*. That is, even when trying to improve the technical aspects of the mix (as opposed to creative choices), there are justifications for almost any placement of the effects. Having said that, we will look at a variety of options and see if we can establish some guidelines.

First, you need to consider your concerns. Are you starting with a noisy signal, or are you worried about the noise introduced by the effects? What is each effect intended to do and what sort of input signal does each effect expect? Do some effects counteract other effects? And of course, what is your goal in manipulating the sound, regardless of the effects and the placement that you might choose in order to achieve this result?

Noise Gates

If the audio effects are producing significant noise (or amplifying existing noise), then a noise gate could be placed toward the end of the audio effects chain. Placing the noise gate after those other effects ensures that the noise is not heard when the signal should be silent. The exception to this is if the effects applied to a signal involve reverberation or a delay line effect. In which case, having a noise gate at the end of the effect chain may eliminate the ends of slowly decaying sounds. If instead a gate is applied before a delay or before reverberation, then even if it abruptly cuts off the sound, the sustain produced by the delay line effect will partly disguise this. Delay-based effects help ensure that the sound will decay naturally rather than having a sudden silence.

Compressors and Noise Gates

Noise gates can be applied to give a relatively clean signal, on which various effects can act. So the noise gate would appear first in the change. Figure 13.10 presents an audio signal with some low-level noise on top, and the same signal after a dynamic range compressor has been applied on bottom. In the top waveform, a noise gate can be applied to remove most of the noise because it can be distinguished from the high-amplitude signal. However, the compressed signal has a smaller dynamic range than the original. Compression results in the noise being boosted until it is comparable in level to the wanted signal, making it very difficult to establish the threshold for the noise gate. Hence, compression is often applied after noise gating. Keep in mind though, that the noise gate is not noise reduction, and there may still be unwanted noise when the source is active and the gate is not in operation.

If gates are not applied, another option is to put compression first in the chain. This is appropriate if the compressor will be combined with other effects that may introduce noise and artifacts. When the compressor is active (reducing output peak dynamic range) and the output level is increased by the compressor's make-up gain, the noise will be amplified along with the instrument's sound. Other audio effects can introduce more noise into the system. If the compressor is placed after those effects, it may amplify that noise as well.

Care must be taken, however, to consider all the editing, mixing, and mastering that is done. Though the compressor has a make-up gain, additional gain may be applied at any point, or even in a later mastering stage. It may be that the gain applied during compression is small in comparison to later gain stages.

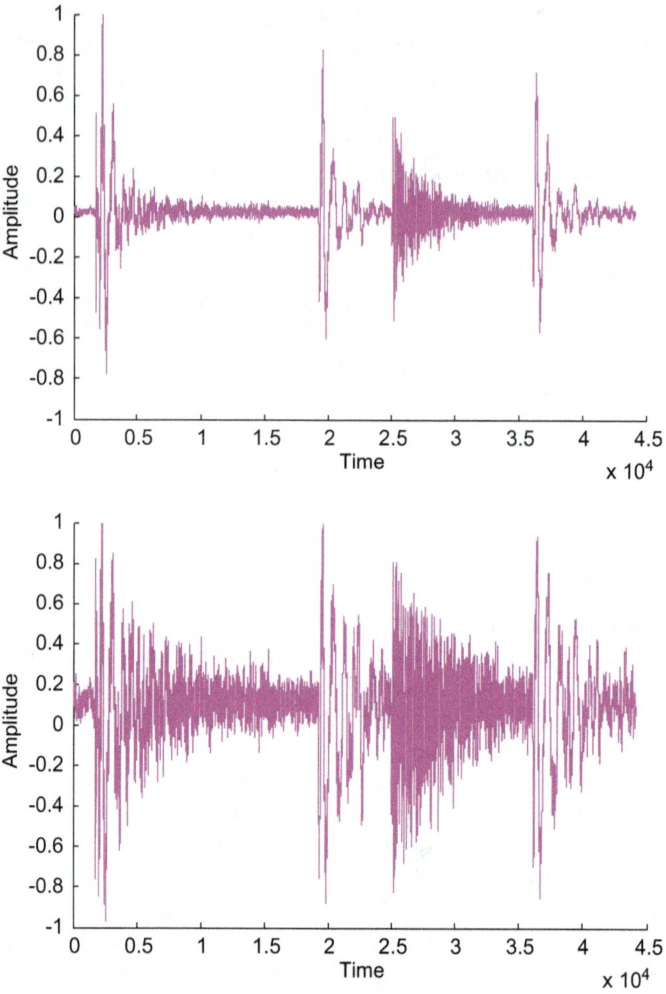

FIGURE 13.10
A gate can be applied to reduce the low-level background noise in the top waveform, whereas this noise is more difficult to gate in the compressed waveform depicted at bottom.

Compression and EQ

Whether to put compression before equalization, or vice-versa is a rather tricky question. There is no universal answer for this, partly because compression can serve different purposes. Transients are generally broadband and high level, so both EQ and compression will act on these together.

Suppose there is a single instrument signal with a highly resonant filter sweep. On some notes, the level may become very large when a note's fundamental frequency is very close to the resonant frequency. So a compressor can be used to reduce this. But you may also want to apply an equalizer to boost some broad mid-frequency range. In this case, one could put the compressor first to attenuate the high transient peaks, and then apply equalization on a well-behaved signal.

But now suppose that the resonance is not serious, but a much more significant boost is required on the midrange frequencies. In which case, the compressor may be placed after the equalizer, since the equalization may be causing unnaturally high levels. With a high threshold on the compressor, a midrange boost can still be achieved while at the same time avoiding the most problematic level issues.

Another reason to place an equalizer before a dynamic range compressor is to make the compression more sensitive to frequency content. This approach can also be achieved, with subtle differences, using multi-band compression or a side-chain filter on a single-band compressor.

Reverb and Flanger

Effect ordering can be quite challenging when reverberation is applied. Suppose both reverberation and flanging, which can be quite a dramatic effect, are to be applied on a signal. Should the flanger be placed before or after reverberation?

Keep in mind that the flanger is driven by a low-frequency oscillator. If the flanger is placed before a reverberator, the reflections due to reverberation will break up the low-frequency, periodic nature of the flanger, resulting in a more diffuse sound. The flanger will sound more subtle, like a reverb with shimmer. But if the reverberation is applied before flanging, the late reflections will all be flanged, producing the flanger's characteristic 'whooshy' effect.

Reverb and Vibrato

With vibrato but without reverb, what we hear resembles the direct sound of the instrument. However, we are used to hearing instruments from a distance, and in a room that provides reverberation. In other words, normal listening includes the direct sound coming from the instrument added to the sound that is reflected off walls, floor, and ceiling. When no vibrato is used, this reverberation makes relatively little difference. The frequencies of all the reflections are the same, so they all add up to make a relatively simple spectrum. By contrast, when playing with vibrato, the delayed sounds may have different frequencies, and that difference changes with time. This gives rise to complex interference effects, producing a richer, livelier sound than a note played without vibrato.

Delay Line Effects

Delay line effects (and here we include reverb) are very often placed near the end of an effect chain, to give a natural sounding decay. So the real questions become when to break this rule, and how to order multiple delay line effects placed in series.

Consider again the flanger, but this time used with a delay block. If the flanger is placed before a delay with feedback, then the delay effect will produce delayed repeats of the flanging sweep each repeat will be, in effect, an image of the same section of the flanger sweep. These 'flanging echoes' may overlap, especially for sustained sounds. The result will be to overlay a flanger sweep with several more sweeps, decaying in level, but starting at different times depending on the delay time setting. The flanging effect may become less prominent, but the sound will become more complex and interesting. However unlike applying reverb after flanging, due to the sparsity of delays, some impression of flange sweep or movement should remain.

Putting the flanger last in the effect chain will impart the flanger's sweeping sound onto the original signal and every other effect that was applied.

Chorus is sometimes applied before a reverb effect to add depth and richness to the signal. This often works well on more ethereal sounds or pianos. Chorus can also be placed prior to delay or echo, to heighten the impression of distance and space.

Let's return to 'flanger before delay.' A small amount of reverb could be added after a delay (and after other possible effects), because then each discrete delay would then be transformed into a diffuse set of decaying delays over a short duration, rather than just a repetition of the signal. This makes the delay line effects seem more natural and increases the sense of space.

Distortion

Distortion will often be placed at the beginning of the effect chain so that any later effects will apply to the new harmonics that were introduced by the distortion. For example, a flanger applied to a distorted signal can sound quite dramatic, because it produces a sweep on a rich set of harmonics.

Because distortion often applies a high gain, and also introduces harmonics of background sounds, the effect can introduce unwanted noise.

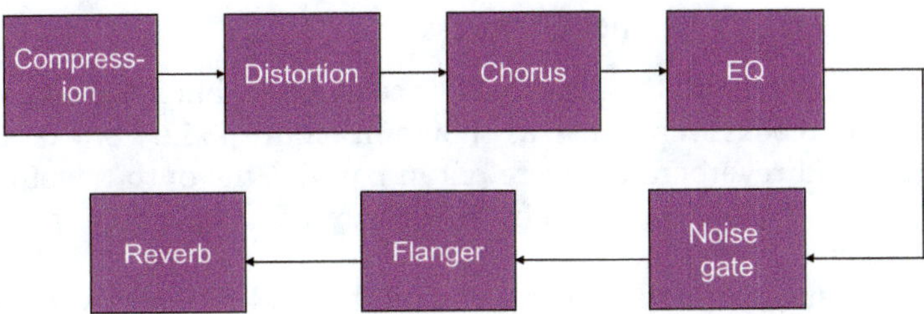

FIGURE 13.11
Just one approach to audio effect ordering, where noise from other effects is gated toward the end of the chain, but delay effects have a natural, ungated behaviour.

So noise gates are often applied in conjunction with distortion. These may be placed directly after the distortion block (or after any filtering applied to the distortion).

An example ordering based on some of the above justifications is shown in Figure 13.11. In this case, the ordering was made with the goal of minimizing the accumulation of noise due to the signal processing in each effect.

Order Summary

Here, we list some of the suggestions for effect ordering.

- Distortion and other nonlinear effects should come right at the front of the signal chain so that any following effects blocks can work on the new harmonics introduced by the distortion.
- Gates come before compressors, so that the make-up gain in compression does not boost significant noise.
- Delay or reverb at the end of the chain to create a natural sounding signal decay.

Combinations of Audio Effects

Many of the effects that have been described can be combined. In fact, common, advanced implementations will often present a single effect that has combined the functionality of multiple effects. In this section, we will take a look at some more ways that effects can be combined together.

Parallel Effects and Parallel Compression

So far we have been talking about series connection, but it is also possible to put effect blocks in parallel, as shown in Figure 13.12. For example, if distortion and reverberation were put in parallel, the output would be a mix of two distinctly separate effects: a distorted direct sound and clean reflections.

Running audio effects in parallel actually occurs whenever there is a wet mix, which involves adding the original signal back in with the modified signal. In which case, we place an audio effect in parallel with an empty effect that just passes the signal without modification. Recall the distinction between effects and processors (as mentioned, this is awkward terminology); processors replace the input with the processed version, but effects add the modified version to the original sound. So, simply by adding a dry/wet mix, a processor can be made to act as an effect. This is the case with *parallel compression*. Consider a dynamic range compressor with extreme settings (high ratio, short attack and release). By adding this output to the input, we have an interesting result. When the signal level is below the threshold, the sum is twice the original, resulting in a 6 dB increase. If the input signal level is far above the threshold, then most of the signal will be compressed. So the summed output is almost entirely due to the input signal on the direct (unprocessed) path, and there is almost no level change. Thus, simply by putting compression in parallel with the direct path, the compressor, without a make-up gain, will apply a boost to the quiet parts of a signal while leaving the loud parts relatively unaffected. This is a form of *upward compression*.

Putting a reverb and a delay, or just multiple reverbs, in parallel is also widely used since it gives the impression of rich acoustics. Complex sounds can also be introduced by putting an audio effect in the feedback loop of another, or by feeding the output signals from audio effects acting in parallel back into the inputs of the other, parallel effects.

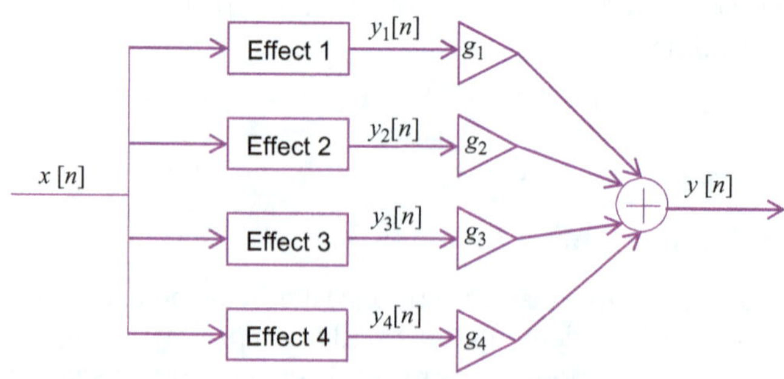

FIGURE 13.12
Parallel combination of audio effects.

Sidechaining

As mentioned in Chapter 7, the sidechain refers to a path within an audio effect other than the main path that produces the output. In effects (as opposed to processors), it often refers to the signal path that generates the affected signal to be added to the original. *Sidechaining* involves feeding an additional signal into an audio effect, where the effect is applied to some other signal. Although frequently used on compressors, gates, limiters, and expanders, it can also be found on vocoders, synthesizers, and other effects. It allows one to modify one signal depending on the characteristics of another signal. For instance, it can be used to prevent multiple sources in the same frequency range from clashing in a mix. As an example of this, the instrumental tracks in a multitrack music recording can be put into a bus, which then feeds the sidechain of a dynamic range compressor that has a large make-up gain. With this sidechain compression, the vocal level can be raised whenever the sum of the background tracks becomes particularly loud.

Ducking

In some cases, it is useful to have a signal's level controlled by a different signal so that when one signal level is high, the other signal is attenuated, as in Figure 13.13. This is known as *ducking* or *cross-limiting*, where one signal ducks under the other one, and is a form of side-chain compression. Ordinarily, dynamic range compression will reduce the gain based on the volume of the signal going through it. But a ducker reduces the gain based on a different signal. The most common application would be a radio or podcast DJ. If music is playing, when the DJ talks into the microphone, the level of the music will drop in order to ensure speech intelligibility. So the microphone signal is the input to the side chain, but the control acts on the music signal. A ducker may also be used to emphasize certain elements in a mix. Hitting the kick drum could lower the other tracks, increasing its presence. Ducking is also often used in dance music to achieve a characteristic "pumping" sound, as discussed in Chapter 7. Though pumping

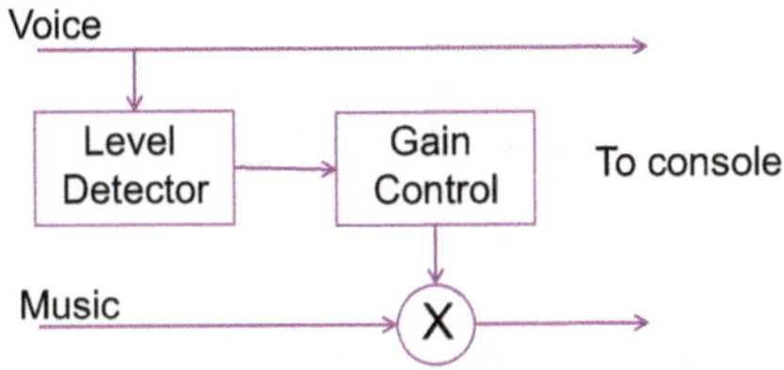

FIGURE 13.13
A 'ducker' or cross-limiting setup.

may be seen in general as an artifact of compression, it can also be a means by which the rest of the mix reacts to the presence of a particular source.

De-esser

The compressor forms the basis for another audio effect, the de-esser. A de-esser is used for reducing sibilant sounds in speech and singing, such as "s", "sh", and "ch." These sibilants often have highly exaggerated high-frequency response in recordings. Instead of monitoring the overall level of the input signal, it is possible to monitor and modify only a certain frequency range. This is what a *de-esser* does to attenuate the 'ess' sounds.

There are two common forms of de-esser, as shown in Figure 13.14. The split-band de-esser is essentially a form of multi-band compressor, where compression is only applied to the frequency range that produces the problematic sibilance. In the broadband de-esser, a broadband compressor is applied that uses an estimate of the signal level based on a filtered version of the input signal. That is, the side-chain of the compressor applies a filter to boost the frequency range where sibilance occurs. This filtered signal is used to trigger the compressor, but the gain reduction still applies to the original signal.

Side-Chain Compression for Mastering

Compressing a mixdown, as is often done in the mastering stage of audio production, is different from compressing a single track because every

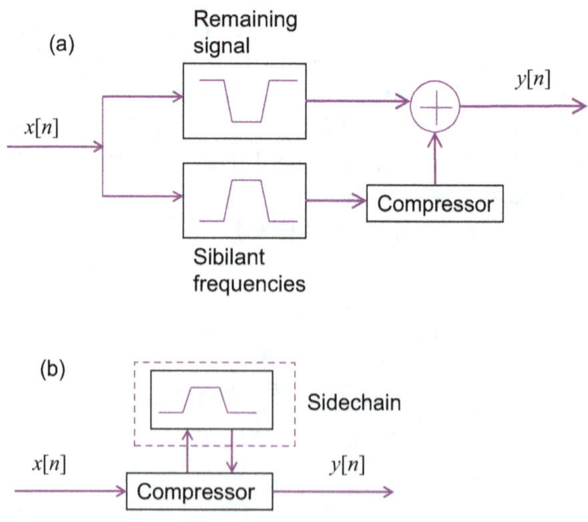

FIGURE 13.14
Split-band (a) and broadband (b) de-essing.

instrument in the mix will get compressed the same amount. In practice, this often means that the bass instruments cause the whole mix to get over-compressed, which makes it sound like the bass instruments are "punching out" the other instruments. This can be compensated for by feeding the side-chain of a compressor with a filtered signal. Similar to the de-esser, high pass filtering can be applied in order to minimize the compression due to low-frequency content.

Multiband Compression

Multiband compression is a versatile and popular tool that combines a filter bank and a dynamic range compressor. Like loudspeaker crossover, discussed in Chapter 4, low pass, band pass, and high pass filters are first used to separate the input signal into several frequency bands. Each band is then passed to a dynamic range compressor, which can be controlled independently of the other compressors. These parallel paths are then summed together to produce the output signal. A block diagram of a multiband compressor is shown in Figure 13.15.

Multiband compression is a particularly useful effect when mastering. For instance, the mixdown of a multitrack session may include an overly loud kick drum as well as vocals with wide dynamic range. If a normal, single-band compressor is applied, it may be triggered by both

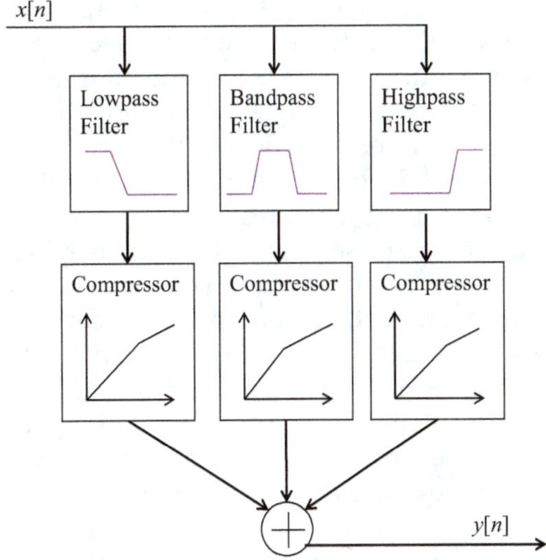

FIGURE 13.15
A multiband dynamic range compressor, with three frequency bands

the offending kick drum and the wanted vocals. By applying multiband compression, we can ensure that loud low-frequency components, such as from the kick drum, only result in dynamic range compression over the low frequencies. Multiband compressors are also often used in broadcast, and the particular implementation and settings used can often give a radio station a characteristic sound. Multiband compression can also be used as an alternative to side-chain compression, often with better results, and is highly effective as a de-esser.

However, the dynamic range compressor is a nonlinear effect with a large set of standard parameters that interact in complicated ways. The multiband implementation greatly increases the complexity of the effect. The number of adjustable parameters is usually on the order of the number of bands multiplied by the number of parameters in a single band compressor. Thus, great care should be taken in using the effect. It is often considered good practice to start with the compressors on each band configured with identical settings, and then selectively make small adjustments to address known issues or achieve predefined goals.

Dynamic Equalization

In dynamic equalization, the frequency response of an equalizer depends on the signal level. So, for instance, a notch filter could be applied such that the amount of attenuation at the center frequency of the notch increases as the signal level increases. The equalizer is no longer a linear filter, and the transfer function $H(\omega)$ must now include level dependence, $H(\omega,L)$. Simplified versions of such functions are known as *describing functions*, with output dependent on both frequency and amplitude. However, level dependence in a dynamic equalizer is usually based on some estimation of the signal envelope, and would include time constants such as are usually found in a dynamic range compressor.

Dynamic equalizers are closely related to multiband compressors and can be used to address many of the same audio production tasks. In multiband compression, different input-output level curves are applied for different input frequencies, whereas in dynamic equalizers, different frequency response curves are applied for different input levels. Thus, in both cases, the output signal level is a function of input level and input frequency. This relationship between the two effects, as well as the overall framework in which they fit, is described in [166].

Dynamic equalization is not a widely used effect, though there are a few commercial implementations, and some analogue equalizers and digital emulations of analogue equalizers implicitly have dependence on signal level.

Combining LFOs with Other Effects

Many of the effects that we have encountered are driven by low-frequency oscillators, such as phasing and chorus. The LFO can also be used with other effects to create interesting combinations. For example, stereo panning can be modulated via an LFO, or even by the envelope of the sound being processed. This latter option needs an effects unit that includes an envelope follower. Filter sweeps, similar to those found in synthesizers, that follow the input signal level can be created. Another option is to use the envelope of the input signal to control a gain placed before a reverberator. This technique, known as reverb ducking, can be used to increase the reverb during quiet sections, but emphasize the direct sound when the signal level is high. It is a useful tool for applying reverb when dealing with a dense mix.

Another tool that can be used is the sample and hold LFO, which can be seen in many classic analogue synthesizers. The source has a variable frequency like any LFO, but it generates a random series of steps rather than a periodic signal. So rather than creating a continuous filter sweep, it will produce regular steps at random frequencies, which can produce interesting 'electronic' sounds when used to control filter frequencies. Other creative techniques include using the sample and hold waveform to trigger an autopan so that the sound will jump to random positions, or to use the input envelope follower to control the stereo positioning, so that a source's stereo position is a function of its loudness.

The outcome is possibly unstable, and generally less predictable, when complicated feedback loops are used. For example, pitch shifters can include both delay and feedback options. So if a pitch shift is used with a feedback delay, each repeat produced by the delay will be frequency shifted further than the preceding one, until it becomes inaudible.

Discussion

In this chapter, we have provided a taster of how the audio effects are used, both in terms of their application within audio devices such as mixing consoles and Digital Audio Workstations, and in terms of how they are ordered and combined in the editing and mixing of audio devices. Audio production (and audio mixing in particular) is of course, a vast field, combining many technical and creative challenges. Even an understanding of the theory of audio effects, along with the skills to create them, does not give someone the knowledge required to effectively use them. For this, both critical listening skills and in-depth understanding of production and perception are necessary.

However, there are conceptual approaches that can be of great assistance. David Gibson [70] suggests that the challenges in mixing audio can be conceptualized by positioning each audio track on four axes: the volume, spatial position, frequency contour, and depth placement of the track. This visual approach can be used to assist in applying effects to tracks so as to organize the tracks along the axes and avoid conflicts in a mix. Others take a more bottom-up approach [17,35], beginning with an understanding of the tools at one's disposal (including those described in previous chapters), and how these tools may be used to address mixing goals.

A note of caution should be made here. Although talented, experienced engineers have provided their own guidance and wisdom regarding approaches to mixing, 'best practices' have, with few exceptions [75,167–169], not been formally established. Most of the 'rules' that can be found in the literature have not been subject to formal study and evaluation. In [170,171], researchers found that many of the assumptions often made by practicing engineers or researchers regarding recommended approaches for audio production were unfounded, or did not agree with the actual approaches that were taken [172]. This suggests that formal study of the best practices in and psychoacoustics of audio production is ripe for further investigation.

Further Reading

There are a huge number of high-quality texts on audio production, many of which we have already mentioned [17,43,70,71]. But since the first edition of this book, a new field has emerged. Intelligent Audio Production [173] is concerned with emerging technologies that assist with or automate much of the audio production process. Developments in the related subfield, Automatic Mixing, up to 2017, are nicely summarized in [174]. But the field took a huge turn after that point with the emergence of deep learning techniques. For intelligent systems that automatically mix multitrack audio, rule-based approaches still (as of 2025) outperform multilayer neural networks. This is primarily due to the high-dimensional nature of the problem, the lack of training data (ideally, studio multitracks), and our extensive knowledge of best practices on which rule-based approaches are based. But nevertheless, deep learning approaches have been very effective for style transfer in audio production tasks [175–177]. In some ways, these Intelligent Audio Production tools represent the culmination of many ideas in this text, since they aim to control the combination of a wide range of audio effects, and their construction requires an understanding of the theory, implementation and application of those effects.

Problems

1. Where are the compressor and noise gate usually placed in the effects chain, and why?
2. In Figure 13.7, 'Signal flow for auxiliary sends,' what operations are performed on Aux 1? Is this mono or stereo?
3. Compare and contrast dynamic EQ and multiband compression.
4. Draw a block diagram of a *de-esser*. Explain how it works and its relationship to a compressor.
5. Draw a block diagram of a *ducker*. Explain how it works and its relationship to a compressor.
6. What sort of effects or processing might be applied in the mastering stage during post-production, and which ones might be applied in mixing but rarely in mastering? Which effects might rarely be used in live sound?
7. What audio effects, or combinations of audio effects, might you use and why to deal with the following problems? You may wish to refer to the effects discussed in other chapters.
   i. A singer is slightly out of tune and off-pitch.
   ii. There is an intermittent loud, low-frequency noise in the background.
   iii. A 35 s piece of music needs to fit into a 30 s TV advertisement and catch the listener's attention.
   iv. An electronic, synthesized sound needs to be mixed with other tracks that were all recorded in a large venue.

14

Advanced Concepts in Audio Programming

This book has focused on the theory behind audio effects, how to implement them in software and how they are typically used by musicians and producers. We have given comparatively little attention to the computational hardware that actually implements these audio effects. To be able to overlook how hardware works is partially a luxury of writing plug-ins for computers in the present era, when even the most inexpensive laptop has ample resources to implement any of the effects described in this book. However, writing efficient code is always a good practice, regardless of what hardware the code runs on.

This chapter will provide a brief overview of some computational considerations on how to implement audio effects safely and efficiently, avoiding crashes or glitches like underruns, when a computation finishes too late to be useful. Most of the considerations in this chapter apply to any computing system, but we will devote special attention to embedded hardware systems like microcontrollers or single-board computers, which, compared to computers or mobile phones, are often far more constrained in their computational resources. As we will see, changes in how an effect is implemented can have a substantial impact on its performance and even whether it runs properly without crashing!

Components of a Real-Time Audio System

Every digital audio system will contain the following three components:

- A *processor* that performs the mathematical calculations in an audio effect;
- *Memory* (RAM) to hold audio signals, effect parameters, and other data; and
- *Input/output* (I/O) interfaces to communicate with the outside world.

Depending on the effect and on the hardware implementing it, any or all of these three components could become the gating factor to its performance.

DOI: 10.1201/9781003593942-14

Processor and RAM

The *processor* on desktop computers and mobile phones will be a central processing unit (CPU) whose speed and architecture will depend on the device. Most desktop and mobile CPUs, at the time of this book's publication, have multiple *cores,* each of which is an independent processor capable of running its own set of instructions while sharing RAM, cache, and I/O interfaces with the other cores. Computers and phones typically have at least several gigabytes of RAM, vastly more than what is required for a typical audio effect.

On embedded systems, the processor might instead be a *microcontroller,* typically a low-power, low-cost processor that is much slower than the CPUs found in computers. Microcontrollers are designed for integration with lower-level electrical systems, so the same chip typically contains many different I/O peripherals on the same chip, such as serial interfaces, USB interfaces, or analog-to-digital converters. Compared to computers, microcontrollers have much lower clock speeds (typically in the tens or hundreds of megahertz) and far more limited amounts of memory (RAM), typically measured in megabytes or even in kilobytes. Memory can become a substantial constraint on the implementation of certain audio effects, as we will discuss later in this chapter.

Single-board computers, like Raspberry Pi or Bela, will typically have similar resources to a low-end mobile phone: considerably more computing power and much more RAM than a microcontroller, but with more constrained resources than a desktop computer. Compared to desktop computers, the advantages of single-board computers are low cost, low-power consumption, and the ability to embed them within a physical object while still retaining the flexibility and connectivity of a full operating system like Linux.

Other forms of processors exist. Graphics processing units (GPUs) are powerful, highly parallel processors, which are usually used for graphics or for machine learning but can be deployed for audio calculations [178, 179]. Specialized digital signal processor (DSP) chips are optimized for working with real-time signals, though these have been increasingly supplanted by microcontrollers and single-board computers in the audio industry. Field programmable gate arrays (FPGAs) allow the programmer to design their own digital logic, which can be extremely efficient and powerful but lacks flexibility of a CPU or microcontroller. We will not discuss the other forms of processors in this chapter.

Managing I/O

Every computing system, from the largest server farm to the smallest microcontroller, needs a way of getting data to and from the outside world. Chapter 1 discussed how audio is converted between the analog and digital domains using ADCs and DACs. At an electrical level, ADCs and DACs are usually implemented as specialized integrated circuits, though in some microcontrollers and single-board computers, they can even be integrated into the same chip as the processor. In all cases, a digital interface is needed to send data between the processor and the audio converters. The computer will also need many other forms of I/O, for example, to connect to displays, input devices, network, and storage.

Hardware implementation is beyond the scope of this book, but how audio gets from our plug-in to the electrical interface to the ADC or DAC has consequences for our code. On any desktop computer or mobile device, our audio code will not talk to the electrical interface directly nor will the DAW that hosts our audio plug-ins. Instead, access is managed by an *operating system* that shares access to the hardware across many different programs and processes. The operating system seeks to ensure that each program receives an adequate share of CPU time, memory, and I/O resources. As we will see later in the next section, this resource-sharing arrangement has important effects we should be aware of in implementing audio effects.

Microcontrollers, unlike larger computers, often run with a very simple operating system (real-time operating system or RTOS), which manages only a few basic tasks, or they may run with no operating system at all (so-called *bare metal* programming). These configurations place greater responsibility on the programmer to manage all of the I/O peripherals directly but often allow more precise control of timing, since no other programs are competing for computing resources.

In the next sections, we will survey some of the limitations and constraints that audio effect programmers often face and identify ways that implementing code differently can help work within these constraints.

Latency

Latency in digital signal processing refers to a time delay between input and output. It is a fundamental property of every digital system (see Chapter 13 for more details).

Audio effects are usually expected to run in real time, processing audio signals as they come in. For a signal processing system to be able to run in real time, it must meet two criteria (see Figure 14.1):

1. On average, the amount of time it takes to compute an output sample must be no longer than one sampling period.
2. For every sample, there must be some maximum latency between the input and output, which is never exceeded.

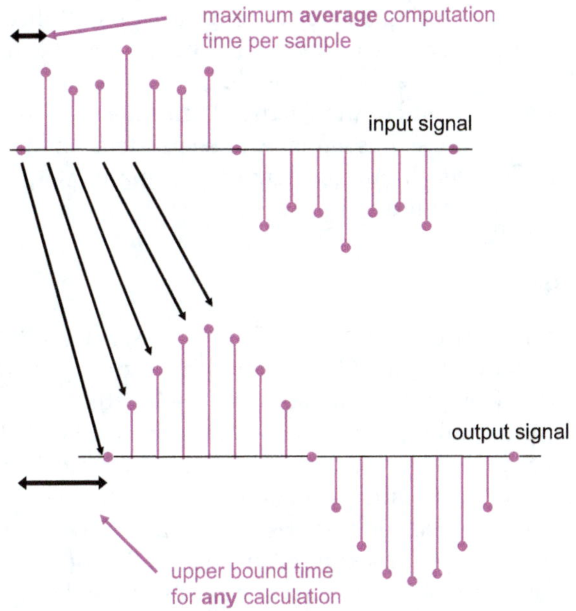

FIGURE 14.1
Calculating audio in real time places limits on the maximum average computation time and the upper bound for any computation. The upper bound determines the minimum usable latency.

For example, at 44.1 kHz, one sample lasts 22.7 μs. Averaged over time, processing one sample of audio cannot take any longer than this, or delay would accumulate without bound. However, for reasons we will soon see, it would be very unusual for the output to lag only 22.7 μs behind the input. A typical digital audio system might have a latency of 5–10 ms or more—more than a thousand times larger! The key is to choose the lowest possible latency, which we guarantee will *never* be exceeded for any audio sample.

Consequences of Latency

Before discussing the causes of latency, first consider its effects. Long latencies on the order of 50ms or more may be audible as an obvious delay. This can be distracting and annoying for musicians and can make it hard to synchronize with other musicians or recorded tracks.

Even latencies that are short enough not to sound like a delay can degrade the experience of playing a musical instrument. Experiments have suggested that total action-to-sound latency should be no more than 10ms to preserve the perceived quality of an instrument [180]. Latency of any length also translates to phase shift, and when multiple audio sources are mixed together, differing latency can produce phase cancellations that affect the frequency response of the track. This property is used deliberately in the *flanger*, discussed in Chapter 3.

It is important to remember that latency is cumulative across every element of the audio chain, so it is valuable to keep latency in our audio code as low as possible, even if our code by itself remains within acceptable latency limits for the application.

Causes of Latency

Latency has many sources in digital audio systems, a detailed breakdown of which can be found in [181]. One of the biggest latency sources on most computers comes from processing audio in *blocks* rather than one sample at a time. The audio callback function in a plug-in is not called every time a single new sample needs to be calculated; rather, it is called only when a sufficiently large block (buffer) of audio samples has accumulated, and then the function is expected to process them all at the same time.

The reason audio is processed in blocks has to do with CPU and operating system factors: when the processor needs to juggle many different tasks at the same time, it is not possible to respond to every single audio sample at the moment it arrives. Accumulating samples into blocks allows for more timing variation of when the audio callback runs. Block-based audio processing typically incurs a latency of at least twice the block size (see Figure 14.2). For example, at 44.1 kHz sample rate, a block size

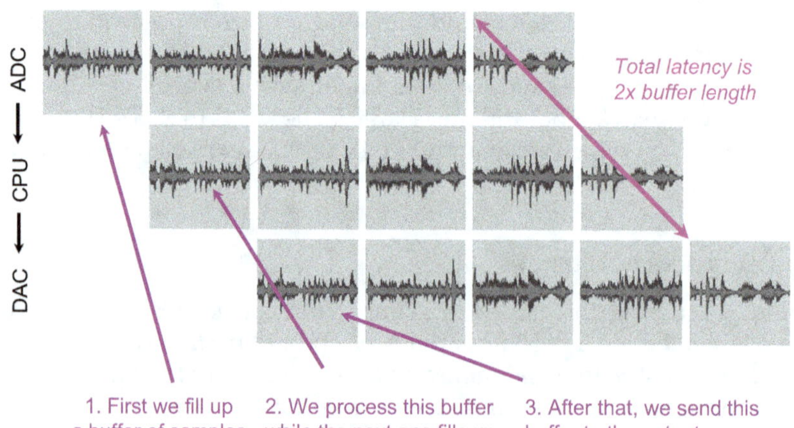

FIGURE 14.2
In a real-time system, three things generally happen at the same time: the ADC fills up a buffer of samples; the CPU processes a buffer of samples; and the DAC plays back a buffer of samples. The total buffering latency is equal to twice the buffer size.

of 256 samples would incur a buffering latency of 512 samples or 11.6 ms (on top of any latency in the audio plug-in itself or in the ADC and DAC). The tradeoff, therefore, is between larger block sizes (more robustness to timing uncertainty) and lower latency (with greater possibility of glitches from timing uncertainty).

Block size is typically set by the audio software, often selectable by the user. The programmer of an audio plug-in will have no control over its value. However, there are other sources of latency within a plugin that a developer should be aware of:

1. Some types of algorithms, like the phase vocoder, operate on blocks or windows of samples at a time. The *window size* and *hop size* of the phase vocoder need not be the same as the system audio block size. Phase vocoder processing will add its own latency of at least one window size plus one hop size [182].
2. Audio filters incur latency. In particular, FIR filters (Chapter 1) are often designed to be *linear phase*, which means that every frequency component of the signal is delayed by the same amount. Linear phase FIR filters incur a latency of $(N-1)/2$ samples where N is the filter length. Longer filters typically allow sharper frequency selectivity at the cost of higher latency.
3. IIR filters also introduce latency, but the amount varies by frequency. This is more commonly known as *group delay* and tends to be the most pronounced around the critical frequency of the filter. Further discussion of FIR and IIR filter design can be found in Chapter 4.

It is important to note that computation time alone does not incur latency, as long as the system is capable of running in real time. In other words, running the sample algorithm with the same audio block size on a faster CPU will not reduce the latency. Latency (group delay) is a mathematical property of different algorithms and an inevitable result of buffering, regardless of the speed of the CPU.

For the audio programmer seeking to minimize latency, the key point is to choose algorithms or filters with the lowest possible latency. On embedded systems or other environments where the programmer can choose the block size, seek the smallest block size needed to avoid underruns or other audible artefacts. In turn, this might require attention to the operating system or even the hardware specifications to make sure that the processor can respond quickly enough to each new audio block.

Computational Complexity

This book has focused on writing audio effects in C++. C++ code must be compiled before it can be run. The compiler converts the code to the native machine language of the processor. Thus, ultimately, the mathematical formulas in the code are implemented by a sequence of specific low-level processor instructions. The amount of time it takes to run a particular audio algorithm depends on the number and type of instructions that the compiler generates.

Precisely estimating the runtime of an algorithm can become quite involved and often depends on optimizations within the compiler, but a few heuristics are still widely used. One of these is to estimate the number of *multiplications per audio sample*. The idea is that a multiply instruction (or *multiply-and-accumulate* (MAC) instruction, which combines multiplication and addition in one instruction) is relatively time-consuming for most processors, so the fewer of these that are needed for each audio sample, the better.

When implementing filters, there is typically one multiply-and-accumulate for each filter coefficient (multiplying a sample by a coefficient and adding it to the others). Hence, the computational complexity of a filter scales linearly with the filter length. This means that long FIR filters, like those representing a room impulse response, can require tens of thousands of multiplies for every audio sample! Fortunately, as Chapter 10 discusses, there are mathematical workarounds using Fast Fourier Transforms (FFTs) to more efficiently implement long FIR filters.

For the FFT, which is also used in the phase vocoder (Chapter 9), the number of multiplies scales on the order of log(N), where N is the FFT size. On the phase vocoder, a longer N also often means the FFTs are calculated less frequently. Thus, using larger FFTs less frequently is sometimes more efficient than calculating many small FFTs. For phase vocoders, the hop size also affects the computational complexity. The larger the amount of overlap from one hop to the next, the larger the number of multiplies per sample. For the same window size N, a hop size of $N/4$ will require twice the number of multiplies as a hop size of $N/2$.

The number of multiplies is also a factor when considering interpolation strategies in delay buffers and wavetables, as discussed in Chapter 3. Cubic interpolation will usually produce higher-quality results than quadratic interpolation, which in turn produces higher-quality results than linear interpolation, with the principal tradeoff being the number of multiplies per sample.

Other mathematical functions

Computational complexity is classically measured in multiplies per sample, but what about other computations such as addition, subtraction, division, trigonometric functions, or exponentiation? Although processor architectures differ in the relative timing of different instructions, it is common for addition and subtraction operations to be relatively fast. For this reason, addition and subtraction are often not considered in the basic heuristic of multiplies-per-sample. On some architectures, division instructions might take a similar amount of time as multiplication, while on others, division will be substantially slower.

Many higher-level mathematical functions, such as `sin()`, `tan()` or `log()`, will not have an instruction in the processor at all. Such functions tend to be disproportionately expensive to compute. Instead, these functions are emulated or approximated by a computation which might involve high-order polynomials or iterative algorithms. Sometimes, processor or operating system vendors will maintain special code libraries with processor-specific optimizations for these functions; the implementations might also make coarser approximations to the mathematical ideal in order to save computation time. The `ne10` library from ARM is one such example.

Audio programmers may not always know what systems their code will be compiled for. In many cases, implementing expensive mathematical functions as look-up tables may be more efficient than using the standard C++ library function. *Wavetable oscillators* work on this principle; their principles are similar to delay-based audio effects, discussed in Chapter 3.

Single and Double Precision

Audio calculations within an effect are typically done with floating point numbers (i.e., numbers that can have a fractional part), except on the smallest microcontrollers that don't have a hardware floating-point unit (FPU). Floating point typically comes in two different sizes: 32-bit *single precision* (`float` in C++) and 64-bit *double precision* (`double` in C++). Double precision offers much more precise representation at the cost of higher memory footprint and, on many architectures, slower computation time.

For the fastest performance, it is often best to use float for audio except for computations where precision is critical, such as applying IIR filters whose feedback can magnify the effects of small rounding errors. In C++, aside from declaring the relevant variables as float rather than double, the programmer should use single-precision versions of the standard mathematical functions, such as `sinf()` instead of `sin()` or `powf()` instead of `pow()`. For example, consider the following code using the `sin()` function to implement an oscillator:

```
// Parameters of the oscillator
float frequency = 440.0;
float amplitude = 0.5;
```

```
// Get this information from the plug-in host
int numSamples;
float sampleRate;

// Phase should be a global variable to be saved across callbacks
float phase = 0;

for(int i = 0; i < numSamples; i++)
{
    // Increment the phase and wrap when it exceeds 2*pi
    phase += 2.0 * M_PI * frequency / sampleRate;
    if(phase >= 2.0 * M_PI) phase -= 2.0 * M_PI;

    // Calculate the sine wave and store in the output buffer
    channelData[i] = amplitude * sin(phase);
}
```

On some systems, using `sinf()` instead of `sin()` may result in a performance improvement as long as the extra precision is not needed. But a wavetable implementation might be faster still:

```
// Parameters of the oscillator
float frequency = 440.0;
float amplitude = 0.5;

// Get this information from the plug-in host
int numSamples;
float sampleRate;

// Parameters of the wavetable
const int wavetableLength = 512;
float wavetable[wavetableLength];

// Use a read pointer rather than a phase; fractional to allow
// interpolation
float readPointer = 0;

// This code goes in initialisation of audio plugin or device, NOT in
// audio callback:
for(unsigned int n = 0; n < wavetableLength; n++)
{
    // Wavetable should be one complete oscillation of the sine wave
    wavetable[n] = sin(2.0 * M_PI * (float)n / (float)wavetableLength);
}

// This code goes in the audio callback:
for(int i = 0; i < numSamples; i++)
{
    // Increment the location in the wavetable and wrap when it reaches
    // the end
    readPointer += wavetableLength * frequency / sampleRate;
    while(readPointer >= wavetableLength)
        readPointer -= wavetableLength;

    // Simplest version, without interpolation:
    // channelData[i] = amplitude * wavetable[(int)readPointer];
```

```
    // Better quality with linear interpolation:.Pointer takes
    // fractional
    // index. Look for samples either side which are indices we can read
    // into
    // buffer. If we get to the end of the buffer, wrap around to 0.
    int indexBelow = floorf(readPointer);
    int indexAbove = indexBelow + 1;
    if(indexAbove >= wavetableLength) indexAbove = 0;

    // For linear interpolation, need to decide how much to weigh each
    // sample. Closer
    // the index's fractional part is to 0, the more weight we give to
    // "below" sample.
    // The closer the fractional part is to 1, more weight we give to
    // "above" sample.
    float fractionAbove = readPointer - indexBelow;
    float fractionBelow = 1.0 - fractionAbove;

    // Calculate the weighted average of the "below" and "above" samples
    channelData[i] = fractionBelow * wavetable[indexBelow] +
                     fractionAbove * wavetable[indexAbove];
}
```

Even though the second example is many more lines of C++ code than the first, it is likely to be faster on most systems. The key to the performance speed-up is that the expensive `sin()` calculations are all performed once at the beginning of the program rather than every sample in real-time. Instead, the stored values are read from the table (with linear interpolation in this example). For further efficiency, rather than keeping track of the phase between 0 and 2π, we keep track of the position in the table between 0 and $N-1$ where N is the length of the table in samples. Notice the similarities between this code, circular buffering in the phase vocoder (Chapter 9), and delay-based effects (Chapter 3).

Vectorization

Most modern CPUs, and even some high-end microcontrollers, have instructions that operate on vectors (arrays) of numbers at the same time. A vector multiplication instruction might perform two or four independent multiplications at the same time, instead of needing separate instructions for each one.

Using vector computation can substantially speed up the execution of an audio effect, but getting the conditions right to be able to use the instructions can be a major challenge. Not all C++ compilers easily generate vectorized code. Loops (such as a `for()` loop applying coefficients of an FIR filter) may need to be organized in a particular way for the compiler to understand how to vectorize them. In some cases, compiler-specific directives (`#pragma`) may also help. In the most ambitious case, where the specific hardware target is known, the programmer might write critical sections of code in assembly language to be sure to make the fullest use of the vector instructions the processor provides.

A full discussion of vectorization is beyond the scope of this book, and best practices change regularly with each new hardware platform and compiler version. General rules of thumb are: to make sure several computations can be executed entirely independently of one another (without one computation depending on the results of a previous one); to write clear and simple loops; and sometimes to *unroll* loops by hand to demonstrate to the compiler how several independent calculations can be done at the same time.

Audio Rate and Control Rate

When calculating a signal that carries a high computational cost, but the signal changes relatively slowly, a simple trick is just to calculate that signal less frequently. For example, consider the *wah-wah* effect (Chapter 5). The effect is generated by a second-order IIR filter whose coefficients change whenever the user adjusts a control. Applying the filter is inexpensive, requiring only 5 multiplications per sample. However, updating the coefficients involves a complicated formula including trigonometric functions, which require many more multiplications to implement. It is not efficient to recalculate the coefficients of the filter at each sample. Instead, a more efficient approach will recalculate the coefficients only when they are changed by the user, for example, by comparing the current value of the cutoff frequency to a version that was cached the last time the coefficients were updated.

For decades, computer music programming languages have implemented the idea of a separate audio rate and control rate, the latter of which is a fraction of the audio rate. For example, with an audio rate of 44.1 kHz, the control rate might be 1/16 of this, or 2.7 kHz. Control rate calculations would only be made once every 16 samples, reducing their total computational load.

For example, consider the calculation of *low-frequency oscillators* (LFOs). In an *auto-wah* effect, the cutoff frequency of the filter changes continuously over time under the control of an LFO. It might appear that recalculating coefficients at every sample is necessary. However, with most LFOs, the change in value from one audio sample to the next is so subtle that the filter coefficients can be updated at a lower control rate with no audible change in performance. Here is an example, extending the code examples in Chapter 5, of how a control-rate filter update could work:

```
int numSamples;          // Indicates how many audio samples to process
float *channelData;      // Array of audio samples, length numSamples
float coefficients[6];   // Previously calculated filter coefficients
float x1, x2, y1, y2;    // Previous values of the input and output
```

```
const int controlInterval = 16; // Interval of control rate updates, in
samples
int controlCounter = 0;

for (int i = 0; i < numSamples; ++i)
{
    const float in = channelData[i];

    float out = coefficients[0] * in /* b0 */
    + coefficients[1] * x1 /* b1 */
    + coefficients[2] * x2 /* b2 */
    - coefficients[4] * y1 /* a1 */
    - coefficients[5] * y2; /* a2 */

    x2 = x1;
    x1 = in;
    y2 = y1;
    y1 = out;

    channelData[i] = out;

    // Update filter coefficients every control rate frame
    controlCounter++;
    if(controlCounter >= controlInterval) {
        controlCounter = 0;
        // This is the function where more expensive calculations are
        // made
        // to calculate the filter coefficients based on its
        // specifications
        recalculateFilterCoefficients();
    }
}
```

The key lines are at the very end of the example, where a counter increments every audio sample to decide when the next control rate interval has elapsed. Then a function is called, which can be more computationally expensive, since it will run fewer times per second than the main audio processing. If defining a separate audio rate and control rate, it is best practice to fix the control rate interval in your code rather than using the system audio buffer size, as the latter option would make the audio plugin behave differently depending on how the user configures the host audio environment.

Memory Footprint

Beyond CPU time, one of the main resource constraints of audio effects is their *memory footprint*: how much RAM they take to run. Most of the effects in this book take very little memory to implement, to the point that memory usage will not be a major constraint except on the very smallest microcontrollers. The exception is anywhere that long chunks of recorded audio need to be stored.

In delay-based audio effects, the amount of memory required scales with the maximum length of the delay. This is discussed in the Implementation section of Chapter 3, where it was shown that 10 s of delay at 44.1 kHz with 32-bit samples would require approximately 1.7 MB of memory to store. For multichannel audio effects, the memory footprint will also scale linearly with the number of channels.

Similar principles apply wherever *wavetables* are used, as in the case where wavetables are used to avoid the need to calculate trigonometric functions at each audio sample. The length of the wavetable presents a tradeoff: longer wavetables will produce more accurate reconstructions of the original function at the cost of greater memory footprint. Here, too, the effect is only likely to be significant on embedded systems with less than a few megabytes of RAM.

Although most audio effects do not use enough memory to present a strain on a modern computer, this is not the case with audio synthesizers. These synthesizers often use a *sample library*, which may include thousands of audio recordings, each of several seconds or longer, covering different variations in pitch, dynamics and articulation for every instrument in the library. Professional sample libraries are often recorded at high sample rates and bit depths (for example, 24 bits at 96 kHz). The total amount of storage for a sample library can run into the tens or hundreds of gigabytes, considerably more than could be stored in the computer's RAM.

The usual approach is to store all of the samples in non-volatile storage-- hard drives in previous eras, and now flash memory-- and only load into RAM the samples that are likely to be played in the immediate future. It is usually not a good idea to try to play audio samples in real time directly from a storage medium because the latency in accessing storage can be high and unpredictable. This is particularly true if the samples are stored across a network. Moreover, loading samples from storage into RAM should not take place within the audio callback, because of the unpredictable latency in accessing storage. This is discussed further in the next section.

Threading

Most audio plug-ins are multi-threaded. This means that two different methods could be running simultaneously in different threads of execution, with no guarantee that one method finishes before the other begins.

In particular, the thread that runs the audio callback function is often different than the thread that handles changing parameters. This results in the potential for subtle bugs and even outright crashes.

Consider the case of a phase vocoder plug-in that needs to gather samples in a buffer before performing a Fast Fourier Transform. Suppose that the user changes the window size of the effect using one of the graphical controls. When this happens, the thread that controls the user interface will update the value of the associated parameter (for example, an object of type `juce::AudioParameterChoice` or `juce::AudioParameterInt`). The plug-in will then be responsible for reallocating the window buffer to hold a different number of samples. In what thread should this happen?

In a multi-threaded environment, the update to `juce::AudioParameterChoice` might happen at the same time as the callback `processBlock()` is running. If the user interface thread tried to reallocate the window buffer, the audio callback thread would be unaware that the buffer size was about to change. It might then attempt to access an invalid index and cause the plug-in to crash, potentially bringing down the entire DAW. Similar problems may occur on recalculating coefficients for filters and other situations where memory is allocated or deallocated while the effect is running.

One approach to avoiding a crash is to ensure that the parameter cannot change value in the middle of a call to `processBlock()`. In other words, only one of the user interface thread and the audio thread can manipulate the `AudioParameterChoice` object at the same time. If one thread begins working with the object first, the other thread must wait until it finishes before proceeding. This behavior can be achieved using a special variable type called a *mutex* (short for mutually exclusive). When one thread locks the mutex, the other thread must stop and wait (block) until the mutex is unlocked again.

Although using a mutex is a common way of addressing threading issues in other contexts, they are generally discouraged in real-time audio programming, except in carefully controlled situations. The reason is that anything that can stop the audio thread from running, even temporarily, can cause it to miss its real-time deadlines and lead to audio glitches. In particular, if the other thread performs memory allocation or I/O while the mutex is locked, this leaves the timing of the audio thread at the mercy of how long those activities take, and most operating systems do not offer any guarantees on the speed of I/O or memory allocation. Therefore, if a mutex is used, it should cover only a few critical lines of code, changing the values of variables in memory.

In real-time audio, the more common way of dealing with threading issues is to do all of the most critical work in the audio thread and make sure that any changes from the user interface thread always leave the parameters in a consistent state. Sometimes, changing a parameter leads to a task that shouldn't be done in the audio thread, such as allocating new memory buffers or loading samples from storage. In this case, a common approach is to use two copies of any relevant memory buffers. While the audio thread continues processing the first copy, the user interface thread can do any allocation or initialization on the second copy. Once the second copy is fully initialized, the audio thread can swap pointers to work with the second memory buffer instead.

The most thread-critical part of the code is setting and testing the flag `shouldSwapBuffers`. Setting or testing this needs to be fully atomic (i.e., cannot be interrupted by a context switch in the middle). Moreover, most modern CPUs implement some form of out-of-order execution where machine-language instructions can take place in a different order than specified in the program if the CPU believes it will produce the same result more efficiently. Testing variables that synchronize threads must be done strictly in order to preclude unexpected behavior. Every operating system and most microcontroller vendor APIs will offer some form of thread library support (e.g., the commonly used `pthreads` library on UNIX systems) to manage these issues. Semaphores and condition variables are some of the techniques available for synchronizing threads in cases like this.

```
// Declare two audio buffers (or data structures) which have identical
// purposes
const int bufferSize;
float audioBuffer1[bufferSize];
float audioBuffer2[bufferSize];

// At any given time, one buffer will be in use by the audio code while
// the other is free to be reallocated. Pointers indicate which is
// which.
float *activeAudioBuffer = audioBuffer1;
float *inactiveAudioBuffer = audioBuffer2;
bool shouldSwapBuffers = false;

// This function could be called on a separate thread than the audio
// thread
void reallocateBuffers()
{
    // If we have already reallocated the buffers, wait for the audio
    // thread to pick that up before continuing. Can also use a
    // OS-specific
    // threading library to wait on this condition.
    while(shouldSwapBuffers == true)
        delay(1);    // This implementation will vary by platform

    // Do any reinitialisation here, which could include reallocating
    // memory
    // This function is pseudo-code for anything you might want to do:
    reinitialiseBuffer(inactiveAudioBuffer);
```

```
    // Tell the audio thread that the pointers are ready to swap.
    // This might be implementation with an atomic thread-safe operation
    // such as a semaphore.
    shouldSwapBuffers = true;
}

void processBlock(AudioSampleBuffer& buffer, MidiBuffer& midiMessages)
{
    if(shouldSwapBuffers)
    {
        float *tempPointer = activeAudioBuffer;
        activeAudioBuffer = inactiveAudioBuffer;
        inactiveAudioBuffer = tempPointer;
        shouldSwapBuffers = false;
    }

    // Do the rest of the audio processing here
    // Always use activeAudioBuffer only!
}
```

Conclusion

The audio programmer needs many tools in their toolbox. A well-balanced programmer should understand the mathematics behind the effects, the resources and limitations of the hardware that will implement the effects, and the musical use cases. This chapter has provided a brief overview of some of the practical considerations of implementing audio effects on desktop computers and embedded hardware.

Hardware performance and software APIs will change every year, but the basic principles in this chapter, and in the rest of the book, are likely to remain similar for the foreseeable future.

Further Reading

A number of practical tutorials on audio programming can be found online. The video course by co-author McPherson [182] covers audio programming on embedded systems, looking at some of the practical issues that developers for embedded hardware face, including latency, computation time, and threading. Many of the lectures are also applicable to writing plug-ins in JUCE. Multi-threaded programming in particular can be quite subtle and challenging, particularly as the number of threads and the number of shared resources increase. Several good texts exist on the subject, e.g., [183]. Other related work [180,181] has examined sources and effects of latency in digital audio systems.

References

[1] U. Zolzer, *Digital Audio Signal Processing*, 2nd Ed., John Wiley and Sons, Ltd, 2008.

[2] A. V. Oppenheim and R. W. Schafer, *Discrete-Time Signal Processing*, 3rd Ed., Prentice Hall, 2009.

[3] S. J. Orfanidis, *Introduction to Signal Processing*, Prentice Hall, 2010.

[4] J. O. Smith, *Spectral Audio Signal Processing*, W3K Publishing, 2011.

[5] A. Koenig and B. E. Moo, *Accelerated C++*, Addison Wesley, 2000.

[6] S. B. Lippman, J. Lajoie and B. Moo, *C++ Primer*, 5th Ed., Addison Wesley, 2012.

[7] B. Stroustrup, *The C++ Programming Language*, Addison Wesley, 2000.

[8] R. Boulanger, V. Lazzarini, M. Mathews, M. V. Mathews, J. Ffitch and R. W. Dobson, *The Audio Programming Book*, MIT Press, 2010.

[9] W. Pirkle, *Designing Audio Effect Plugins in C++: For AAX, AU, and VST3 with DSP Theory*, Routledge, 2019.

[10] W. C. Pirkle, *Designing Software Synthesizer Plugins in C++: With Audio DSP*, Focal Press, 2021.

[11] M. J. Yee-King, *Build AI-Enhanced Audio Plugins with C++*, CRC Press, 2024.

[12] F. Bianchi, A. Cipriani and M. Giri, *Pure Data: Electronic Music and Sound Design*, Contemponet, 2021.

[13] J. D. Reiss, *Working with the Web Audio API*, Routledge, 2022.

[14] O. Niemitalo, "Polynomial Interpolators for High-Quality Resampling of Oversampled Audio," 2001. https://yehar.com/blog/wp-content/uploads/2009/08/deip.pdf

[15] A. M. Stark, M. D. Plumbley and M. E. P. Davies, "Real-time beat-synchronous audio effects," in *New Interfaces for Musical Expression (NIME)*, New York, 2007.

[16] J. O. Smith, *Physical Audio Signal Processing*, W3K Publishing, 2010.

[17] A. Case, *Mix Smart: Professional Techniques for the Home Studio*, CRC Press, 2012.

[18] P. Pestana and J. Reiss, "User preference on artificial reverberation and delay time parameters," *Journal of the Audio Engineering Society*, vol. 65, no. 1/2, 2017.

[19] V. Valimaki and T. I. Laakso, "Principles of fractional delay filters," in *IEEE International Conference on Acoustics, Speech, and Signal Processing (ICASSP)*, Istanbul, Turkey, 2000.

[20] A. Wright and V. Valimaki, "Neural modeling of phaser and flanging effects," *Journal of the Audio Engineering Society*, vol. 69, no. 7/8, pp. 517–529, 2021.

[21] A. Clifford and J. D. Reiss, "Using delay estimation to reduce comb filtering of arbitrary musical sources," *Journal of the Audio Engineering Society*, vol. 61, no. 11, pp. 917–927, 2013.

[22] E. Vincent, T. Virtanen and S. Gannot, Eds., *Audio Source Separation and Speech Enhancement*, John Wiley & Sons, 2018.
[23] L. Wang, T. K. Hon, J. D. Reiss and A. Cavallaro, "Self-localization of Ad-hoc arrays using time difference of arrivals," *IEEE Transactions on Signal Processing*, vol. 64, no. 4, 2016.
[24] P. A. Naylor and N. D. Gaubitch, Eds., *Speech Dereverberation*, Springer Science & Business Media, 2010.
[25] Y. Li and L. Deng, "An overview of speech dereverberation," in *8th Conference on Sound and Music Technology*, Singapore, 2021.
[26] S. J. Orfanidis, "High-order digital parametric equalizer design," *Journal of the Audio Engineering Society*, vol. 53, pp. 1026–1046, 2005.
[27] J. D. Reiss, "Design of audio parametric equalizer filters directly in the digital domain," *IEEE Transactions on Audio, Speech and Language Processing*, vol. 19, pp. 1843–48, August 2011.
[28] V. Valimaki and J. D. Reiss, "All about audio equalization: Solutions and frontiers," *Applied Sciences, Special Issue on Audio Signal Processing*, vol. 6, no. 5, May 2016.
[29] D. Self, *The Design of Active Crossover*, Focal Press, 2011.
[30] J. A. Moorer, "The manifold joys of conformal mapping: Applications to digital filtering in the studio," *Journal of the Audio Engineering Society*, vol. 31, pp. 826–841, 1983.
[31] G. Massenburg, "Parametric equalization," in *42nd AES Convention*, Los Angeles, CA, USA, 1972.
[32] J. D. Reiss and A. McPherson, *Audio Effects*, 1st Ed., CRC Press, 2014.
[33] J. T. Colonel, C. J. Steinmetz, M. Michelen, and J. D. Reiss, "Direct design of biquad filter cascades with deep learning by sampling random polynomials," in *IEEE International Conference on Acoustics, Speech and Signal Processing (ICASSP)*, Singapore, 2022.
[34] C. Y. Yu, et al., "Differentiable all-pole filters for time-varying audio systems," in *Intl. Conf. Digital Audio Effects (DAFx)*, Surrey, UK, 2024.
[35] R. Izhaki, *Mixing Audio: Concepts, Practices and Tools*, Focal Press, 2008.
[36] E. Perez Gonzalez and J. D. Reiss, "Automatic equalization of multi-channel audio using cross-adaptive methods," in *127th AES Convention*, New York, 2009.
[37] M. Holters and U. Zölzer, "Graphic equalizer design using higher-order recursive filters," in *Digital Audio Effects (DAFx)*, Montreal, 2006.
[38] ISO, "ISO 266, Acoustics– Preferred frequencies for measurements," 1975.
[39] L. Bregitzer, *Secrets of Recording: Professional Tips, Tools & Techniques*, Focal Press, 2009.
[40] Z. Ma, J. D. Reiss and D. Black, "Implementation of an intelligent equalization tool using Yule-Walker for music mixing and mastering," in *134th AES Convention*, Rome, Italy, 2013.
[41] A. Loscos and T. Aussenac, "The wahwactor: A voice controlled wah-wah pedal," in *New Interfaces for Musical Expression (NIME)*, Vancouver, Canada, 2005.
[42] R. G. Keen, "Technology of Wah Pedals," Sept., 1999. http://geofex.com/Article_Folders/wahpedl/wahped.htm

[43] R. Izhaki, *Mixing Audio: Concepts, Practices and Tools*, 4th Ed., Focal Press, 2023.

[44] V. Bruschi, Cecchi, S., A. Nicolini and V. Välimäki, "Real-time implementation of a linear-phase octave graphic equalizer," in *International Conference on Digital Audio Effects (DAFX)*, 2024.

[45] S. Hafezi and J. D. Reiss, "Autonomous multitrack equalisation based on masking reduction," *Journal of the Audio Engineering Society*, vol. 63, no. 5, 2015.

[46] M. A. Martinez Ramirez and J. D. Reiss, "End-to-end equalization with convolutional neural networks," in *Digital Audio Effects (DAFx)*, Aveiro, Portuga, 2018.

[47] A. Carson, C. Valentini-Botinhao, S. King, and S. Bilbao, "Differentiable grey-box modelling of phaser effects using frame-based spectral processing," in *Intl. Conf. Digital Audio Effects (DAFX)*, Copenhagen, Denmark, 2023.

[48] M. A. Martinez Ramirez, E. Benetos and J. D. Reiss, "A general-purpose deep learning approach to model time-varying audio effects," in *22nd International Conference on Digital Audio Effects (DAFx)*, Birmingham, UK, 2019.

[49] R. Parncutt, "A perceptual model of pulse salience and metrical accent in musical rhythm," *Music Perception*, vol. 11, Summer 1994.

[50] H. Fastl and E. Zwicker, *Psychoacoustics: Facts and Models*, 3rd Ed., Springer, 2007.

[51] S. Rosen, "Temporal information in speech: Acoustic, auditory and linguistic aspects," *Philosophical Transactions of the Royal Society B*, vol. 336, June 1992.

[52] P. Dutilleux and U. Zölzer, "Modulators and demodulators," in *DAFX: Digital Audio Effects*, 2nd Ed., 2011.

[53] C. Mitcheltree, C. J. Steinmetz, M. Comunità, J. D. Reiss, "Modulation extraction for LFO-driven audio effects," in *26th Int. Conf. Digital Audio Effects (DAFx)*, Copenhagen, Denmark, 2023.

[54] J. M. Chowning, "The synthesis of complex audio spectra by means of frequency modulation," *Journal of the Audio Engineering Society*, vol. 21, no. 7, pp. 526–534, 1973.

[55] V. Lazzarini and J. Timoney, "Modulation synthesis in digital and analogue computing environments," in *11th Ubiquitous Music Workshop*, Matosinhos, Portugal, 2021.

[56] B. Lachaise and L. Daudet, "Inverting dynamics compression with minimal side information," in *Digital Audio Effects Workshop (DAFx)*, Helsinki, 2008.

[57] D. Giannoulis, M. Massberg and J. D. Reiss, "Digital dynamic range compressor design— A tutorial and analyis," *Journal of the Audio Engineering Society*, vol. 60, June 2012.

[58] B. Rudolf, "1176 Revision history," *Mix Magazine Online*, 1 June 2000. https://mixonline.com/recording/1176-revision-history-372048

[59] B. E. Lobdell and J. B. Allen, "A model of the VU (volume-unit) meter, with speech applications," *Journal of the Acoustical Society of America*, vol. 121, 2007.

[60] D. Barchiesi and J. D. Reiss, "Reverse engineering the mix," *Journal of the Audio Engineering Society*, vol. 58, pp. 563–576, July 2010.

[61] R. J. Cassidy, "Level detection tunings and techniques for the dynamic range compression of audio signals," in *117th AES Convention*, Los Angeles, CA, USA, 2004.
[62] J. S. Abel and D. P. Berners, "On peak-detecting and RMS feedback and feedforward compressors," in *115th AES Convention*, New York, NY, USA, 2003.
[63] P. Dutilleux, K. Dempwolf, M. Holters and U. Zolzer, "Nonlinear processing," in *Dafx: Digital Audio Effects*, Wiley, John & Sons, 2011, p. 554.
[64] J. Bitzer and D. Schmidt, "Parameter estimation of dynamic range compressors: Models, procedures and test signals," in *120th AES Convention*, 2006.
[65] L. Lu, "A digital realization of audio dynamic range control," in *Fourth International Conference on Signal Processing Proceedings (IEEE ICSP)*, 1998.
[66] Sonnox, *Dynamics Plug-in Manual*, Sonnox Oxford Plug-ins, 2007.
[67] M. Zaunschirm, J. D. Reiss and A. Klapuri, "A sub-band approach to musical transient modification," *Computer Music Journal*, vol. 36, 2012.
[68] M. J. Terrell, "Automatic noise gate settings for drum recordings containing bleed from secondary sources," *EURASIP Journal on Advances in Signal Processing*, vol. 2010, pp. 1–9, 2010. Article ID 465417.
[69] S. Gorlow and J. D. Reiss, "Model-based inversion of dynamic range compression," *IEEE Transactions on Audio, Speech, and Language Processing*, vol. 21, 2013.
[70] D. Gibson, *The Art of Mixing*, 3rd Ed., Taylor & Francis, 2018.
[71] M. Senior, *Mixing Secrets for the Small Studio*, 3rd Ed., Focal Press, 2025.
[72] J. T. Colonel and J. Reiss, "Approximating ballistics in a differentiable dynamic range compressor," in *153rd Audio Engineering Society Convention*, New York, NY, USA, 2022.
[73] J. T. Colonel and J. D. Reiss, "Reverse engineering a nonlinear mix of a multitrack recording," *Journal of the Audio Engineering Society*, vol. 71, no. 9, pp. 586–595, 2023.
[74] S. Gorlow, J. D. Reiss and E. Duru, "Restoring the dynamics of clipped audio material by inversion of dynamic range compression," in *IEEE International Symposium on Broadband Multimedia Systems and Broadcasting (BMSB)*, Beijing, China, 2014.
[75] Z. Ma, B. De Man, P. D. Pestana, D. A. A. Black and J. D. Reiss, "Intelligent multitrack dynamic range compression," *Journal of the Audio Engineering Society*, vol. 63, no. 6, 2015.
[76] A. Mason, N. Jillings, Z. Ma, J. D. Reiss and F. Melchior, "Adaptive audio reproduction using personalised compression," in *AES 57th International Conference*, Hollywood, CA, USA, 2015.
[77] D. Giannoulis, M. Massberg and J. D. Reiss, "Parameter automation in a dynamic range compressor," *Journal of the Audio Engineering Society*, vol. 61, no. 10, 2013.
[78] C. J. Steinmetz and J. D. Reiss, "Efficient neural networks for real-time modeling of analog dynamic range compression," in *Audio Engineering Society Convention 152*, 2022.
[79] A. Wright and V. Valimaki, "Grey-box modelling of dynamic range compression," in *25th International Conference on Digital Audio Effects (DAFx20in22)*, Vienna, Austria, 2022.

[80] M. Comunità, C. J. Steinmetz and J. D. Reiss, "Differentiable black-box and gray-box modeling of nonlinear audio effects," *arXiv preprint arXiv:2502.14405*, 2025.

[81] R. M. Corey and A. C. Singer, "Modeling the effects of dynamic range compression on signals in noise," *The Journal of the Acoustical Society of America*, vol. 150, no. 1, pp. 159–170, 2021.

[82] B. D. Man and J. D. Reiss, "An intelligent multiband distortion effect," in *AES 53rd International Conference on Semantic Audio*, London, UK, 2014.

[83] J. T. Colonel, M. Comunità and J. Reiss, "Reverse Engineering Memoryless Distortion Effects with Differentiable Waveshapers," in *153rd Audio Engineering Society Convention*, New York, NY, USA, 2022.

[84] D. Yeh, J. Abel and J. O. Smith, "Simulation of the diode limiter in guitar distortion circuits by numerical solution of ordinary differential equations," in *Digital Audio Effects (DAFX)*, 2007.

[85] V. Välimäki, S. Bilbao, J. O. Smith, J. S. Abel, j. Pakarinen and D. Berners, "Virtual analog effects," in *Dafx:Digital Audio Effects*, 2nd Ed., Wiley, John & Sons, 2011.

[86] D. Yeh, J. S. Abel, A. Vladimirescu and J. O. Smith, "Numerical methods for simulation of guitar distortion circuits," *Computer Music Journal*, vol. 32, pp. 23–42, 2008.

[87] J. Pakarinen and M. Karjalainen, "Enhanced wave digital triode model for real-time tube amplifier emulation," *IEEE Transactions on Audio, Speech, and Language Processing*, vol. 18, pp. 738–746, 2010.

[88] J. Pakarinen and D. Yeh, "A review of digital techniques for modeling vacuum-tube guitar amplifiers," *Computer Music Journal*, vol. 33, pp. 85–100, 2009.

[89] W. M. J. Leach, "SPICE models for vacuum-tube amplifiers," *Journal of the Audio Engineering Society*, vol. 43, pp. 117–126, 1995.

[90] H. Shapiro and C. Glebbeek, *Jimi Hendrix: Electric Gypsy (New and Improved ed.)*, St. Martin's Press, 1995.

[91] M. A. Martinez Ramirez and J. D. Reiss, "Modeling nonlinear audio effects with end-to-end deep neural networks," in *IEEE International Conference on Acoustics, Speech and Signal Processing (ICASSP)*, Brighton, UK, 2019.

[92] M. Comunità, C. J. Steinmetz and J. D. Reiss, "NablAFx: A framework for differentiable black-box and gray-box modeling of audio effects," *arXiv preprint arXiv:2502.11668*, 2025.

[93] J. Imort, G. Fabbro, M. A. Martinez Ramirez, S. Uhlich, Y. Koyama and Y. Mitsufuji, "Distortion audio effects: Learning how to recover the clean signal," in *23rd Int. Society for Music Information Retrieval (ISMIR)*, Bengaluru, India, 2022.

[94] J. L. Flanagan and R. M. Golden, "Phase vocoder," *Bell System Technical Journal*, pp. 1493-1509, Nov. 1966.

[95] M. Portnoff, "Implementation of the digital phase vocoder using the fast Fourier transform," *IEEE Transactions on Acoustics, Speech, and Signal Processing*, vol. 24, pp. 243–248, 1976.

[96] J. Laroche and M. Dolson, "Improved phase vocoder timescale modification of audio," *IEEE Transactions on Speech and Audio Processing*, vol. 7, pp. 323–332, 1999.

[97] M. Dolson, "The phase vocoder: A tutorial," *Computer Music Journal*, vol. 10, pp. 14–27, 1986.
[98] M. Liuni and A. Röbel, "Phase vocoder and beyond," *Musica/Tecnologia*, pp. 73–89, 2013.
[99] M.-H. Serra, "Introducing the phase vocoder," in *Musical Signal Processing*, Routledge, pp. 31–90, 2013.
[100] M. Pilia, S. Mandelli, P. Bestagini and S. Tubaro, "Time scaling detection and estimation in audio recordings," in *IEEE International Workshop on Information Forensics and Security (WIFS)*, Montpelier, France, 2021, pp. 1–6.
[101] M. Scott, *Rise of Auto-Tune*, PublishDrive, 2025.
[102] L. L. Beranek, "Analysis of Sabine and Eyring equations and their application to concert hall audience and chair absorption," *Journal of the Acoustical Society of America*, vol. 120, Sept. 2006.
[103] D. Howard and J. Angus, *Acoustics and Psychoacoustics*, 5th Ed., Routledge, 2017.
[104] D. Davis and E. Patronis, *Sound System Engineering*, 3rd Ed., Elsevier, 2006.
[105] D. Davis, *A More Accurate Way of Calculating Critical Distance*, Altec TL-207, 1971.
[106] M. R. Schroeder and B. F. Logan, "Colorless artificial reverberation," *Journal of the Audio Engineering Society*, vol. 9, 1961.
[107] M. R. Schroeder, "Natural sounding artificial reverberation," *Journal of the Audio Engineering Society*, vol. 10, 1962.
[108] J. A. Moorer, "About this reverberation business," *Computer Music Journal*, vol. 3, pp. 13–18, 1979.
[109] M. Jot and A. Chaigne, "Digital delay networks for designing artificial reverberators," in *90th Audio Engineering Society Convention*, Paris, France, 1991.
[110] J. Stautner and M. Puckette, "Designing multichannel reverberators," *Computer Music Journal*, vol. 6, no. 1, pp. 52–65, 1982.
[111] J. Allen and D. Berkley, "Image method for efficiently simulating small-room acoustics," *Journal of the Acoustical Society of America*, vol. 65, pp. 943–950, 1979.
[112] E. Lehmann and A. Johansson, "Prediction of energy decay in room impulse responses simulated with an image-source model," *Journal of the Acoustical Society of America*, vol. 124, July 2008.
[113] P. Peterson, "Simulating the response of multiple microphones to a single acoustic source in a reverberant room," *Journal of the Acoustical Society of America*, vol. 80, Nov. 1986.
[114] W. G. Gardner, "Efficient convolution without input-output delay," in *97th Convention of the Audio Engineering Society*, San Francisco, CA, 1994.
[115] V. Valimaki, J. D. Parker , L. Savioja and J. O. Smith, "Fifty years of artificial reverberation," *IEEE Transactions on Audio, Speech, and Language Processing*, vol. 20, no. 5, pp. 1421–1448, 2012.
[116] V. Välimäki, J. Parker, L. Savioja, J. O. Smith and J. Abel, "More than 50 years of artificial reverberation," in *60th AES Intl. Conf.: Dereverberation and Reverberation of Audio (DREAMS)*, Leuven, Belgium, 2016.
[117] I. Ibnyahya and J. Reiss, "A Method for matching room impulse responses with feedback delay networks," in *153rd Audio Engineering Society Convention*, New York, NY, USA, 2022.

[118] S. Schlecht, "FDNTB: The feedback delay network toolbox," in *International Conference on Digital Audio Effects. DAFx*, Vienna, Austria, 2020.
[119] S. Siddiq, "Optimization of convolution reverberation," in *23rd International Conference on Digital Audio Effects (DAFx-20)*, Vienna, Austria, 2020.
[120] B. De Man, K. McNally and J. Reiss, "Perceptual evaluation and analysis of reverberation in multitrack music production," *Journal of the Audio Engineering Society*, vol. 65, no. 1/2, 2017.
[121] E. Chourdakis and J. Reiss, "A machine learning approach to design and evaluation of intelligent artificial reverberation," *Journal of the Audio Engineering Society*, vol. 65, no. 1/2, 2017.
[122] S. Lee, H.-S. Choi and K. Lee, "Differentiable artificial reverberation," *IEEE/ACM Transactions on Audio, Speech, and Language Processing*, vol. 30, pp. 2541–2556, 2022.
[123] J. Vilkamo, T. Lokki and V. Pulkki, "Directional audio coding: Virtual microphone-based synthesis and subjective evaluation," *Journal of the Audio Engineering Society*, vol. 57, 2009.
[124] D. Beheng, *Sound Perception*, Deutche Welle, Radio Training Centre (DWRTC), 2002.
[125] B. Bernfeld, "Attempts for better understanding of the directional stereophonic listening mechanism," in *44th Audio Engineering Society Convention*, Rotterdam, Netherlands, 1973.
[126] V. Pulkki, "Virtual sound source positioning using vector base amplitude panning," *Journal of the Audio Engineering Society*, vol. 45, 1997.
[127] D. G. Malham and A. Myatt, "3-D sound spatialization using ambisonic techniques," *Computer Music Journal*, vol. 19, pp. 58–70, 1995.
[128] J. S. Bamford, *An Analysis of Ambisonics Sound Systems of First and Second Order*, University of Waterloo, 1995.
[129] M. A. Poletti, "Three-dimensional surround sound systems based on spherical harmonics," *Journal of the Audio Engineering Society*, vol. 53, Nov. 2005.
[130] P. M. Morse and K. U. Ingard, *Theoretical Acoustics*, McGraw-Hill, 1968.
[131] E. G. Williams, *Fourier Acoustics: Sound Radiation and Nearfield Acoustical Holography*, London: Academic Press, 1999.
[132] M. Gerzon, "Design of ambisonic decoders for multispeaker surround sound," in *58th Audio Egineering Society Convention*, New York, NY, USA, 1977.
[133] E. N. G. Verheijen, "Sound Reproduction by Wave Field Synthesis (PhD thesis, TU Delft)," 1998.
[134] J. Blauert, *Spatial Hearing: The Psychophysics of Human Sound Localization*, MIT Press, 1997.
[135] F. Rumsey, *Spatial Audio*, Focal Press, 2001.
[136] J. W. Strutt, "On our perception of sound direction," *Philosophical Magazine*, vol. 13, pp. 214–232, 1907.
[137] V. R. Algazi, R. O. Duda and D. M. Thompson, "Motion-tracked binaural sound," *Journal of the Audio Engineering Society*, vol. 52, Nov. 2004.
[138] C. P. Brown and R. O. Duda, "A structural model for binaural sound synthesis," *IEEE Trans. Speech and Audio Processing*, vol. 6, pp. 476–488, 1998.
[139] R. O. Duda, "3-D Audio for HCI," 2000.

[140] V. Pulkki, "Spatial sound generation and perception by amplitude panning techniques (PhD thesis)," Helsinki, Finland, 2001.

[141] D. de Vries, E. W. Start and V. G. Valstar, "The wave field synthesis concept applied to sound reinforcement: Restrictions and solutions," in *96th Convention of the Audio Engineering Society*, Amsterdam, 1994.

[142] A. Farina, "Simultaneous measurement of impulse response and distortion with a swept-sine technique," in *108th AES Convention*, Paris, France, 2000.

[143] M. A. Gerzon, "Periphony: With-height sound reproduction," *Journal of the Audio Engineering Society*, vol. 21, no. 1, pp. 2–10, 1973.

[144] M. Morimoto and Y. Ando, "On the simulation of sound localization," *Journal of the Acoustical Society of Japan*, vol. 1, no. 3, pp. 167–174, 1980.

[145] R. Alexander, *The Inventor of Stereo: The Life and Works of Alan Dower Blumlein*, Routledge, 2013.

[146] R. C. Alexander, *Michael Gerzon: Beyond Psychoacoustics*, Dora Media Productions, 2008.

[147] V. R. Algazi, et al., "The CIPIC HRTF database," in *IEEE Workshop on the Applications of Signal Processing to Audio and Acoustics (WASPAA)*, New Paltz, NY, USA, 2001.

[148] P. Majdak, F. Brinkmann, J. De Muynke, M. Mihocic and M. Noisternig, "Spatially oriented format for acoustics 2.1: Introduction and recent advances," *Journal of the Audio Engineering Society*, vol. 70, pp. 565–584, 2022.

[149] V. Pulkki, "Spatial sound reproduction with directional audio coding," *Journal of the Audio Engineering Society*, vol. 55, no. 6, pp. 503–516, 2007.

[150] J. Breebaart, J. Engdegård, C. Falch, O. Hellmuth, J. Hilpert, A. Hoelzer, J. Koppens, W. Oomen, B. Resch, E. Schuijers and L. Terentiev, "Spatial audio object coding (SAOC)-The upcoming MPEG standard on parametric object based audio," in *124th Audio Engineering Society Convention*, Amsterdam, Netherlands, 2008.

[151] B. Rafaely, et al., "Spatial audio signal processing for binaural reproduction of recorded acoustic scenes–review and challenges," *Acta Acustica*, vol. 6, p. 47, 2022.

[152] S. Agrawal, et al., "Defining immersion: Literature review and implications for research on immersive audiovisual experiences," *Journal of Audio Engineering Society*, vol. 68, no. 6, pp. 404–417, 2019.

[153] A. Pierce, *Acoustics: An Introduction to Its Physical Principles and Applications*, American Institute of Physics, 1989.

[154] J. O. Smith, "Doppler simulation and the Leslie," in *5th International Conference on Digital Audio Effects (DafX)*, Hamburg, 2002.

[155] B. G. Quinn, "Doppler speed and range estimation using frequency and amplitude estimates," *The Journal of the Acoustical Society of America*, vol. 98, pp. 2560–2566, 1995.

[156] A. R. Gondeck, "Doppler time mapping," *The Journal of the Acoustical Society of America*, vol. 73, pp. 1863–1864, 1983.

[157] M. J. Morrell and J. D. Reiss, "Inherent Doppler properties of spatial audio," in *129th AES Convention*, San Francisco, CA, USA, 2010.

[158] D. Rocchesso, "Fractionally addressed delay lines," *IEEE Transactions on Speech and Audio Processing*, vol. 8, Nov. 2000.

[159] C. Kyriakakis, "Fundamental and technological limitations of immersive audio systems," *Proceedings of the IEEE*, vol. 86, no. 5, pp. 941–951, 1998.

[160] X. Sun, "Immersive audio, capture, transport, and rendering: A review," *APSIPA Transactions on Signal and Information Processing*, vol. 10, p. e13, 2013.

[161] J. Herre and S. Disch, "MPEG-I immersive audio—Reference model for the virtual/augmented reality audio standard," *Journal of the Audio Engineering Society*, vol. 71, no. 5, pp. 229–240, 2023.

[162] N. Robson, A. McPherson and N. Bryan-Kinns, "Being with the waves: An ultrasonic art installation enabling rich interaction without sensors," in *New Interfaces for Musical Expression (NIME)*, 2022.

[163] S. Wilkie, "The effect of audio cues and sound source stimuli on looming perception (PhD thesis)," Queen Mary University of London, London, UK, 2015.

[164] D. Menexopoulos, P. Pestana and J. Reiss, "The state of the art in procedural audio," *Journal of the Audio Engineering Society*, vol. 71, no. 12, pp. 826–848, 2023.

[165] D. Ronan, H. Gunes and J. D. Reiss, "Analysis of the Subgrouping Practices of Professional Mix Engineers", 142nd AES Convention, Berlin, May 20-23, 2017.

[166] D. K. Wise, "Concept, design, and implementation of a general dynamic parametric equalizer," *Journal of the Audio Engineering Society*, vol. 57, no. 1/2, pp. 16–28, Jan. 2009.

[167] A. Tom, J. D. Reiss and P. Depalle, "An automatic mixing system for multitrack spatialization for stereo based on unmasking and best panning practices," in *Audio Engineering Society Convention 146*, 2019.

[168] J. P. Wakefield and Dewey, Christopher, "An investigation into the efficacy of methods commonly employed by mix engineers to reduce frequency masking in the mixing of multitrack musical recordings," in *138th AES Convention*, Warsaw, Poland, 2015.

[169] N. Jillings, "Automating the production of the balance mix in music production (PhD thesis)," 2023.

[170] P. D. Pestana, "Automatic mixing systems using adaptive audio effects (PhD thesis)," Universidade Catolica Portuguesa, Lisbon, Portugal, 2013.

[171] P. D. Pestana and J. D. Reiss, "Intelligent Audio Production Strategies Informed by Best Practices," in *AES 53rd International Conference on Semantic Audio*, London, UK, 2014.

[172] A. Pras, K. Turner, T. Bol and E. Olivier, "Production processes of pop music arrangers in Barnako, Mali," *153rd Audio Engineering Society Convention*, New York, NY, USA, 2019.

[173] B. De Man, R. Stables and J. D. Reiss, *Intelligent Music Production*, Focal Press, 2019.

[174] B. De Man, J. Reiss and R. Stables, *Ten Years of Automatic Mixing*, Salford, UK, 15 September 2017.

[175] C. J. Steinmetz, N. J. Bryan and J. D. Reiss, "Style transfer of audio effects with differentiable signal processing," *Journal of the Audio Engineering Society*, vol. 70, no. 9, pp. 708–721, 2022.

[176] S. S. Vanka, C. Steinmetz, J. B. Rolland, J. Reiss and G. Fazekas, "Diff-MST: Differentiable mixing style transfer," in *25th International Society for Music Information Retrieval (ISMIR)*, San Francisco, CA, USA, 2024.

[177] C. J. Steinmetz, S. Singh, M. Comunità, I. Ibnyahya, S. Yuan, E. Benetos and J. D. Reiss, "St-ito: Controlling audio effects for style transfer with inference-time optimization," in *25th International Society for Music Information Retrieval (ISMIR)*, San Francisco, CA, USA, 2024.

[178] N. Jillings, R. Stables and J. D. Reiss, "Zero-delay large signal convolution using multiple processor architectures," in *IEEE WASPAA*, Mohonk, New York, 2017.

[179] L. Savioja, V. Välimäki and J. O. Smith, "Audio signal processing using graphics processing units," *Journal of the Audio Engineering Society*, vol. 59, no. 1/2, pp. 3–19, 2011.

[180] R. H. Jack, A. Mehrabi, T. Stockman and A. McPherson, "Action-sound latency and the perceived quality of digital musical instruments: Comparing professional percussionists and amateur musicians," *Music Perception*, vol. 36, no. 1, 2018.

[181] A. McPherson, R. H. Jack and G. Moro, "Action-sound latency: Are our tools fast enough?," in *International Conference on New Interfaces for Musical Expression (NIME)*, Brisbane, Australia, 2016.

[182] A. McPherson, "C++ real-time audio programming with Bela," 2021. [Online]. Available: https://learn.bela.io/tutorials/c-plus-plus-for-real-time-audio-programming/course-introduction/.

[183] M. Walmsley, *Multi-Threaded Programming in C++*, Springer, 1999.

Index

For Product Safety Concerns and Information please contact our EU representative GPSR@taylorandfrancis.com
Taylor & Francis Verlag GmbH, Kaufingerstraße 24, 80331 München, Germany

www.ingramcontent.com/pod-product-compliance
Lightning Source LLC
LaVergne TN
LVHW020604110826
845149LV00002B/371